Lithium–Sulfur Batteries

Lithium–Sulfur Batteries

Edited by

Mark Wild
OXIS Energy, E1 Culham Science Centre, Abingdon, UK

Gregory J. Offer
Department of Mechanical Engineering, Imperial College London, London, UK

This edition first published 2019
© 2019 John Wiley & Sons Ltd

The right of Mark Wild and Gregory J. Offer to be identified as the authors of the editorial material in this work has been asserted in accordance with law.

Registered Offices
John Wiley & Sons, Inc., 111 River Street, Hoboken, NJ 07030, USA
John Wiley & Sons Ltd, The Atrium, Southern Gate, Chichester, West Sussex, PO19 8SQ, UK

Editorial Office
9600 Garsington Road, Oxford, OX4 2DQ, UK

For details of our global editorial offices, customer services, and more information about Wiley products visit us at www.wiley.com.

Wiley also publishes its books in a variety of electronic formats and by print-on-demand. Some content that appears in standard print versions of this book may not be available in other formats.

Library of Congress Cataloging-in-Publication Data

Names: Wild, Mark, 1974- editor. | Offer, Gregory J., 1978- editor.
Title: Lithium–Sulfur batteries / edited by Mark Wild, OXIS Energy, E1 Culham
    Science Centre, Abingdon, UK, Gregory J. Offer, Mechanical Engineering, Imperial
    College London, London, UK.
Description: First edition. | Hoboken, NJ, USA : John Wiley & Sons, Inc.,
    2019. | Includes bibliographical references and index. |
Identifiers: LCCN 2018041630 (print) | LCCN 2018043007 (ebook) | ISBN
    9781119297857 (Adobe PDF) | ISBN 9781119297901 (ePub) | ISBN 9781119297864
    (hardcover)
Subjects: LCSH: Lithium–Sulfur batteries.
Classification: LCC TK2945.L58 (ebook) | LCC TK2945.L58 L57 2019 (print) |
    DDC 621.31/2424–dc23
LC record available at https://lccn.loc.gov/2018041630

Cover design: Wiley
Cover image: Courtesy of OXIS Energy

Set in 10/12pt WarnockPro by SPi Global, Chennai, India
Printed and bound in Singapore by Markono Print Media Pte Ltd

10 9 8 7 6 5 4 3 2 1

Contents

Preface

In 2014, a team from industry and academia came together to develop a Revolutionary Electric Vehicle Battery and Energy Management System (REVB) funded by the EPSRC and Innovate UK. The project brought together material scientists, electrochemists, physicists, mathematicians, and engineers from OXIS Energy Ltd, Imperial College London, Cranfield University, Lotus Engineering, and Ricardo PLC. In 2016, the team held the first lithium–sulfur conference in the United Kingdom in the Faraday lecture theatre of the Royal Institution in London – a place where scientists, artists, authors, and politicians have shared ideas for over 200 years with the aim of diffusing science for the common purpose of life. The conference was named LiS–M^3 and continues annually. M^3 stood for materials, mechanisms, and modeling; we also held a fourth session for applications of Li–S technology. Following that first conference, the two conference chairs, Dr. Gregory J. Offer from Imperial College and Dr. Mark Wild from OXIS Energy, were approached to edit this book. The chapters have been provided by those that gave presentations, were invited that day, or were part of the REVB team in the Faraday lecture theatre.

The organization of this book follows the structure of the conference and has the same aim of educating a diverse scientific community about the most promising next-generation Li–S battery technology, enabling applications requiring batteries with superior gravimetric energy density. Lithium–sulfur batteries are game changers in the world of lightweight energy storage with a theoretical gravimetric energy density of $\sim$2600 Wh kg^{-1}. Yet, there are challenges, and today the practical energy density target is 500 Wh kg^{-1}.

Materials. In Part I we start with basic electrochemical theory to understand the challenges and complexity of lithium–sulfur batteries, and then focus on the approaches by material scientists to overcome those challenges. It soon becomes clear that there are no silver bullets, but that a systems approach is required to increase the areal loading of sulfur in a stable cathode, to increase sulfur utilization through the electrolyte/cathode interface and to reduce degradation at the electrolyte/anode interface. It is also evident that lithium–sulfur technology has reached a point in its development that it can now be tailored to meet the needs of commercial markets such as aviation, marine, or automotive.

Mechanisms. Part II considers the current understanding of the complex mechanisms in a lithium–sulfur cell. There remains an incomplete understanding of the mechanism from materials research, and analytical studies only see part of the picture. Elucidating the mechanism of a lithium–sulfur cell is complex and intriguing. There are many studies that have opened windows onto the association and disassociation reactions of the lithium polysulfides and the precipitation and dissolution of solid products at the end of charge and discharge. These underlying mechanisms lead to the unique discharge and charge characteristics and degradation pathways. Included are chapters on polysulfide reactivity and an enlightening look at the lithium–sulfur cell from the perspective of its insoluble end product, lithium sulfide.

Modeling. Part III starts by looking broadly at physics-based models that mimic and predict the performance and degradation of a lithium–sulfur cell under operational conditions. The section concludes with control models used to predict state of charge and state of health in real-time battery management systems. Modeling requires knowledge of the mechanism (Part II) and performance characteristics of the technology (Part I) and is used to develop the control algorithms and working models required by engineers developing applications (Part IV).

Applications. Part IV addresses the commercial application of lithium–sulfur battery technology. It starts with a market analysis, takes in key differences that battery engineers must be aware of in the design of a lithium–sulfur battery, and concludes with the first real-world application of lithium–sulfur batteries, the high-altitude long-endurance unmanned aerial vehicle (HALE–UAV).

As a guide to access the book, each part begins with its own introduction and also each chapter. If you are looking for a good overview, then start with the part introductions. If you are a material scientist, then start with Part I and continue with mechanisms in Part II; Chapter 10 may also be of interest to identify a target market. If you are an applications engineer you might like to start with modeling in Part III and move to applications in Part IV. If you are interested in modeling lithium–sulfur cells then start with Chapter 2 of Part I and then move to Parts II and III.

Mark Wild
OXIS Energy, Abingdon, UK

Part I

Materials

Lithium–sulfur cells utilize a very similar architecture as today's Li ion pouch cells. Double side coated cathodes (sulfur) are assembled with layers of separators and anodes (lithium foil) either through winding or electrode stacking and subsequently vacuum packaged into a pouch (aluminum/polymer laminate foil). In Li–S cells, cathodes are typically assembled in the charged state with lithium metal as anode.

At first glance, the combination of the lightest, most electropositive metal (lithium) with a safe, abundant (and reasonably light) nonmetal (sulfur) makes good sense as a prospective battery. However, while the lithium–sulfur battery offers a very high theoretical specific energy ($\sim$2600 Wh kg^{-1}) the actual performance delivered is proving to be limited and today a gravimetric energy density target of 500 Wh kg^{-1} is thought to be an achievable step change in battery performance with this technology.

Materials research lies at the heart of lithium–sulfur cell development and relies on a good understanding of the underlying mechanisms (Part II). The game changer is to achieve a lightweight battery with sufficient cycle life and power performance for relevant applications (Part IV) where the weight of large Li ion battery systems hinders product performance, e.g. aircraft and large vehicles. The goal is to increase the ratio of active sulfur to inactive, yet functional, materials in the cell and to make the best use of this sulfur by achieving the highest sulfur utilization cycle on cycle.

In Chapter 1, we begin with a grounding in basic electrochemical theory. We explore how basic theory translates to a more complex electrochemical system such as a lithium–sulfur cell. The chapter concludes with a theoretical explanation of the main challenges faced by materials research scientists developing commercial lithium–sulfur products.

In Chapter 2, we move on to a discussion of the sulfur cathode, where due to its nonconductive nature, sulfur is most often combined with carbons, additives, and binders to be coated onto a primed aluminum current collector. Even when optimized for high areal sulfur loading, cathode materials contribute to reduced gravimetric energy density and release reactive polysulfides into the electrolyte, leading to degradation.

In Chapter 3, we continue with a discussion of electrolytes and it will become apparent that there is an intimate relationship between sulfur loading and electrolyte loading. Stability of the electrolyte components toward both lithium and polysulfides is also

Lithium–Sulfur Batteries, First Edition. Edited by Mark Wild and Gregory J. Offer.
© 2019 John Wiley & Sons Ltd. Published 2019 by John Wiley & Sons Ltd.

critical to optimizing sulfur utilization and cycle life. A balance is to be struck between trapping polysulfides within the cathode and dissolution of polysulfides into the electrolyte to achieve acceptable energy density, cycle life, and power.

In Chapter 4, we briefly summarize the electrolyte anode interface and the key challenges. This is an area that is poorly covered by the academic literature but is a vital area of research to improve the cycle life of a lithium sulfur cell in tandem with other approaches. Throughout all chapters and in Chapter 7 we make reference to the role of anode in relation to degradation and reduced cycle life. Primarily efforts have included the use of a range of barrier layers either at the cathode surface, as a modification to the separator or as a polymer, or at the ceramic coating on the lithium itself in addition to optimization of cathode and electrolyte formulations.

Not to be lost is the shift in lithium–sulfur cell development to choose the thinnest and lightest components to reduce gravimetric energy density. Such materials require special consideration during scale-up activities when compared to handling procedures in standard lithium ion manufacture.

1

Electrochemical Theory and Physics

Geraint Minton

OXIS Energy, E1 Culham Science Centre, Abingdon, Oxfordshire OX14 3DB, UK

1.1 Overview of a LiS cell

On discharge, the overall process occurring in a lithium–sulfur (LiS) cell is the reaction of lithium and sulfur to form lithium sulfide, Li_2S, according to the reaction shown in Eq. (1.1).

$$16Li + S_8 \rightleftharpoons 8Li_2S \tag{1.1}$$

Although both reactants are present in the cell, its design, shown in Figure 1.1, prevents the reaction from taking place directly. The cell comprises a lithium metal electrode and a mixed carbon/sulfur (C/S) electrode. The latter is composed of a mix of highly porous carbon, which provides both electronic conductivity and an electrochemically active surface; sulfur, which is the active material in this electrode; and binder, which holds the structure together. The two electrodes are divided by a separator material, which stops the active materials from making direct contact and also prevents electrons passing internally between the electrodes. Contact between the active materials in each electrode is made indirectly, via an electrolyte, which is in contact with the lithium electrode and permeates the separator and C/S electrode. The electrolyte is composed of a solvent in which a lithium salt, plus any additives, has been dissolved. Adjacent to the C/S electrode is a current collector material to facilitate the flow of electrons to and from and external circuit, a task which is performed by the lithium metal on the other side of the cell.

By preventing the direct contact of lithium and sulfur, the design of the cell means that the reaction in Eq. (1.1) takes place indirectly: one or both species have to enter the electrolyte in order to react, a process which consumes electrons from (or releases them to) the electrode surfaces. These electrochemical reactions are the fundamental process which must occur in any electrochemical cell, since without them charge would not be replenished on the electrodes when an external current flows, causing the voltage to rapidly decrease to zero. Considering the electrochemical steps occurring in a LiS cell, the overall cell reaction in Eq. (1.1) can be split into the following overall half-reactions occurring at each electrode:

$$\text{Lithium electrode} \quad Li^+ + e^- \rightleftharpoons Li_{(s)} \tag{1.2a}$$

$$\text{C/S electrode} \quad S_{8(s)} + 16e^- \rightleftharpoons 8S^{2-} \tag{1.2b}$$

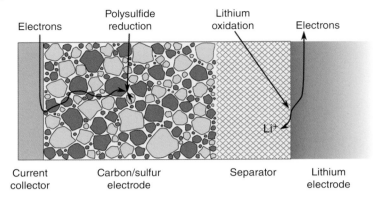

Electrons   Polysulfide   Lithium   Electrons
             reduction   oxidation

Li⁺

Current   Carbon/sulfur   Separator   Lithium
collector   electrode   electrode

Figure 1.1 Structure of a LiS cell, indicating the two electrodes, separator, and C/S electrode current collector. Also indicated are the reactions at each electrode and the electron pathway on discharge.

where the equations are written in the standard form $Ox + ne^- \rightleftharpoons Re$, in which Ox is the oxidized (more positively charged) form of the species, Re is the reduced (more negatively charged) form, and e^- is an electron. In order for the overall cell reaction to be satisfied, on discharge the lithium reaction must run to the left (electrochemical oxidation), with ions and electrons being formed from the lithium metal, while the C/S reaction must run to the right (electrochemical reduction), with electrons being consumed and ions being formed by the reaction of solid-phase sulfur. In separating out the reactions in this manner, it is possible to see how the reactions generate the electronic charge in the electrodes which flows as the external current when the electrodes are connected.

The cell half-reactions might indicate that only electrochemical reactions occur in the cell, but this is not the case. The initial state of the cell includes solid-phase sulfur, so a dissolution step is required; and to form the precipitated Li_2S, a chemical reaction is required to combine the lithium and sulfur ions:

$$\text{Sulfur dissolution} \quad S_{8(s)} \rightleftharpoons S_{8(d)} \tag{1.3a}$$

$$\text{Li}_2\text{S precipitation} \quad 2Li^+ + S^{2-} \rightleftharpoons Li_2S \tag{1.3b}$$

Furthermore, as with the phase change processes at either end of the reaction mechanism, the reactions occurring at the C/S electrode do not appear to take place in a single step, as Eq. (1.2b) may suggest. Instead, a host of intermediate species are produced through both chemical and electrochemical elementary steps [1]. These intermediate species consist of sulfur anions of varying chain lengths, collectively known as polysulfides. They are commonly split into two groups: "high-order" species are those with chain lengths of five to eight atoms, and "low-order" species are those with chain lengths of one to four atoms.

Since the exact reaction mechanism is currently unknown, and may even vary under different operational conditions [2–5] or electrolyte compositions, this chapter does not attempt to describe how each individual process affects a LiS cell. Instead, the focus is on how different types of reaction processes contribute to (or detract from) the electrochemical performance of a cell, using a LiS as a reference point.

Before continuing, it will be useful to define how the electrodes are referred to throughout this chapter. The lithium reaction has to generate electrons on discharge, while the polysulfide species have to accept them. Since current flow is in the direction opposite to the flow of electrons, electronic current will flow from the C/S electrode to the lithium electrode, making the lithium electrode the anode on discharge. However, the direction of the current is reversed on charge, which would make the C/S electrode the anode; and if the cell is at rest, then there is no anode. In order to simplify the terminology, and because it is commonly the discharge which is of more interest, the lithium electrode will herein be referred to as the anode, regardless of the direction (or presence) of a current, and the C/S electrode will be referred to as the cathode.

1.2 The Development of the Cell Voltage

The purpose of an electrochemical cell is to drive an electric current through a circuit which is connected across the terminals of the cell. This flow of current occurs spontaneously: naively electrons move from the electrode with the lower electric potential to the one with the higher electric potential, causing the difference between the two electrode potentials to decrease. Thus, maintaining a current requires electrons to be spontaneously generated at the lower potential electrode and consumed at the higher potential electrode; otherwise, the electrode potentials would equilibrate and the flow of electrons would stop.

Thermodynamically, a process will occur spontaneously if it lowers the free energy of a system, where the free energy is the internal energy of the system minus that part of the internal energy which can do no useful work. In the case of an electrochemical cell, the system is the electrochemical cell and the external circuit, and a process is anything occurring within the system, for example the conversion of one species to another via a reaction, or the trend for a species to move in one direction or another. A useful aspect to remember when discussing processes is that we are always considering the net outcome of a very large number of individual random events, some of which increase the free energy and some of which lower it. The "direction" represents how this stochastic process is biased, and this bias is always in the direction which minimizes the free energy, because it is more likely that a particle in a higher energy state will move to a lower energy state than the reverse occurring.

Coupled with the notion of the events occurring at random is the fact that the processes do not stop when equilibrium is reached – for example, individual particles in a gas do not stop moving just because there is no concentration gradient. Instead, equilibrium implies that the bias to the process has been removed – the number of particles in the gas moving to the left is now equal to the number moving to the right, so there is no net change in the concentration.

There are two forms of the free energy commonly used to describe processes in an electrochemical system: the Gibbs free energy G, typically used when discussing the reaction processes, and the Helmholtz free energy F, commonly used when discussing the electrolyte composition [6–9] and species transport. How they are related and defined is beyond the scope of this chapter and, since we will not consider volume or pressure changes in the cell, the two are essentially equivalent. Related to the free energy is the electrochemical potential μ_i of the species in the system, which is also often more convenient to work with [10, 11]. This term represents the change that occurs in the free energy of a system when a particle of type i is added to a point in the system from a point outside of it, and can be used to derive expressions for the system behavior. Under the assumption that there are a large number of particles in the system [12], the electrochemical potential is determined by the following relationships:

$$\mu_i = \left(\frac{dG}{dN_i} \right)_{(p,T,N_{j\neq i})} = \left(\frac{dF}{dN_i} \right)_{(V,T,N_{j\neq i})} \tag{1.4}$$

where N_i is the number of molecules of type i and the subscripts indicate the properties held constant during the differentiation: p is the pressure, T is the temperature, V is the volume, and $N_{j\neq i}$ is the quantity of all species except species i. The notion of how the electrochemical potential relates to the overall direction a process occurs is summarized in the following two examples:

1) Particle movement can be thought of as the process of taking a particle of type 1 from point A inside the system, moving it outside of the system, and then placing it back in the system at a different point, B. The removal process changes the free energy of the system by μ_1^A and the addition step changes it by μ_1^B. If $\mu_1^A > \mu_1^B$, the free energy is reduced by the particle's movement, so the bias for species movement is for particles to move from A to B.
2) A reaction process effectively removes a particle of type 1 from point A in the system and replaces it with a particle of type 2 (or vice versa). If $\mu_1^A > \mu_2^A$, the free energy of the system is reduced by the formation of particle 2 and so this direction is preferred. If not, the reverse process is preferred and particle 1 tends to be formed from particle 2.

There are two types of particle that we are typically interested in when considering an electrochemical system: electrons in the electrodes and particles in the electrolyte, itself comprising ions, neutral species, and solvent molecules. It is common practice when modeling electrochemical systems to treat the solvent molecules as a continuum dielectric background through which the remaining particles in the electrolyte move and interact [13–16], an approach also taken throughout this discussion. The electrochemical potentials of the electrons and the solvated mobile species can be written as [17, 18]

$$\mu_{e^-}(\boldsymbol{r}) = E_f(\boldsymbol{r}) - e_0\phi(\boldsymbol{r}) \tag{1.5a}$$

$$\mu_i(\boldsymbol{r}) = \mu_i^\ominus + k_B T \ln\left(\frac{c_i(\boldsymbol{r})}{c^\ominus} \right) + \mu_i^{ex}(\boldsymbol{r}) + z_i e_0 \phi(\boldsymbol{r}) \tag{1.5b}$$

where E_f is the energy of the thermodynamic Fermi level of the electrode, discussed subsequently, e_0 is the elementary charge, k_B is the Boltzmann constant, T is the temperature and, for species i, c_i is its concentration, μ_i^{ex} is its excess chemical potential, and z_i is its valence. $\mu_i^\ominus$ is the chemical potential of species i in the standard state, defined as an ideal solution of that species at the standard concentration $c^\ominus=1$ M, at a temperature of 298 K and a pressure of 1 bar. Finally, ϕ is the local electrostatic potential, defined relative to some fixed reference. For the species in solution, the logarithmic term represents the ideal, or entropic, contribution to the electrochemical potential, the excess chemical potential describes all non-electrostatic contributions, and the final term describes the contribution from the electrostatic potential.

1.2.1 Using the Electrochemical Potential

To give an example of how the electrochemical potential is the quantity of interest in an electrochemical system, we first look at the behavior of conductive materials. The energy of an electron in a material is represented by the energy of the Fermi level E^f, which is the energy level within the band structure of the material which has a 50% probability of occupancy by electrons. The Fermi level is defined by the kinetic Fermi energy E^k, which is a function of the effective electron masses and their number density, and which sits at a constant position relative to the conduction band edge [17, 19]. Because of the differing band structures of different materials, their Fermi level energies will differ.

The other quantity of interest for a material is its work function, which is the amount of energy required to excite an electron from the Fermi level to the vacuum, defined as a point just outside the material where the electric field is zero, such that the electron ends the process with zero kinetic energy. The electric potential at this point is referred to as the vacuum potential. The work function depends on factors such as the crystal face through which the electron exits the material and defects in the surface structure, meaning that its value is different for different surfaces of the material.

In the bulk material, the Fermi level is constant, meaning that an electron added to the material from a vacuum state will have the same final energy, regardless of which surface it enters through. Since the final state must be the same in both cases but the change in energy to get there (the negative of the work function) differs, the implication is that the vacuum potentials at different surfaces are not equivalent. This is represented in Figure 1.2a, showing the positions of the conduction band and Fermi level and the positions of the vacuum potential at two surfaces. Also indicated is the internal potential of the material in the bulk near each surface. We assume a conductive material, which implies that $\phi_1 = \phi_2$.

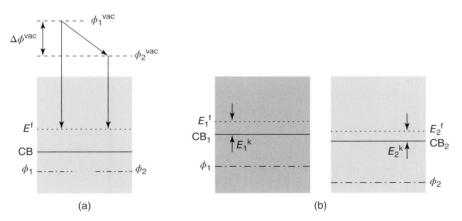

(a) (b)

Figure 1.2 Scheme for the relevant energy levels in a conductive material. (a) The Fermi level energy lies a fixed distance above the conduction band edge, CB, but differences in the work functions across different surfaces mean that the vacuum potential is not constant with respect to E^f. Choosing one vacuum potential as the reference point, the energy required to move a charge to the Fermi Level is always the same because of the work which has to be done against the external potential difference between the vacuum potentials, $\Delta\phi^{vac}$. (b) The CB, Fermi level energy, and electrostatic potentials in two disconnected conductors may all be different.

An electron at surface 1, with potential ϕ_1^{vac}, will lose energy upon moving to the Fermi level, while an electron at surface 2, with potential ϕ_2^{vac}, will lose a different amount of energy to reach the same final state. Since the final internal states are identical, this means that an electron which starts at surface 1 but enters the material via surface 2 must undergo a change of energy as it moves externally around the material. This change in energy is associated with working against the electric field between the two surfaces, an electric field which must exist because of the differences between the two vacuum potentials.

We now consider the change in the electrochemical potential associated with moving an electron from surface 1 to the Fermi level via each route. Defining the reference electrostatic potential as the vacuum potential at surface 1, the change in the electrochemical potential when the electron moves directly to the Fermi level is $\mu_1 = E_f - e_0(\phi_1 - 0)$. Alternatively, if the electron takes the indirect path, the electrochemical potential is $\mu_2 = E_f - e_0(\phi_2 - \phi_2^{vac}) - e_0(\phi_2^{vac} - 0)$. Since the Fermi level is constant, and there are no internal variations in the electric potential, the final electrochemical potential in both cases can be seen to be the same and the two routes are equivalent, overall.

We now consider the case of connecting two conductive materials together. Prior to contact, as indicated by Figure 1.2b, the different materials will have different conduction band edges and kinetic Fermi energies E^k, so the Fermi levels will differ. Also, because there may be an arbitrary electric field between the two phases, the internal electrostatic potentials may also be different, and so the electrochemical potential of the electrons in the two materials are not the same. When placed into contact (Figure 1.3), if material 1 has a larger Fermi energy, then there will be electron transfer from material 1 to material 2 (Figure 1.3a). As a result, material 2 will acquire a negative charge and material 1 will be left positively charged.

This separation of charge generates an electric field $\vec{E}$, which grows as more charge is transferred. The direction of the field is such that it resists the continued transfer of electrons, so the transfer process is self-limiting: eventually, the field grows large enough to prevent further electron transfer and an equilibrium state is reached. In this state, the field is such that the potential varies smoothly from ϕ_1, the value in the bulk of material 1, to ϕ_2, the value in the bulk of material 2 (Figure 1.3b). The total potential difference, which can now be explicitly defined in terms of the electric field, is known as the Galvani potential, denoted here as $\Delta\phi^G$.

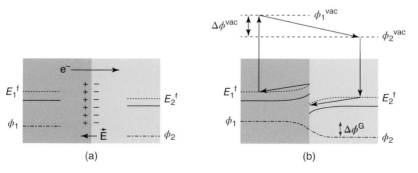

Figure 1.3 (a) Connecting two conductors allows electrons to move from the material with the larger Fermi level energy to the one with the lower Fermi level energy, resulting in the development of a charge-separated layer and electric field at the interface. (b) The field defines the Galvani potential difference $\Delta\phi^G$ between the materials and causes the conduction band edges to bend near the interface. At equilibrium, no net work is done moving a charge along the path indicated.

The other implication of the system being at equilibrium is that an electron moving from material 1 to material 2 should do no overall work. Note that the Fermi levels of the two bulk materials have not changed, so this element still favors the transfer of electrons between materials; it is just that this is now counterbalanced by the electric field. Any energy an electron would be able to lose by moving to the material with the lower Fermi level would be offset by the necessary energy gain required to move up the electric field at the interface.

To state this in terms of the electrochemical potentials of the two phases, a reference point is still required for the electrostatic potential, and here we choose the vacuum potential of one of the external surfaces of material 1. The work function at that surface, being far from the interface region, is largely unchanged by the interfacial field, so the electrochemical potential of an electron in the bulk of material 1 remains as $\mu_1 = E_f^1 - e_0(\phi_1 - 0)$. For material 2, however, the electrochemical potential is now $\mu_2 = E_f^2 - e_0(\phi_2 - \phi_2^{vac}) - e_0(\phi_2^{vac} - 0)$, where ϕ_2 can be written in terms of ϕ_1 as $\phi_2 = \phi_1 - \Delta\phi^G$. Since the electrochemical potential must be spatially invariant at equilibrium, the previous expressions can be equated:

$$\mu_2 = \mu_1$$
$$E_f^2 - e_0(\phi_2 - \Delta\phi^G - \phi_2^{vac}) - e_0(\phi_2^{vac} - 0) = E_f^1 - e_0(\phi_1 - 0)$$
$$E_f^2 + e_0\Delta\phi^G = E_f^1$$

From this, we find that the electrostatic potential difference across the interface equals the difference in the Fermi energies of the two materials.

There are some complications in the interfacial region, where the electric potential varies spatially. In order for the electrochemical potential to be invariant in this region, the definition of the electrochemical potential implies that the Fermi level energies E^f of the two materials must also change. However, E^f is not affected by the field directly. Instead, the electric field changes the shape of the conduction band, causing it to curve toward the vacuum potential on the positively charged side and away from it on the negatively charged side. The Fermi level, which is separated from the conduction band edge by an amount equal to the kinetic Fermi energy (which is unchanged by the electric field), therefore effectively curves in the same direction as the conduction band. The resulting smooth variation in the Fermi level counterbalances the electric potential at all positions. This spatially dependent Fermi energy has been referred to as the thermodynamic Fermi energy [19].

A second observation is that while the electrostatic potential must be continuous across the interface, the changes to the Fermi levels of the two materials do not necessarily cause these to become continuous. The discontinuity is due to the work which has to be done in crossing from the crystal lattice in one material to the crystal lattice in the other, and can be demonstrated by considering the movement of an electron along the path indicated in Figure 1.3b, which must be the equivalent of an electron taking the direct route across the interface.

The potential difference between the phases is unmeasurable in all but a few situations. A voltmeter, for example, is essentially a long conducting wire, and connecting this between conductive materials would simply lead to electron rearrangement at the interface of each end of the wire with the material it is in contact with. The electrochemical potential would equilibrate throughout the entire system, which comprises material one, the wire, and material two, with a Galvani potential developing at each interface, but there would be no continuous net flow of electrons, since the free energy would quickly be minimized. However, a voltmeter requires some flow of electrons, since this does the work to turn the needle of the meter and give a reading. Consequently, no voltage would be measured in the example case.

This lack of a voltage might make us question the use of a voltmeter to measure a potential difference, but what we are generally interested in when measuring a voltage is the propensity of two phases to drive a current between them. In this case, the voltmeter is exactly the device we want; but rather than the difference in the electrostatic potential, what we are actually measuring is a difference in the electrochemical potential across the voltmeter's probes. Such a difference in electrochemical potential implies that the free energy of the system is not minimized, meaning the system is out of equilibrium. The flow of electrons through the voltmeter occurs spontaneously as the system attempts to minimize its free energy, providing the energy to turn the voltmeter's needle while also causing the electrochemical potentials to equilibrate.

Since the driving force for the current is proportional to the difference between the electrochemical potentials in the two materials, connecting the voltmeter will cause the voltage to decrease over time as the materials equilibrate. From the converse of this, we can see that if a constant voltage is measured, processes must be occurring to maintain the electrochemical potential difference across the voltmeter's probes. How the electrochemical potential difference is maintained is at the heart of the function of an electrochemical cell, as discussed in the following sections.

1.2.2 Electrochemical Reactions

The process by which an electrochemical cell is able to develop and maintain a measureable voltage is at the core of what the cell does. As with the behavior of two or more connected metals, the electrochemical potential provides a reasonably intuitive route to understanding the processes driving the cell. However, there is more scope for complexity, even while trying to keep descriptions of individual processes as simple as possible. To begin, we start by considering what happens at the single interface when a conductive electrode material is placed in contact with an electrolyte (Figure 1.4).

At the microscopic level, the interfacial region between the phases is highly complex even before any electrochemical reactions have taken place – a vast number of short- and long-range interactions take place between the mobile molecules which form

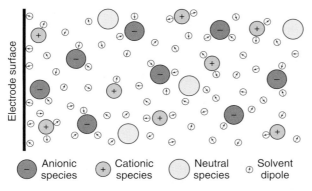

Anionic species	**Cationic species**	**Neutral species**	**Solvent dipole**

Figure 1.4 Diagram of the simplified electrode–electrolyte system. Mobile species in the electrolyte move through a sea of solvent dipoles and may approach the flat, smooth electrode surface.

the electrolyte and between these molecules and the surface, as well as other effects like image–charge interactions, surface adsorption, or the simple fact that the surface is not necessarily microscopically flat. In order to help build a simpler description of the system, we make a number of assumptions.

First, we assume that the only explicit surface interaction which occurs is the electrochemical reaction, and that the excess chemical potential in Eq. (1.5b) implicitly describes all other interactions that a particle experiences. Second, we assume that the surface is smooth and flat. Third, we define the reference point for the electric potential as being in the bulk electrolyte. Finally, for simplicity, we assume that the work function of the electrode surface facing the electrolyte is the same as the work function for the electrolyte surface facing the electrode. This means that the potential in the electrode, relative to the bulk electrolyte, is zero. In reality, the ions in the electrolyte would respond to the electric dipole on the electrode surface, compensating for any electric potential difference; but since this potential difference does not contribute to the operation of the cell, it can be ignored.

The electrolyte is formed by the dissolution of the general salts AB and CD into a solvent, forming a homogeneous mix of completely dissociated ions A^+, B^-, C^+, and D^-. Also present is a quantity of species E, which has no net charge and is again uniformly distributed. The electrochemical potentials of all species in the electrolyte are given by Eq. (1.5b), with $z_{A^+} = z_{C^+} = +1$, $z_{B^-} = z_{D^-} = -1$, and $z_E = 0$. The species in solution and electrons in the electrode are homogeneously distributed, so there is no initial electric field and the electric potential is zero throughout the system. From Eq. (1.5a), this means that the electrochemical potential of the electrons in the electrode is initially equal to the Fermi level energy of the electrode material, E_f.

As previously stated, the standard form of an electrochemical reaction is

$$Ox + n_e e^- \rightleftharpoons Re \tag{1.6}$$

where n_e is the number of electrons involved. In the example system, we assume that ions C^+ and D^- are inert, so there are two possible reactions of the standard type:

$$A^+ + e^- \rightleftharpoons E \tag{1.7a}$$
$$E + e^- \rightleftharpoons B^- \tag{1.7b}$$

For both reactions, reduction makes the electrode positively charged while increasing the amount of negative charge in the electrolyte, either by replacing positive ions with neutral species or replacing neutral species with negative ions. Conversely, oxidation, the process shown in Figure 1.5a, makes the electrode negatively charged while increasing the amount of positive charge in the electrolyte. Note that charge is conserved at the reaction-site, which means that the overall electrochemical cell remains charge neutral throughout its operation. However, local charge densities do arise either side of the electrode–electrolyte interface: the electrons have to exist in the electrode phase and ions have to exist in the electrolyte phase, so the reaction creates a local separation of charge across the interface.

As outlined at the start of this section, a reaction process alters the free energy of the system by removing one or more particles and replacing them with one or more and others; and while the reaction is free to occur in either direction, the direction which minimizes the free energy is dominant. The change in the free energy associated with the

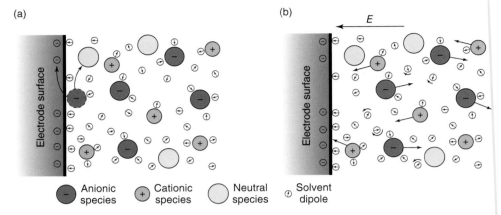

(a)

(b)

⊖ Anionic species	⊕ Cationic species
◯ Neutral species	⊘ Solvent dipole

Figure 1.5 (a) Example of an oxidation process, in which an anionic species in the solution releases an electron to the electrode surface and becomes a neutral molecule. (b) The response of the ions in solution to the electric field generated by the surface charge on the electrode.

electrochemical reaction, ΔG_{RXN}, is the difference between the electrochemical potentials of the active species:

$$\Delta G_{RXN} = \mu_{Re} - \mu_{Ox} - \mu_{e^-}$$
$$= \left[\mu_{Re}^{\ominus} + k_B T \, \ln \left(\frac{c_{Re}}{c^{\ominus}} \right) + \mu_{Re}^{ex} + z_{Re} e_0 \phi \right]$$
$$- \left[\mu_{Ox}^{\ominus} + k_B T \, \ln \left(\frac{c_{Ox}}{c^{\ominus}} \right) + \mu_{Ox}^{ex} + z_{Ox} e_0 \phi \right] - [E_f - e_0 \phi] \tag{1.8}$$

where the terms in the second equality are grouped by species. This thermodynamic expression tells us the overall direction in which the reaction will spontaneously occur: if $\Delta G_{RXN} > 0$, the oxidation process will occur, making the electrode more negatively charged; while if $\Delta G_{RXN} < 0$, the reduction process occurs, making the electrode more positively charged. As with all thermodynamic expressions, however, it does not tell us the rate of a reaction, although a range of methods are available for determining this [20–23]. For our purposes, it is sufficient to know that the reaction rate increases with the magnitude of ΔG_{RXN}. Essentially, the further the system is from equilibrium, the faster it tries to move toward it.

As well as determining the direction that the reaction spontaneously occurs, it can be seen that the magnitude of ΔG_{RXN} will always tend to zero as the reaction proceeds, because the reaction reduces μ_i for the species being consumed while increasing μ_i for the species being produced, regardless of the direction. When ΔG_{RXN} reaches zero, the reaction reaches equilibrium. At this point the net reaction rate is zero, although this is not because no reaction events occur, but because the frequency of reaction events in either direction is the same.

While the reaction is taking place, the excess of the reaction product and deficit of the reactant near the surface, relative to the bulk, ordinarily drives a species flux due to diffusion: the product moves away from the surface to the bulk, where the concentration is smaller, and the reactant moves toward the surface from the bulk, where the concentration is larger. While the reaction is occurring, the diffusion component of the flux acts to prevent the accumulation of product or depletion of reactant at the surface, which would otherwise quickly cause ΔG_{RXN} to shrink at the surface, significantly slowing the reaction.

The diffusion flux is countered by a migration flux caused by the electric field generated by the separation of charge at the interface. The electric field acts on all charged particles in the system, causing cations to move down the field (to lower potentials) and anions to move up the field, as shown in Figure 1.5b. It also has a number of other effects, including the alignment of solvent dipoles against the field, which, together with the presence of the ions themselves, alters the permittivity/dielectric constant of the electrolyte and therefore how the field propagates [24–27], although we do not consider this further here. For a given surface charge, ions of the opposite sign in solution are termed "counterions" and those with the same sign are "co-ions." Counterions are always attracted to a charged surface and co-ions are repelled. Since the surface charge is generated by an electrochemical reaction, the electric field it generates always acts either to draw the product toward the surface or repel the remaining reactant in solution, depending on which of those species is charged. The migration flux therefore always acts to reduce ΔG_{RXN} by decreasing $\mu_{Re} - \mu_{Ox}$.

Finally, changes in the concentrations of species near the surface also alter the excess chemical potential μ^{ex} of all species in that region. We do not deal too much with this term, but one of the main contributing factors is the excluded volume interaction (EVI), which represents the fact that two particles cannot occupy the same volume of space. As the concentration of any species increases at a point, the EVI energy increases, so, for example, it prevents the counterion concentrations becoming unphysically large. It also means that a concentration gradient in one species will induce a gradient in μ^{ex} for all others, leading to a displacement flux, for example.

Overall, the structure of the electrolyte near the interface quickly becomes complicated, so to get a better understanding of the processes at play and how they interact, we temporarily halt the reaction and allow the electrolyte to equilibrate spatially.

1.2.3   The Electric Double Layer

With the reaction paused, the quantity of each species in the electrolyte is fixed, as is the electrode surface charge density. In the electrolyte, if μ_i is not homogeneous, then the species will spontaneously rearrange themselves until it is, at which point the free energy will be minimized and the system reaches equilibrium.

We know that there is an electric field between the electrode surface and the bulk electrolyte because of the separation of charge caused by the reaction. As indicated by Figure 1.5b, this field attracts counterions to the surface and repels co-ions, regardless of their involvement in the reaction. All ion concentrations near the surface therefore differ from their bulk values: there is an increase in the concentration of counter-ions and a decrease in the concentration of co-ions (Figure 1.6a). The accumulation of counter-charge alters the electric field, causing it to decrease to zero (the defined value in the bulk) with distance from the surface at the same time as the ion concentrations return to their bulk values, where the net charge is zero (Figure 1.6b).

In order to express the equilibrium state in terms of the species' electrochemical potentials, the relationship between the accumulation of charge and the electric field,

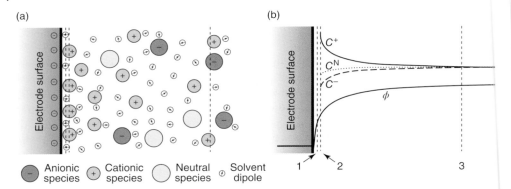

Figure 1.6 (a) Diagram of the structure of the double layer and (b) representative plot of the species concentrations, c, and electrostatic potential, ϕ. Near the electrode, a layer of adsorbed solvent molecules defines the inner Helmholtz plane (line 1) and the minimum approach distance defined by the finite size of the ions is known as the Stern layer (line 2). Beyond a certain distance from the surface (line 3), the bulk is effectively shielded from the surface charge on the electrode, with the species concentrations and the electrostatic potential converging on their bulk values.

which defines the electrostatic potential, is needed. To determine this, we again look to the free energy of the system, which is always minimized with respect to changes in the electrostatic potential ϕ:

$$\left(\frac{\mathrm{d}F}{\mathrm{d}\phi}\right)_{(T,V,N_i)} = 0 \tag{1.9}$$

We will again not dwell on the specific form of the ϕ-dependence of the free energy, but the simplest model assumes that the electric field at a point in space is defined by the average field generated by all other charges in the system – the so-called mean-field approximation. This leads to the electrostatic potential being defined by the Poisson equation

$$-\frac{1}{4\pi}\nabla \cdot [\epsilon_0\epsilon_r\nabla\phi] = \rho \tag{1.10}$$

where ϵ_r is the permittivity of the electrolyte relative to the permittivity of free space, ϵ_0, and ρ is the charge density, which is a function of the species concentrations

$$\rho = \sum_i z_i e_0 c_i \tag{1.11}$$

The electric field causes the migration of counter-charge to the surface, where it accumulates. Since the sign of the charge changes across the interface, the electric field is strongest at the surface itself. However, the Poisson equation shows that the field becomes weaker as it passes through the accumulated counter-charge in solution, which means that its ability to attract more counterions toward the surface also diminishes. Therefore, with distance from the surface, both the electric field and the amount of counter-charge decrease, although the rate of this decrease slows, such that there is a convergence on the bulk values. Over the distance of this convergence, the concentrations of all species return to their bulk values and the electric potential converges on zero (the reference potential). The region in which the charge density in the electrolyte is non-zero is called the electric double layer.

The exact structure of the double layer, and the variation of the electrostatic potential within it, depends on how the counter-charge accumulates. The most energetically favorable arrangement would be for all of the counter-charge to be on the surface, so that the electric field is very large but tightly confined to the surface. However, the large concentration gradients in this arrangement would induce diffusion, causing the counter-charge to spread out from the surface, reducing the field at the surface slightly while causing it to extend further into the electrolyte before becoming zero. Furthermore, because μ^{ex} also affects how counter-charge accumulates, this too will alter the electric field and therefore the structure of the double layer. The way that species interactions play a crucial role in the formation of the double layer make understanding its structure in detail complicated, even in equilibrium, although many approaches of varying complexity are possible, ranging from the most commonly applied Gouy–Chapman model, through various density functional theory methods, to Monte Carlo and molecular dynamics techniques [28–33].

The fact that the double layer causes the electric field to fall to zero with distance from the surface means that it has the effect of shielding the bulk electrolyte from the surface charge on the electrode. With the bulk electrolyte being homogeneous and there being no electric field, it will be at equilibrium, because each component of the electrochemical potential will be constant. In the double-layer region, the variation in the potential is offset by the variations in the species concentrations (plus contributions from the μ^{ex} terms), with the total μ for each species at every point being equal to the value in the bulk.

From the perspective of the processes which cause species transport, the equilibrium structure of the double layer can be thought of as being the result of the diffusion, migration, and displacement fluxes competing with one another until they balance each other out. If the field increases, ions will migrate to the surface (or away from it) until the concentration gradient is such that the diffusive flux rebalances the migration flux.

1.2.4 Reaction Equilibrium

Having arbitrarily stopped the electrochemical reaction and allowed the electrolyte to equilibrate, we now allow the reaction to restart and for the system to reach a full equilibrium. While the reaction takes place, ΔG_{RXN} tends toward zero, causing the reaction rate to decrease to zero. In parallel to this, the electrochemical potential gradients tend to zero, with a double layer present at the interface to confine the electric field to a boundary layer and shield the bulk from the excess surface charge on the electrode.

In the equilibrium state, the magnitude of the potential difference across the double layer is defined by the amount by which the electrochemical potential of the electrons has changed during the reaction process, which is equal to the difference between the electrochemical potentials of the chemical species:

$$\Delta\mu_{e^-} = (E_f - e_0\phi) - E_f = -e_0\phi = \mu_{Re} - \mu_{Ox} \tag{1.12}$$

Although it differs from how we have defined it in this work, the final term is termed the Gibbs free energy of the reaction, which leads us to the commonly stated relationship between this form of ΔG_{RXN} and the electrode potential

$$\Delta G_{RXN} = -n_e e_0\phi \tag{1.13}$$

where the n_e term has been added to account for multi-electron transfer reactions. The electrostatic potential is clearly linked to the difference between the electrochemical potentials of the oxidant and reductant in solution, and the potential difference across the double layer is the reduction potential of the reaction. Furthermore, because the electrochemical potential of a species at equilibrium does not vary spatially, the preceding equations hold regardless of where the electrochemical potential is determined: the bulk and the surface are equivalent in this regard, meaning that the properties of the bulk can be used to determine something which is very much a surface property.

The dependence of the reduction potential on μ_{Ox} and μ_{Re} also implies that its value is not fixed for a particular reaction, but is determined by the state of the system. Something which alters one of the electrochemical potentials will ultimately alter the electrostatic potential. This is the essence of the Nernst equation, which is a fundamental thermodynamic relationship between the reduction potential of a particular reaction and the reduction potential when all active species are in their standard states. By writing

$$-e_0\phi - e_0\phi^\ominus = (\mu_{Re} - \mu_{Ox}) - (\mu_{Re}^\ominus - \mu_{Ox}^\ominus) \tag{1.14}$$

and substituting in Eqs. (1.5a) and (1.5b), it is straightforward to derive the expression

$$-e_0(\phi - \phi^\ominus) = k_B T \ln \left(\frac{c_{Re} \exp(\mu_{Re}^{ex})}{c_{Ox} \exp(\mu_{Ox}^{ex})} \right) + e_0(z_{Re} - z_{Ox})\phi \tag{1.15}$$

The term on the left of Eq. (1.15) is a surface term, but because the terms on the right came from the chemical species' electrochemical potentials, which is homogeneous at equilibrium, the values can be taken from anywhere in solution. Choosing the bulk, where the electrostatic potential is defined as zero, the preceding expression becomes the Nernst equation:

$$\phi = \phi^\ominus - \frac{k_B T}{n_e e_0} \ln \frac{a_{Re}}{a_{Ox}} \tag{1.16}$$

where we have introduced the activity of the species, defined as

$$a_i = \exp \left(\frac{\mu_i - \mu^\ominus}{k_B T} \right) \tag{1.17}$$

Values of the standard state reduction potentials for many reactions are measured and commonly tabulated, although for reasons discussed subsequently single-electrode potentials cannot be measured. Because of this, they are quoted relative to a second reaction potential, typically that of a standard hydrogen electrode. From these values, the reduction potential in any system can be determined, provided the activities are known.

The Nernst equation is a fundamental thermodynamic relationship and must always be obeyed, so any kinetic model for the reaction rate must reduce to this expression at equilibrium. The equation also starts to hint at some of the complexity which arises in a multicomponent system like LiS. For example, the presence of a species which interacts differently with the two active species will alter their relative activities, and therefore the equilibrium electrode potential. Similarly, a change of solvent may alter the stability of one active species in the electrolyte, again altering the equilibrium state. However, these factors are often difficult to account for in a way that does not significantly hinder the time taken to run a model, meaning that simplifications are always made.

1.2.5   A Finite Electrolyte

The single-electrode system discussed so far assumes that the bulk electrolyte is large enough that the reaction does not affect the species concentrations; but in a real system, the quantity of each species is finite. When a reaction converts one species into another, the activities of those species change throughout the system: one increases at the expense of the other. If the reaction is allowed to continue, such as happens when an electrochemical cell is discharged, the accumulation of the product and the removal of the reactant in the bulk will cause the equilibrium reduction potential to shrink, ultimately becoming zero. Such a situation will arise when enough reactant has been converted to product that their electrochemical potentials are equal, at which point there is no driving force for the reaction, no further charge added to the electrode, no double layer, and so no potential difference between the electrode and the electrolyte.

This decrease in the equilibrium cell voltage is characteristic of a typical electrochemical cell, and is indicative of the fact that there is a smaller amount of free energy remaining within the cell as the cell is discharged. As a result, the cell voltage is usually a good indicator of the state of charge of a cell. However, in one of the many challenges facing LiS cell developers, the equilibrium voltage of a LiS cell is constant at approximately 2.15 V over a wide range of the capacity of the cell. As well as posing a problem for the developers of battery management systems, this indicates some complexity in the reaction mechanism, with the exact cause for the flat equilibrium voltage not yet being fully understood.

1.2.6   The Need for a Second Electrode

While a single electrode in an electrolyte can be studied in isolation theoretically, no information can be extracted from such a system experimentally. Taking the example of measuring the voltage across the interface, doing so would require the use of a voltmeter, which measures the difference in the electrochemical potential of the electrons between its two probes. If we first connect one of the probes to the electrode material, this causes a brief rearrangement of the electrons at the interface as the electrochemical potentials of the electrode and the probe materials equilibrate. This leads to the development of a Galvani potential between the two phases but does not lead to a current flow, because there is no sustained electrochemical potential gradient driving electron flow. Since there is no current flow, no voltage is measured.

If the second probe of the voltmeter is immersed into the electrolyte (Figure 1.7a), and if it is electrochemically inactive, there will be no change in μ_{e^-} in the second probe Consequently there will still be no current flow and, again, no voltage measured. However, the situation will be very different if we insert a second, electrochemically active, electrode into the electrolyte and connect the second probe of the voltmeter to that (Figure 1.7b).

Because the second electrode is electrochemically active, the reaction taking place at its surface will cause the same series of changes to take place as happened at the first electrode, culminating in an accumulation (or deficit) of electrons on the electrode surface, the formation of a double layer, and an electrostatic potential difference between the electrode and the bulk electrolyte. When the second probe of the voltmeter is connected to the second electrode, the system will spontaneously try and reduce the free energy

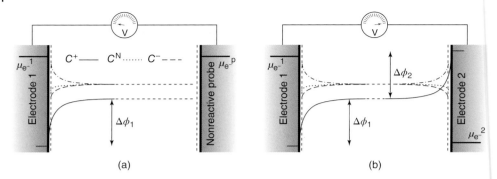

Figure 1.7 (a) Assuming the probes of the voltmeter are electrochemically inert, connecting one probe to an electrode and inserting the other into an electrolyte will yield a reading of zero volts, because no electrochemical potential difference is maintained between the two probes. (b) If the second probe is connected to a second, electrochemically active, electrode, a voltage will be measured, but its value will be defined by the sum of the two double-layer potentials.

by equilibrating the electrochemical potentials of the electrons in the three conductive phases. However, doing so alters μ_{e^-} in both electrodes, placing the reactions out of equilibrium. This causes them to restart, which adds or removes electrons at each electrode as appropriate in order to maintain $\mu_{e^-} = \mu_{Re} - \mu_{ox}$, which then maintains the electrochemical potential difference across the probes of the voltmeter. Because of this, a continuous current will flow, and the voltmeter will register a voltage. Assuming that the voltmeter only passes a small current, the reaction will not move too far from equilibrium, μ_{e^-} in the two electrodes will not change by much, and so the measured voltage is effectively the difference between the electrochemical potentials of the electrodes at equilibrium.

Within the electronically conducting phases, there will still be a variation in the electrostatic potential, because the material's differing Fermi level energies lead to the generation of the Galvani potential. But these do not make a contribution to the flow of current, which is only maintained by the extent to which the electrochemical reactions maintain an electrochemical potential difference between two electrode surfaces. However, the total difference is the sum of two separate processes, neither of which can be measured independently because the flow of current is required to make the measurement.

Although we can say from a theoretical perspective that the total measured potential difference must be the sum of the two double-layer potentials, the fact that we can experimentally only ever measure pairs of potentials poses a problem, because it means that the properties of one electrochemical reaction can only ever be measured with respect to a second, and we have no way of determining what proportion of the total is due to each.

To resolve this, a particular reaction, namely, the H/H^+ redox couple in the standard hydrogen electrode is defined as being as having zero potential. All other reaction potentials are then measured relative to this, with the entire measured potential difference being assigned to the second reaction. These are the values which are tabulated as the standard reduction potentials of the reactions. With regard to the Nernst equation, this still holds because although we might change the activities of the species involved in the second reaction, the activities of the species in the standard hydrogen electrode are assumed to remain such that its potential remains at the defined value of zero.

Finally, it should be possible to see that unless the electrodes are close enough that double layers overlap, the electrode separation should have no direct effect on their equilibrium potential difference, because the formation of the two double layers effectively occurs independently. In a closed system, however, the changing concentrations of the active species in the bulk due to one reaction may alter the electrochemical potentials of the species involved in the other reaction, for example through species interactions. In this case, the equilibrium positions, and therefore the double-layer potentials, of both reactions may change.

1.3 Allowing a Current to Flow

Rather than just driving a trickle current through a voltmeter, the purpose of building an electrochemical cell is to drive larger currents which power an electronic device. As indicated by the example of a voltmeter, this can be achieved simply by connecting the cell electrodes across an electric circuit, which permits a spontaneous flow of electrons as the system attempts to remove the gradient in μ_{e^-} between the two electrode surfaces. However, this flow of current, being much larger than that which passes through a voltmeter, significantly disturbs μ_{e^-} in the electrodes, leading to a decrease in the cell voltage.

The current I passing through a wire depends on its resistance R and the voltage V across it according to Ohm's law

$$V = IR \tag{1.18}$$

If the resistance is fixed, then the fact that the voltage varies means that the current also changes, which makes discussing the electrode processes slightly more complex. To simplify this (and to match how cells are typically tested experimentally), we assume that the external circuit contains a variable resistance which is varied in such a way that the current remains constant. In this way, the rate at which the current alters μ_{e^-} in the electrodes is constant.

Electrons move from where their electrochemical potential is largest to where it is smallest or, equivalently, from the electrode with the more negative electrostatic potential to the one with the more positive electrostatic potential, i.e. from the anode to the cathode. Note that this does not necessarily tell us the sign of the net charge on the electrodes, only that the electrochemical potential is different in each.

This flow of electrons (at a constant rate) causes μ_{e^-} at the electrodes to change, decreasing at the anode and increasing at the cathode. As can be seen from Eq. (1.8), this alters ΔG_{RXN} for both reactions, making it positive at the anode and negative at the cathode, placing both reactions out of equilibrium. In response, the reactions restart, and they do so in such a way that they try and maintain μ_{e^-} on their respective electrodes. As with the connection of the voltmeter, this reaction process allows for the continual flow of electrons. However, if the reactions are unable to supply or remove electrons from the electrode surfaces at the same rate as the current moves electrons between them, the difference between the electrochemical potentials of the electrodes will decrease, i.e. the cell voltage will decrease as the current flows.

The extent of the decrease in the cell voltage depends on a number of physical factors, but all affect the rate of the reaction through a limited number of mechanisms, predominantly by altering the actual conversion rate or by altering the rate of transport of the active species. The total reduction in cell voltage during the passage of a current is called the overpotential, and the two main contributors are therefore called the reaction overpotential and the transport overpotential.

1.3.1 The Reaction Overpotential

Although the passage of a current alters ΔG_{RXN}, causing one of the reaction directions to be favored, there remains an intrinsic energy barrier to the reaction. This can be described by transition state theory, in which the active species in a reaction are connected via an unstable intermediate transition state species with a larger electrochemical potential than either active species, as shown in Figure 1.8. Any active species particle which gains enough energy may enter the transition state, but because it is unstable it will decay back into one of the reagents.

The rate of reduction depends on the temperature-dependent frequency with which the oxidant state attempts to enter the transition state, together with the energy difference between the two states, $\Delta G_1^{\ddagger}$. Similarly, the rate of oxidation depends on the frequency with which the reductant state attempts to enter the transition state as well as its energy difference $\Delta G_2^{\ddagger}$. Since the transition state energy is the same in both cases, a difference between $\Delta G_1^{\ddagger}$ and $\Delta G_2^{\ddagger}$ implies that one species is likely to form the transition state more frequently than the other (in the figure, this would be the oxidant state), ultimately leading to the reaction favoring the formation of the species with the larger $\Delta G^{\ddagger}$.

As the reaction depicted proceeds, $\Delta G_1^{\ddagger}$ increases due to the loss of active species and $\Delta G_2^{\ddagger}$ increases due to the formation of reductant. When the two values are equal, the forward and reverse rates are the same, so the net reaction stops. This coincides with ΔG_{RXN} becoming zero, which we already know to be the equilibrium requirement for the reaction. If the cell is already in equilibrium when a current is drawn, the flow of electrons alters the total electrochemical potential of the oxidant species, changing $\Delta G_1^{\ddagger}$ and biasing the direction of the reaction depending on whether $\Delta G_1^{\ddagger}$ shrinks or grows. Because the current is drawn continuously, this bias will continue to grow until $\Delta G_1^{\ddagger}$ is sufficiently large that the reaction occurs at the same rate as the electron flow, at which point there will be no further change in μ_{e^-}.

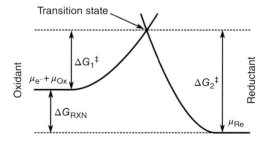

Figure 1.8 According to transition state theory, the oxidant or reductant species must acquire enough energy to enter the transition state before being converted to the other species. The difference between the energy of the two states and the transition state ($\Delta G_i^{\ddagger}$) determines the likelihood of this occurring, and the difference between the two $\Delta G_i^{\ddagger}$ values determines the net reaction direction.

The extent by which $\Delta G_1^{\ddagger}$ has to change depends on how large the energy of the transition state is relative to the oxidant and reductant states: the larger it is, the larger $\Delta G_1^{\ddagger}$ will have to become before the reaction rate matches current. If we assume that the electrochemical potentials of the chemical species remain constant at the surface, then all of the change in $\Delta G_1^{\ddagger}$ comes from a change in μ_{e^-}. The respective changes in μ_{e^-} at the two electrodes combine to reduce the measured cell voltage.

1.3.2 The Transport Overpotential

As with the change in μ_{e^-} due to the consumption and formation of electrons, when the reactions are biased in response to the current, μ_{Ox} and μ_{Re} also change because the reactions consume one species and produce the other. On their own, these changes act to decrease ΔG_{RXN} as the reaction proceeds, countering the effect of the changing value of μ_{e^-} and slowing the reaction rate.

At one extreme, if the reactant is not replenished and the product not removed, then the changes to μ_{Ox} and μ_{Re} reduce ΔG_{RXN} toward zero, slowing the rate to zero. Because the lack of a reaction rate does not stop the current from passing between the electrodes, the μ_{e^-} values are driven ever further from their equilibrium positions and the voltage falls rapidly to zero. At the other extreme, if the reactant is instantaneously supplied to the surface and the product instantaneously removed, then μ_{Ox} and μ_{Re} do not change from their bulk values and only the reaction overpotential contributes to the voltage losses in a cell under load (although there would still be a decrease in the cell voltage with time, because the Nernst potential still decreases).

The reality lies between these two limiting cases, with some contribution to the overpotential caused by the need for the species' electrochemical potentials to be out of equilibrium in order to feed the reactions. For steady-state transport, the flux J_i of a species i in solution is defined by its electrochemical potential gradient according to the Maxwell–Stefan equation

$$\nabla \mu_i = k_B T \sum_{j \neq i} \frac{c_i c_j}{c^2 D_{ij}} \left(\frac{J_j}{c_j} - \frac{J_i}{c_i} \right) \tag{1.19}$$

where D_{ij} is the Maxwell–Stefan diffusion coefficient – the extent to which a flux of one species affects another. Within this formalism, the flux of a species is dependent on the fluxes of all other species in the system; but the complexity of solving this equation means that the simpler Fick's law, which removes the interspecies flux dependencies, is more commonly used:

$$J_i = -D_i c_i \nabla \mu_i \tag{1.20}$$

The current will continue moving the electrochemical potentials of the electrons further away from the reaction equilibrium until the reaction rate is the same as the rate of electron flow. This requires the transport of the reactant to the surface and the product away from it at the same rate. From Fick's law, it can be seen that this requires an electrochemical potential gradient, which means that, relative to the bulk, there has to be a certain amount of product accumulation and reactant deficit at the surface for transport to occur. The changes to μ_{Ox} and μ_{Re} at the surface contribute a change to ΔG_{RXN} which offsets the effect of the change in μ_{e^-}, slowing the reaction rate. Since the external current is still constant, it drives the reaction further from equilibrium by altering

μ_{e^-} until its rate is sufficient to match the current. This change in μ_{e^-} is measured as a further decrease in the cell potential, and this contribution to the total overpotential is the transport overpotential.

1.3.3 General Comments on the Overpotentials

The voltage of a cell under load depends on how effectively the electrochemical reactions are able to maintain the electrochemical potentials of the electrons in the two electrodes. When a constant current passes between the electrodes, the electrochemical potential in the anode will decrease toward the value in the cathode, while the value in the cathode will increase toward that in the anode. This happens independently of the reactions, which are driven out of equilibrium by the changes to μ_{e^-}. The further a reaction is from equilibrium, the faster it occurs, so the cell voltage when a load is applied is simply determined by how far the current drives the system out of equilibrium before all other processes occur quickly enough to match the current.

1.4 Additional Processes Which Define the Behavior of a LiS Cell

So far in this chapter we have developed the basic picture of a simple electrochemical cell, describing how a reaction at one of each of the electrodes maintains an electrochemical potential difference of the electrons in two electrodes, permitting a continuous flow of current. Even within this simple system, the relationship between the current, the state of the cell, and the voltage is not straightforward; but when we start to look at a more complex system like the LiS cell, the interactions between processes as well as morphological changes can also affect the cell performance. In this final section, we consider some of these effects and the mechanisms by which they impact the cell voltage.

1.4.1 Multiple Electrochemical Reactions at One Surface

The overall sulfur reduction process stated in Eq. (1.2b) does not occur as a single step, but rather involves a number of elementary reaction steps in which the higher-order polysulfides are reduced to lower-order species. Furthermore, because species of different orders may be present simultaneously in the electrolyte, it is possible for one or more electrochemical steps to occur at the same time. However, because all electrochemical reactions share the same pool of electrons on the cathode surface, which means that μ_{e^-} must be the same in each process, the reactions alter each other's behavior.

Considering the simpler example case of the following two simultaneous reactions

$$A + e^- \rightleftharpoons A^- \tag{1.21}$$

$$B + e^- \rightleftharpoons B^- \tag{1.22}$$

under the assumption that initially $\Delta G_{A \to A^-} < \Delta G_{B \to B^-} < 0$, so both reactions favor reduction, but the first is further from its equilibrium state. Both reactions initially run in the direction of reduction, removing electrons from the electrode and decreasing μ_{e^-}. Consequently, ΔG_{RXN} for both reactions increases toward zero. Because the second reaction initially starts closer to equilibrium, $\Delta G_{B \to B^-}$ reaches zero first, meaning its rate becomes zero. However, with $\Delta G_{A \to A^-}$ still being negative, the first reaction continues removing electrons from the surface, further decreasing μ_{e^-}. This makes $\Delta G_{B \to B^-}$ positive, pushing the second reaction out of equilibrium in such a way that it begins to oxidize B^-, adding electrons to the surface.

The two reactions now work in competition with each other, with the second replacing some of the electrons consumed by the first, suppressing the continued decrease of μ_{e^-}. At the same time, the difference $(\mu_{A^-} - \mu_A)$ continues to shrink, because A is still being converted to A^-, while $(\mu_{B^-} - \mu_B)$ grows as it is driven in reverse. Eventually, when the two difference terms are both equal to μ_{e^-}, the reactions will both reach equilibrium. However, that equilibrium point will be in a different place to where each reaction individually would have it. Specifically, much more A will have been converted to A^- than if that reaction had occurred alone and, conversely, much less B will have been converted to B^-. At the same time, μ_{e^-} will lie somewhere between the values it would have if each reaction occurred alone, so the electrode potential for this mixed reaction lies somewhere between the values for the two reactions.

The response of an electrode with a mixed potential to the passage of current can be difficult to predict. The current at the cathode increases μ_{e^-} by the same amount for the two reactions, and the reaction with the most facile kinetics and smallest transport limitation will provide the majority (but not all) of the reaction current. This could not only be either reaction; but the dominant reaction will change as the passage of current continues, due to the changing state of the electrolyte.

The reaction mechanism at the cathode in a LiS cell is also complicated by the fact that reactions not only share electrons but also one of the chemical species. Since it is the same high-order polysulfides being reduced to low-order species, the products of one reaction step are the reactants in a later step (although there may be an intermediate chemical step, so the connection may not be direct). Such a link places an additional constraint on the progression of the reactions, because now the progression of the first reaction after the second has equilibrated causes μ_{e^-} to decrease but μ_{Ox} to increase, so the effect on the second reaction is less clear. Determining how the reactions interact becomes increasingly difficult as the complexity of the reaction mechanism grows, often meaning that modeling tools are typically required to understand how they behave.

1.4.2 Chemical Reactions

The electrochemical reactions occurring at the cathode side of a LiS cell are not the only processes involved in the reaction mechanism. As well as the precipitation and dissolution steps at either end, chemical association and dissociation reactions play a role [1, 34, 35], and while these reactions do not directly alter the electrochemical potential of the electrons, they will alter the reaction equilibria and dynamics by changing the electrochemical potentials of the other species involved.

The driving force behind a chemical reaction is the same as that for an electrochemical reaction, being the difference between the electrochemical potentials of two or more active species which are able to interconvert. For a general reaction of the form

$$\sum_i a_i A_i \rightleftharpoons \sum_j b_j B_j \tag{1.23}$$

where the upper case letters represent chemical species and lower case letters are their stoichiometric coefficients, the free energy of the reaction has a form similar to that of an electrochemical reaction:

$$\Delta G_{RXN} = \sum_j b_j \mu_j^B - \sum_i a_i \mu_i^A \tag{1.24}$$

If this value is positive, the reaction spontaneously runs to the left; while if it is negative, the reaction runs to the right. The consumption of one set of species and the production of the others cause ΔG_{RXN} to decrease until it reaches zero, at which point the reaction is in equilibrium. Chemical reactions do not involve free electrons, and so are not limited to occurring at the surface, although electron transfer may occur, for example, when redox mediators are involved.

If a chemical reaction is coupled to an electrochemical reaction through a shared active species, it places a constraint on the equilibrium electrochemical potential of that species while also affecting the transient response of the electrode to a current. Take, for example, the initial steps in the sulfur reduction process in which the S_8^{2-} ion is formed, which might occur as the following two-step process

$$S_{8(s)} \rightleftharpoons S_{8(d)} \tag{1.25a}$$
$$S_{8(d)} + 2e^- \rightleftharpoons S_8^{2-} \tag{1.25b}$$

Starting with the equilibrium state, the first reaction stops when $\mu_{S_{8(d)}}$ is equal to that of the solid phase. This is more commonly thought of in terms of the species activities, Eq. (1.17), where the solid phase is always in its standard state and so has unit activity. The second reaction stops when $\mu_{S_{8(d)}} + \mu_{e^-} = \mu_{S_8^{2-}}$; but with the first reaction fixing $\mu_{S_{8(d)}}$, the second compensates by varying how far it proceeds. Since this alters μ_{e^-}, there will be a change in the equilibrium electrode potential.

The effect of the dissolution step on the electrochemical reaction rate is perhaps more pronounced. As discussed, an applied current will continue altering μ_{e^-} until the electrochemical reaction rate matches the electron flow rate. Ignoring the kinetic and transport limitations, the mechanism for how the dissolution reaction alters the cell potential is that the reaction causes a deficit of the solute at the surface which is not replaced quickly enough by the dissolution reaction. The current continues increasing μ_{e^-} (because we are at the cathode), pushing the reaction further from equilibrium, decreasing the cell voltage, and causing a larger deficit of solute in the electrolyte. Eventually, the solute electrochemical potential becomes small enough that the dissolution reaction replenishes the solute as quickly as it is consumed by the electrochemical reaction, at which point μ_{e^-}, and therefore the voltage, stabilises.

While the example given relates to a dissolution step, a similar behavior results for all chemical steps. By fixing the electrochemical potentials of the electrochemically active species relative to other species in the solution, they alter the equilibrium position of those reactions, altering the cell potential. Under load, if the chemical reaction rate is intrinsically slow, the cell voltage decreases because the system has to move further from equilibrium in order to make the reaction happen sufficiently quickly to supply the reactants to the electrochemical reactions or to remove their products.

One point in the LiS discharge curve where the effect of a chemical reaction step may be evident is in the transition between the two voltage plateaus. There is commonly a small recovery in the voltage at this point, which has been associated with the onset of Li_2S precipitation [36, 37]. In order for precipitation to occur, nucleation of the precipitate species onto the carbon structure must occur, which has a larger energy barrier than precipitation onto existing precipitate. Because nucleation is slow, it acts as a chemical reaction bottleneck which the system has to move further away from equilibrium to overcome, suppressing the voltage. Once nucleation has occurred, however, the bottleneck is removed and the reactions become more facile. This allows the system to move back toward the equilibrium position, leading to a slight recovery in the cell voltage.

1.4.3 Species Solubility and Indirect Reaction Effects

The extent to which a particular species can be dissolved in a solvent depends on the extent to which the dissolved species are stabilized by the solvent molecules, which depends on the species–solvent interaction. However, it also depends on the presence of other species in the electrolyte, which may have a preferential interaction with the solvent molecules or interact more or less favorably with the dissolved species, allowing more or less of the dissolved phase to be formed.

All of these interactions alter the electrochemical potentials of the active species in a particular reaction, regardless of the reaction type. Because of this, not only do the adjacent reactions in the LiS reaction mechanism directly affect each other but the presence of the species formed anywhere in the mechanism will also affect the electrochemical potentials of all other species in the mechanism, changing their reaction rate in the process.

Perhaps the easiest way to see this is by considering what happens when the first two reaction steps stated in Eqs. (1.25a) and (1.25b) are extended to the creation of the S_4^{2-} ion, which we assume occurs as a single step

$$S_8^{2-} + 2e^- \rightleftharpoons 2S_4^{2-} \tag{1.26}$$

Although this reaction removes S_8^{2-} from the system, in theory allowing Reaction (1.25b) to progress further, it does not necessarily reduce the electrochemical potential of the $S_{8(d)}$ species because, for instance, there may be an unfavorable interaction between that species and the S_4^{2-} ion. One example for this may simply be that two of the lower-order polysulfide ions occupy more space than the one higher-order ion, so μ^{ex} increases for all species due to the EVI. As a result, $S_{8(s)}$ becomes less able to dissolve into solution – its solubility is reduced by a reaction farther down the reaction mechanism.

1.4.4 Transport Limitations in the Cathode

So far, we have assumed that the current which flows externally to the cell is spread evenly across the electrode–electrolyte interface, rendering all parts of the interface equivalent and meaning that all processes occur at the same rates everywhere. The structure of the LiS cell indicated in Figure 1.1 should, however, suggest that this is not the case. The fact that the cathode is a porous structure extending away from the anode means there are variations in the distances species have to travel to reach active sites, as indicated by Figure 1.9. This means that the transport overpotential changes throughout

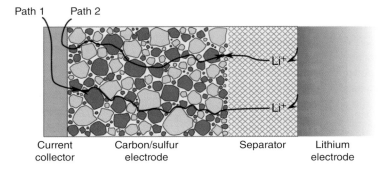

Path 1   Path 2

Current   Carbon/sulfur   Separator   Lithium
collector   electrode   electrode

Figure 1.9 Indication of two paths for the internal flow of current through a cell. The higher conductivity of electrons in the carbon phase than of ions in the electrolyte phase means that path 2 has a lower overall resistance than path 1. Provided the necessary species are present for electrochemical conversion on path 2, this will be the favored route of charge movement through the cell.

the cathode, being smaller for reactions taking place near the anode and larger for reactions occurring at the other side of the cathode (the distance the electrons have to travel and the resistance of the carbon phase introduces a transport overpotential for these too; but this is significantly smaller than that for the chemical species, so we ignore it here).

The transport overpotential of path 2 is initially smaller than that of path 1, meaning that path 2 represents the preferential site for reactions to take place. Because of this, the separator side of the cathode works harder than the current collector side. However, as the polysulfide reactants are depleted near the separator, the dominant region for reactions has to move ever deeper into the cathode, leading to an increasingly large transport overpotential as the cell discharges.

Because the cathode is a porous structure, the rate of species transport is lower than that in a bulk electrolyte, being affected by the porosity – the fraction of pore space in the cathode volume – and the tortuosity – how direct the route between two points in the structure is. In a less porous or a more tortuous structure, net species transport is slower and so the transport overpotential is larger, corresponding to a lower operational cell voltage.

1.4.5 The Active Surface Area

The total current being drawn from a cell is distributed across the active surface area of the electrode. The higher the surface area of the electrode, the lower the required current density at each point, and therefore the slower the reactions at each point have to be. Since the reactions are able to take place more slowly, they operate closer to their equilibrium positions, and the cell operates at a higher voltage than it would if the surface area was smaller. This is part of the reason for using a high surface area carbon to build a cell.

However, one of the main problems at the end of the discharge of the cell is the formation of Li_2S, which precipitates onto the surfaces of the carbon. As described previously, the rate of this reaction affects the reaction mechanism because it affects the electrochemical potentials of the active species in solution; but it also has a second effect, which is that the precipitate covers up the electrochemically active surface. In doing so, it limits the rates of the electrochemical reactions, preventing those regions of the cathode with deposits from contributing to the total required current.

Because some parts of the cathode are now inactive, the total reaction rate slows, allowing the current to cause μ_{e^-} to increase. This forces the remaining active surfaces further from their equilibrium positions, increasing the reaction rates to compensate at the expense of a lower cell voltage. The formation of precipitate therefore not only actively reduces the cell voltage during a single cycle but also likely contributes to the faster degradation of the remaining surfaces, because they have to work harder to compensate.

1.4.6 Precipitate Accumulation

As well as altering the active surface area, the accumulation of precipitates alters the volume of the pore space in the cathode, changing the porosity during the operation of a cell. Furthermore, if enough precipitate builds up, pores may become entirely blocked, disabling pathways through the cathode and increasing the tortuosity, and potentially rendering regions of the cathode inaccessible to active species. The effect of the increased porosity and tortuosity is to increase the transport overpotential, leading to a loss in the cell voltage; and if sections of the cathode become inaccessible, there will be step changes in the available active surface area, having the same end effect.

1.4.7 Electrolyte Viscosity, Conductivity, and Species Transport

From Eq. (1.19) or (1.20), it can be seen that the diffusion coefficient of the active species plays a role in the transport limitation, since it partly determines the electrochemical potential gradient required to generate the necessary species flux to supply the reactions. The diffusion coefficient is related to the viscosity η; and while the relationship is often complex, for simple spherical particles it is of the form

$$D = \frac{k_B T}{8\pi\eta r^3} \tag{1.27}$$

Unfortunately, for multicomponent electrolytes and nonspherical particles it is difficult to predict the viscosity, which depends on how the species in the electrolyte interact and evolve in time. For a LiS cell, determining the electrolyte viscosity presents an even more challenging problem because the composition of the electrolyte varies during the operation of the cell. On the one hand, we do not necessarily know the true composition during the operation; and on the other, it is almost impossible to determine the viscosity

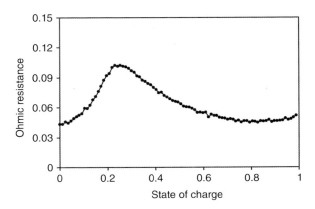

Figure 1.10 Representative resistance variation of a LiS cell as a function of the state of discharge. During the first quarter of the discharge, corresponding to the high plateau, the resistance of the cell increases, before decreasing during the low plateau.

of the particular composition anyway, because of the same polysulfide interconversion which prevents individual reduction potentials being determined.

However, one indication that the electrolyte viscosity changes during the operation of a cell comes from the series resistance of a cell, an example plot of which is shown in Figure 1.10. This can be determined from high-frequency electrochemical impedance spectroscopy data or by the galvanostatic intermittent titration technique, and is an indication of the electrolyte resistance. In both cases, the resistance of a cell is seen to increase up to the transition between the high and low plateaus and then to generally decrease during the low plateau.

The variation in the electrolyte resistance indicates that the viscosity is changing, increasing the transport limitation on the electrochemical reactions. The cause for this changing resistance is likely to be a steady increase in the total concentration of the electrolyte, driven by the reactions converting the fewer higher-order polysulfides into greater numbers of the lower-order polysulfides. This total concentration increase takes place until nucleation of the final precipitate species occurs, at which point the bottleneck on the reaction mechanism is removed and the species clogging up the electrolyte can be removed by precipitating out of the solution.

1.4.8 Side Reactions and SEI Formation at the Anode

Although we have focused mostly on the processes which affect the cathode, since this is where much of the reaction mechanism takes place and is where many of the factors affecting the cell behavior occur, there are a number of processes at the anode which are worth consideration from a fundamental perspective.

The primary anode reaction, and the only one which is essential as far as the mechanism is concerned, is an electroplating/electrodissolution reaction in which the reductant species is lithium from the solid phase of the anode itself. When this is reduced, a lithium cation is released into the solution and an electron released to the remaining electrode, so the electrode itself forms the phase in which the free electrons reside.

Lithium is highly reactive, which is part of the reason that it is a desirable electrode material because it contributes significantly to the high voltage of a lithium–metal cell. However, it also means that it will react with many other species which come into contact with it, typically forming unwanted side products. As well as not contributing to the electrochemical behavior of the cell, these side reactions may be irreversible, consuming lithium and other components of the cell, leading to capacity fade with cycling, cell degradation, or both.

One of these side reactions is between the solvent and the lithium, and the loss of solvent causes severe problems for the cell because it increases the electrolyte viscosity and decreases the quantity of electrochemically active species which can be in the solution phase at any one time. However, the lithium–solvent reaction typically forms a solid–electrolyte interface (SEI) on the anode surface, slowing the side reaction by covering up the lithium. At the same time, the primary electrochemical reaction now has to occur through the SEI, with lithium ions passing through the layer on their way to or from the reaction site. This adds a transport limitation to the reaction, potentially slowing the kinetics and lowering the cell voltage.

The SEI is essential to minimize the loss of the solvent, but the properties of the SEI are critical to the performance of the anode. Because of this, much research is undertaken trying to develop either solvents which form good SEIs or manufacturing artificial SEIs by a range of methods.

1.4.9 Anode Morphological Changes

As well as SEI formation at the anode, the stripping and plating of the electrode during operation leads to morphological changes taking place at the anode surface. This is because microscopic initial variations in the surface structure of the anode makes some places more favorable for stripping or plating than others, acting as seed points for both processes.

A detailed consideration of how and where these processes occur is beyond the scope of this chapter; but in the context of the effects on the reaction rates, it is worth noting that these changes alter the surface area of the anode and introduce some porosity to its structure, particularly if the plating leads to mossy lithium growth. Although this may help from the perspective of increasing the reactive surface area, the same increased surface area will encourage additional SEI formation and lead to structural weakening of the lithium, including the possibility of disconnected regions which cease to be electrochemically active. This can result in the remaining surface having to compensate for the inactive surfaces, leading to more rapid degradation.

1.4.10 Polysulfide Shuttle

As with the lithium being highly reactive at the anode, the high electrochemical potential of electrons on the anode surface means that the lithium anode is able to readily reduce any high-order polysulfide species which find their way to the anode surface. This unwanted process represents one of the major challenges facing the commercialization of a LiS cell, and while it is discussed in more detail elsewhere in this book, here we consider the fundamental mechanism which drives the process.

The problem with the reaction taking place is most easily seen when the cell is charged. Toward the end of charge, the cell produces increasingly large quantities of high-order polysulfide species; but because these species have to accumulate in solution before the solid-phase sulfur precipitates out again, they will diffuse away from the cathode, into the separator, and to the anode surface. Because of the small μ_{e^-} on the anode, these high-order polysulfides are reduced before diffusing back to the cathode to be reoxidized. This whole process generates an internal parasitic current, in which the polysulfide species transport electrons from the anode to the cathode inside the cell. This process is known as the polysulfide shuttle and is the characteristic unwanted side reaction in a LiS cell.

The effects of this process are numerous. First, it can prevent complete charging of the cell: if the shuttle rate matches the charging rate, the charging current is neutralized by the shuttle current. Second, it causes a mixed potential at the anode, contributing to a lowering of the cell voltage. Third, it causes significant internal heating in the cell, which can cause major problems for cell operation, particularly at the pack and module level. Fourth, if the cell is left in a charged state, it contributes to self-discharge of the cell, making it unfavorable even for cell storage. All of these effects of the polysulfide shuttle on the cell behavior are unwanted, so eliminating it from the cell is of paramount importance.

1.5 Summary

The LiS cell functions because of the fundamental tendency of a system to spontaneously move to its lowest available free energy state. For this system, this involves the conversion of higher energy chemical species to lower energy species, releasing electrons to one of the electrodes and consuming them from the other. When no current is drawn from the cell, these processes reach an equilibrium state in which the free energy is minimized, but there is an electrochemical potential difference between the electrons in the terminals of the cell, measured as the cell voltage.

The existence of the cell voltage implies that the free energy is not at a global minimum, which is further evidenced by the flow of electrons when the terminals of the cell are connected. In allowing electronic current to flow, lower free energy states become available by the further conversion of species along the reaction mechanism, and so the system spontaneously moves in that direction. These reactions act to maintain the electrochemical potentials of the electrons in the electrodes, and so a continuous current flows.

For a given flow of current, the voltage of the cell will be reduced, relative to the equilibrium state. These voltage losses occur because the electrons lost or gained at each electrode cannot be replaced or removed instantaneously. The rates of the reactions are dependent on how far the system is from equilibrium and how fast the reaction and species transport processes occur, and it is the specifics of how these processes are affected by the active species, electrolyte composition, and structure of the cell that the specific behavior of a particular cell is determined.

In the case of a LiS cell, a large number of factors contribute to the complexity of the system, mostly arising from the large number of active species and the types of reaction they undergo. The interplay between these processes and the changes to the cell structure upon cycling mean that a wide variety of approaches are necessary to build up a full picture of what is taking place within the cell, how each process affects the overall performance, and, ultimately, how to use this information to improve the performance of the technology.

References

1 Wild, M., O'Neill, L., Zhang, T. et al. (2015). Lithium sulfur batteries, a mechanistic review. *Energy & Environmental Science* 8: 3477.

2 Gorlin, Y., Siebel, A., Piana, M. et al. (2015). Operando characterization of intermediates produced in a lithium–sulfur battery. *Journal of the Electrochemical Society* 162 (7): A1146–A1155.

3 Wang, Q., Jianming, Z., Wailter, E. et al. (2015). Direct observation of sulfur radicals as reaction media in lithium sulfur batteries. *Journal of the Electrochemical Society* 162: A474–A478.

4 Cuisnier, M., Cabelguen, P.E., Evers, S. et al. (2013). Sulfur speciation in Li–S batteries determined by operando X-ray absorption spectroscopy. *Journal of Physical Chemistry Letters* 4: 3227–3232.

5 Kawase, A., Shirai, S., Yamoto, Y. et al. (2014). Electrochemical reactions of lithium–sulfur batteries: an analytical study using the organic conversion technique. *Physical Chemistry Chemical Physics* 16: 9344–9350.

6 Roth, R. Fundamental measure theory for hard-sphere mixtures: a review. *Journal of Physics: Condensed Matter* 22 (6): 063102.

7 Hansen, J.-P. and McDonald, I.R. (2013). Inhomogeneous Fluids Chapter 6. In: *In Theory of Simple Liquids*, 4e, 203–264. Oxford: Academic Press.

8 Archer, A.J. (2009). Dynamical density functional theory for molecular and colloidal fluids: a microscopic approach to fluid mechanics. *Journal of Chemical Physics* 130 (1): 014509.

9 Goddard, B.D., Goddard, B.D., Nold, A. et al. (2012). General dynamical density functional theory for classical fluids. *Physical Review Letters* 109: 120603.

10 Paz-Garcia, J.M., Johannesson, B., Ottosen, L.M. et al. (2014). Modeling of electric double-layers including chemical reaction effects. *Electrochimica Acta* 150: 263–268.

11 Minton, G., Purkayastha, R., and Lue, L. (2017). A non-electroneutral model for complex reaction-diffusion systems incorporating species interactions. *Journal of the Electrochemical Society* 164 (11): E3276.

12 Kaplan, T.A. The chemical potential. *Journal of Statistical Physics* 122 (6): 1237–1260.

13 Ross MacDonald, J. (1987). Comparison and discussion of some theories of the equilibrium electrical double layer in liquid electrolytes. *Journal of Electroanalytical Chemistry and Interfacial Electrochemistry* 223 (1): 1–23.

14 Paz-García, J.M., Johannesson, B., Ottosen, L.M. et al. (2011). Modelling of electrokinetic processes by finite element integration of the Nernst-Planck-Poisson system of equations. *Separation and Purification Technology* 79 (2): 183–192.

15 Wang, H., Thiele, A., and Pilon, L. (2013). Simulations of cyclic voltammetry for electric double layers in asymmetric electrolytes: a generalized modified Poisson–Nernst–Planck model. *Journal of Physical Chemistry C* 117 (36): 18286–18297.

16 Yochelis, A. (2014). Spatial structure of electrical diffuse layers in highly concentrated electrolytes: a modified Poisson–Nernst–Planck approach. *Journal of Physical Chemistry C* 118 (11): 5716–5724.

17 Ashcroft, N.W. and Mermin, N.D. (1976). *Solid State Physics*. Saunders College Publishing.

18 Khan, S.U.M., Kainthla, R.C., and Bockris, J.O.M. (1987). The redox potential and the Fermi level in solution. *The Journal of Physical Chemistry* 91 (23): 5974–5977.

19 Bockris, J.O.'M. and Khan, S., U.M. (1993). *Surface Electrochemistry, A Molecular Level Approach*, Chapter 5. Springer US.

20 Newman, J. and Thomas-Alyea, K.E. (2004). *Electrochemical Systems*, 3), Chapter 8e. Wiley Inter-Science.

21 Hush, N. (1958). Adiabatic rate processes at electrodes. I. Energy-charge relationships. *The Journal of Chemical Physics* 28: 962–972.

22 Hush, N. (1968). Homogeneous and heterogeneous optical and thermal electron transfer. *Electrochimica Acta* 13: 1005–1023.

23 Marcus, R.A. (1956). On the theory of oxidation–reduction reactions involving electron transfer. I. *The Journal of Chemical Physics* 24: 966–978.

24 Booth, F. (1951). The dielectric constant of water and the saturation effect. *The Journal of Chemical Physics* 19 (4): 391–394.

25 Gongadze, E. and Iglič, A. (2012). Decrease of permittivity of an electrolyte solution near a charged surface due to saturation and excluded volume effects. *Bioelectrochemistry* 87: 199–203.

26 Ben-Yaakov, D., Andelman, D., and Podgornik, R. (2011). Dielectric decrement as a source of ion-specific effects. *The Journal of Chemical Physics* 134 (7): 074705.

27 Hatlo, M.M., van Roij, R., and Lue, L. (2012). The electric double layer at high surface potentials: the influence of excess ion polarizability. *Europhysics Letters* 97 (2): 28010.

28 Gouy, M. (1910). Constitution of the electric charge at the surface of an electrolyte. *Journal of Theoretical and Applied Physics* 9: 457–468.

29 Chapman, D.L. (1913). LI. A contribution to the theory of electrocapillarity. *Philosophical Magazine* 6 (25): 475–481.

30 Wu, J. and Li, Z. (2007). Density-functional theory for complex fluids. *Annual Review of Physical Chemistry* 58: 85–112.

31 Gillespie, D., Khair, A.S., Bardhan, J.P., and Pennathur, S. (2011). Efficiently accounting for ion correlations in electrokinetic nanofluidic devices using density functional theory. *Journal of Colloid and Interface Science* 359 (2): 520–529.

32 Martín-Molina, A., Hidalgo-Álvarez, R., and Quesada-Pérez, M. (2009). Additional considerations about the role of ion size in charge reversal. *Journal of Physics. Condensed Matter* 21 (42): 424105.

33 Zarzycki, P., Kerisit, S., and Rosso, K.M. (2010). Molecular dynamics study of the electrical double layer at silver chloride–electrolyte interfaces. *Journal of Physical Chemistry C* 114 (19): 8905–8916.

34 Lu, Y.C., He, Q., and Gasteiger, H.A. (2014). *The Journal of Physical Chemistry* 118: 5733–5741.

35 Wang, Q., Jianming, Z., Wailter, E. et al. (2015). *Journal of the Electrochemical Society* 162: A474–A478.

36 Kumaresan, K., Mikhaylik, Y., and White, R.E. (2008). A mathematical model for a lithium–sulfur cell. *Journal of the Electrochemical Society* 155 (8): A576–A582.

37 Waluś, S., Barchasz, C., Bouchet, R. et al. (2015). Lithium/sulfur batteries upon cycling: structural modifications and species quantification by in situ and operando X-ray diffraction spectroscopy. *Advanced Energy Materials* 5 (16): 1500165.

2

Sulfur Cathodes

Holger Althues, Susanne Dörfler, Sören Thieme, Patrick Strubel and Stefan Kaskel

Fraunhofer-Institut für Werkstoff-und Strahltechnik IWS, Winterbergstraße 28, 01277, Dresden, Germany

2.1 Cathode Design Criteria

2.1.1 Overview of Cathode Components and Composition

The active material layer of a Li–S cathode consists of elemental sulfur, carbon materials, and binders. Typically, aqueous slurries of the components are coated onto a current collector in a reel-to-reel coating process, and the solvent is removed by drying to leave active components. Dry layer thickness may be in the range of 20–200 μm. Similar to Li-ion battery cathodes, thin aluminum foils (thickness: 10–20 μm) are used as current collector and as substrate of the active material layer. Primer coatings, based on carbon-filled polymer films, can be applied on the current collector prior to coating the active materials in order to enhance adhesion or decrease interface resistance to the active layer (Figure 2.1).

Due to the low electrical conductivity of sulfur and the related discharge products (lithium(poly-)sulfide(s)), the conductive carbon materials are key components within the active cathode layer. Carbons are typically applied in fractions of 20–50 wt% within the cathode composition. The particulate carbon materials can be added directly to the cathode slurry or provided as a composite by premixing carbons with sulfur. Polymer binders are added in fractions of 3–20 wt% to enhance adhesion and cohesion of the layer. The sulfur fraction is typically in a range of 45–75 wt%. A statistical lithium–sulfur literature review on the important electrode and cell parameters like the sulfur electrode fraction and electrode thickness is given by Hagen et al. [1].

In the sulfur/carbon composite cathodes, carbon materials fulfill multiple roles:

1) The electrochemical conversion of sulfur species takes place through charge transfer at the conductive carbon surface.
2) The carbon materials, supported by the binder, form a porous and conductive scaffold providing free volume for the uptake of sulfur and electrolyte.
3) This scaffold provides mechanical stability in order to withstand volume changes as well as dissolution and precipitation reactions of sulfur species during charging and discharging the battery cell.
4) Through its morphology or chemical surface properties, the carbon binder scaffold further affects the polysulfide retention hindering the mass transport of polysulfides to the anode.

Lithium–Sulfur Batteries, First Edition. Edited by Mark Wild and Gregory J. Offer.

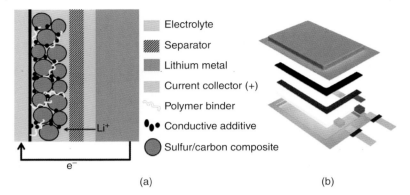

Electrolyte

Separator

Lithium metal

Current collector (+)

Polymer binder

Conductive additive

Sulfur/carbon composite

Li⁺

e⁻

(a)   (b)

Figure 2.1 (a) Scheme of a Li–S cell stack and its components. (b) Scheme of a multilayer Li–S pouch cell.

Through these interactions of the carbon scaffold with the sulfur conversion chemistry, the cathode morphology has a dramatic impact on the Li–S cell performance (including specific energy, rate capability, and cycle life). The influence of the carbon nanostructure as well as the cathode composition and its microstructure on electrochemical properties has been intensively studied over the past decade [2–5] and is discussed in detail in Section 2.1.

2.1.2 Cathode Design: Role of Electrolyte in Sulfur Cathode Chemistry

As the cathode conversion reactions involve dissolution and precipitation of active material species, the electrolyte has a much more active role than just providing ionic transport between anode and cathode [6]. Solvent molecules contribute to solvation of lithium polysulfides (LiPS) and the electrolyte composition determines the solubility of the different polysulfide species. In essence, a quasi-equilibrium between the high surface area carbon scaffold and the polysulfide solution evolves during discharge in which the concentration and species continuously change during discharging [7]. The pore structure and polarity of the scaffold affects the species concentration in solution through complex and dynamic adsorption–desorption processes. In that sense, the electrolyte and its interaction with the carbon scaffold crucially determines the active sulfur mass redistribution within the cathode.

Ether solvents with a high donor number are often used in order to provide a high solvation capability of long chain polysulfides resulting in the typical two-plateau discharge mechanism. A high degree of dissolution of the active material is beneficial for an efficient conversion and high sulfur utilization with this electrolyte concept. Consequently, a rather high amount of electrolyte is required for the cathode chemistry following this mechanism. An additional excess of electrolyte might be required in order to reduce the electrolyte viscosity and enhance the ionic mobility to a certain degree as the dissolution of polysulfides in high concentrations leads to highly viscous fluids. Furthermore, an excess of electrolyte will increase the cycle life of a cell, as the depletion of electrolyte is the most relevant degradation mechanism [8]. However, the electrolyte takes the highest weight and volume fraction of all cell components in a Li–S cell and therefore has a huge impact on specific energy and energy density at cell level. Four milliliter per gram sulfur is estimated to be the maximum electrolyte volume for achieving a competitive specific energy at the cell level, and there is little literature on values below that amount [1]. In order to accommodate the required electrolyte volume, the solid cathode scaffold needs to be designed accordingly to provide a high porosity or swellability. Consequently, reducing the electrolyte content would require an adaption of cathode porosity.

Alternative electrolyte concepts are under investigation, including solid electrolytes [9, 10] and liquid electrolytes with very low polysulfide solubility [11, 12]. These concepts force the active material into the solid state, thereby changing the full conversion mechanism ((quasi)solid–solid conversion). Other cathode design criteria apply for these concepts and the limitations concerning minimum electrolyte content might be shifted to lower values [13]. Similarly, solid-state conversion of sulfur has been demonstrated for cathodes with sulfur molecules immobilized in microporous carbons [14] or through covalent bonding to polyacrylonitrile [15, 16]. Both concepts enable the use of carbonate electrolytes, which are otherwise known to be unstable in the presence of dissolved nucleophilic polysulfides.

As a consequence, the cathode design criteria will strongly depend on the chosen electrolyte concept. In the following, the considerations predominately refer to the widely applied and established ether electrolyte concept, a binary mixture of 1,3-dioxolane (DIOX) and 1,2-dimethoxyethane (DME) as it has been used quite frequently today [6].

2.1.3 Cathode Design: Impact on Energy Density on Cell Level

One of the main motivations to study Li–S cells results from the high theoretical specific energy of 2550 Wh kg^{-1}. Even though the first Li–S prototype cells already exceed the specific energy of Li-ion technology by 30–40%, the recent record value of 425 Wh kg^{-1} is still far away from this theoretical limit.

In the graph in Figure 2.2, the composition and weight distribution of a typical Li–S cell stack is illustrated. While anode and separator take a minor fraction, cathode + electrolyte contribute more than 75 wt% of the cell stack weight. On the other hand, the cathode active material, sulfur, only adds 12 wt%. Increasing the sulfur active material content and its utilization are the main objectives for increasing specific energy.

Decreasing the amount of electrolyte obviously would have the highest impact. However, for a given electrolyte concept, a minimum amount of electrolyte will be required (see Section 2.1.2). Here, a fixed amount of 3 µl mg^{-1} sulfur is assumed.

The sulfur-to-carbon ratio has been studied for various cathode concepts. Maximizing the active material content is a major target in many studies, and sulfur fractions up to 80% have been reported with reasonable utilization. However, as the carbon weight fraction at the cell level is typically below 10%, its impact on the specific energy is rather low. Adding carbon to the cell stack mentioned in Figure 2.2 from 35% cathode fraction to a level of 50%, would decrease the specific energy by less than 4%.

For a given current collector (here, a 15-µm aluminum foil) the weight fraction of that component is mainly determined by the active material loading. Furthermore, separators and, more importantly, anode current collectors (not considered here) also add significantly to the cell mass and their weight fraction is also highly dependent on the active material loading.

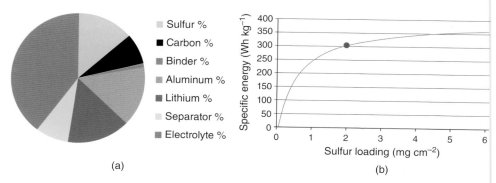

Figure 2.2 (a) Weight distribution of cell components on stack level (estimations based on 2 mAh cm^{-2}, 60% sulfur content in cathode active layer, 35% carbon, 5% binder, 1100 mAh g^{-1}, 4× excess Li, 12-μm polypropylene (PP) separator, 15-μm aluminum current collector). (b) Effect of active material loading on specific capacity. The red spot marks the loading/specific energy estimated for the represented cell stack on the left.

Therefore, depending on the cell design, the active material loading might have a significant impact on the specific energy at the cell level.

Unfortunately, several challenges are related to high active material loadings. From a processing perspective, cathode coating and drying processes are problematic for cathodes with a thickness greater than 100 μm. In addition, high loadings translate into high areal currents at a given C-rate. At the same time, ionic and electronic resistance increases with increasing electrode thickness. As a consequence, cells with high active material loadings will suffer from lowered rate capability.

On the other hand, the weight fraction of current collectors can be significantly lowered by alternative concepts. Lithium metal anodes might be applied without an additional current collector at all and 15-μm aluminum foils might be replaced by even thinner foils or by carbon current collectors in the future. Based on these concepts, high-energy cells with low active material loadings might be envisaged.

In conclusion, optimization of specific energy is only meaningful at the full cell level considering all components. The cell weight is still dominated by inactive materials. Reducing their impact will be the main strategy for future cell development. Overcoming today's limitations concerning the minimum electrolyte content seems to be crucial, but so also is reducing the impact of aluminum current collector and, to a lesser extent, the amount of carbon and binder have the potential to enhance the specific energy. Furthermore, improving the active material utilization could strongly impact cell performance.

2.1.4 Cathode Design: Impact on Cycle Life and Self-discharge

Today, lithium–sulfur chemistry suffers from a relatively low cycle life. High-energy prototype cells typically achieve less than 50 cycles before the capacity drops to 80% of its initial value.

The main degradation phenomena are related to the anode surface reactions, resulting in electrolyte and lithium depletion [8]. However, active material loss and structural degradation have also been observed on the cathode side. Several degradation phenomena can be attributed to the dissolution and precipitation reactions related to sulfur conversion. Through these reactions, sulfur species are redistributed in the cell and precipitates may block the cathode surface or the pores of carbon particles [17–20]. The porous carbon scaffold may collapse through the dissolution of solid sulfur species. Furthermore, dissolved polysulfides may diffuse through the separator to the anode and take part in reductive surface reactions with lithium metal [21, 22]. This results in the so-called redox shuttle or in the formation of solid lithium sulfide or polymeric precipitates [23, 24].

Consequently, capacity loss, low coulombic efficiencies, and self-discharge are observed in cells without any provision against polysulfide diffusion and the shuttle phenomena.

The main concepts to improve cathode stability are as follows:

- Encapsulation or anchoring of polysulfides
- Providing stable, hierarchical porous and conductive scaffolds
- Blocking polysulfide diffusion through interlayers.

These concepts may be based on tailored carbon and binder materials. Examples of various materials and electrochemical results are discussed in detail in Section 2.1.

2.1.5 Cathode Design: Impact on Rate Capability

The rate capability of a Li–S cell depends on various aspects. Ionic and electronic transport as well as charge-transfer phenomena at the anode and cathode contribute to cell resistance and limit the maximum current for charge and discharge. While a high active material load helps increase the specific energy on cell level, the rate capability suffers from high loadings. Increasing the areal loading results in increased areal current (for a given C-rate) and increased resistance due to long and tortuous pathways for the ionic and electronic transport.

A major contribution results from electrolyte resistance which may vary dependent on the state of charge, as dissolved polysulfides significantly increase the viscosity and decrease ionic mobility. Thus, an excess of electrolyte may increase the rate capability of a cell (through diluting and decreasing viscosity), but at the same time it lowers the specific energy through added weight [25]. In this case, charge transfer at the cathode surface is the rate-limiting phenomena, and increasing the carbon surface area or carbon content will positively affect the cell power capability [26].

In conclusion, cell performance of lithium–sulfur cells can be tailored over a broad range through selection of electrolytes and adaption of cathode components. Some aspects in cell design, such as increased active material loading, will favor specific energy at the cost of power capability, while others, such as increasing the accessible, conductive surface area, may increase specific capacity as well as rate performance at the same time.

2.2 Cathode Materials

2.2.1 Properties of Sulfur

Sulfur is a by-product from the oil and coal refinery and available at low costs and at a large scale. Elemental sulfur was extracted from underground deposits by the Frasch process, but nowadays most of the available sulfur stems from natural gas and crude oil desulfurization by the Claus process. Elemental sulfur forms many allotropic and polymorphic structures, including a metastable polymer (plastic sulfur), but its most stable form is the cyclic S_8 molecule. Unlike oxygen, sulfur tends to form single bonds with itself rather than double bonds. This tendency, which leads to the formation of extended rings and chains (catenation) arises due to the relative strengths of p–p σ-bonding and p–p π-bonding. As a result, sulfur aggregates into extended structures and larger molecules.

The commonest and most stable yellow orthorhombic polymorph, α-S_8, consists of crownlike eight-membered rings. Commercial roll sulfur, flowers of sulfur, and precipitated milk of sulfur are all of this form. Orthorhombic α-sulfur has a density of $2.069\,\mathrm{g\,cm^{-3}}$ and close to room temperature [27] is a good electrical insulator 5×10^{-30} S cm^{-1} and excellent thermal insulator. The solubility ranges from 35.5 g sulfur per 100 g for CS_2 to 0.065 g sulfur per 100 g for EtOH and increases with rising temperature. At about 95.3 °C, α-sulfur becomes unstable, the packing of the S_8 rings changes, and monoclinic β-sulfur is formed. This results in a lower density (1.94–2.01 g cm^{-3}), but the dimensions of the S_8 rings in the two allotropes are very similar. The transition is slow even above 100 °C, and this enables a melting point of meta-stable single crystals of α-S_8 to be obtained: a value of 112.8 °C is often quoted but microcrystals may melt as high as 115.1 °C. Monoclinic β-sulfur has a melting point often quoted as 119.6 °C, but this can rise to 120.4 °C in microcrystals or may be as low as 114.6 °C. β-sulfur can be prepared by crystallizing liquid sulfur at about 100 °C and then cooling rapidly to room temperature in order to hinder the formation of α-sulfur. Under these conditions, β-S_8 can be kept for several weeks at room temperature, before it reverts to the more stable α-form. When molten sulfur after heating above 150 °C is cooled slowly, monoclinic γ-sulfur forms, This polymorph consists of S_8 rings like the α- and β-phase, resulting in a higher density (2.19 g cm^{-3}).

The saturated vapor pressure above solid and liquid sulfur depends on the respective temperature. It is known that the composition of the vapor contains all molecules S_n with $2 \leq n \leq 10$ including odd-numbered species. The actual concentration depends on temperature and pressure. In the saturated vapor up to 600 °C, S_6 is the most common species followed by S_6 and S_7 (green color). Between 620 and 720 °C, S_7 and S_6 are slightly more prevalent than S_8. However, the concentration of all three species falls rapidly with respect to those of S_2, S_3, and S_4. Above 720 °C, S_2 is the predominant species, especially at lower pressures. S_3 is a cherry-red, angular molecule-like ozone. The more stable gaseous species is the violet S_2 molecule that has both σ- and π-bonding (violet vapor).

The isotope ^{33}S has a nuclear spin quantum number, $I = 3/2$ and is useful in nuclear magnetic resonance (NMR) experiments. The resonance was first observed in 1951, but the low natural abundance of ^{33}S (0.75%) and the quadrupole broadening of the signals has so far restricted the amount of significant scientific work. However, more results are expected now that pulsed Fourier transform techniques have become available.

Numerous higher polysulfides of electropositive elements (Na, K, Ba, Li) have been characterized. They all contain catenated S_n^{2-} ions and are yellow at room temperature, turn dark on heating, and may be thought of as salts of the polysulfanes. The radical anion S_3^- occurs in the mineral ultramarine, where it is found with Na$^+$ in a cavity formed by coordinated SiO_4 and AlO_4 tetrahedra. The S_3^{2-} ion is bent and isoelectronic with

SCl_2. The S_4^{2-} ion has a twofold symmetry, essentially tetrahedral bond angles, and a dihedral angle of 97.8°. The S_5^{2-} ion also has an approximately twofold symmetry. The S_6^{2-} ion has alternating S–S distances, and a mean bond angle of 108.8° [28, 29].

2.2.2 Porous and Nanostructured Carbons as Conductive Cathode Scaffolds

2.2.2.1 Graphite-Like Carbons

The system of carbon allotropes has a great range of different structures and properties: diamond, for example, has a 3D structure, graphite a 2D structure, fullerene a 0D structure, and carbon nanotubes (CNTs) have a 1D structure. The 1D structure of CNTs can be understood as a hybrid between the structures of fullerene and graphite. Furthermore, there are some linear carbyne compounds in which carbon atoms are bound by alternating single and triple bonds (polyyne) or by continuous double bonds (cumulene). Graphite-like carbons are often used as additives in electrodes as they combine several positive properties such as high electrical conductivity, an acceptable chemical durability, and low material costs. Yet, under the right circumstances, carbon can also act as an electrocatalyst for electrochemical reactions. Graphite has a layered structure, and each layer consists of carbon hexagons which are formed by interconnected sp^2-hybridized carbon atoms with conjugated double bonds with an average length of 0.142 nm. Those individual layers are called graphene and they usually stack up to parallel ordered crystallites (crystallite size >1000 nm). In most cases, they form an ABAB sequence (hexagonal graphite) with a layer distance d_{002} of 0.3354 nm [30, 31].

Shear forces can cause a layer formation with an ABCABC sequence (rhombohedral graphite), which will usually result in a stacking fault within the crystallite. Besides these regular formations, the so-called turbostratic structures are formed (Figure 2.3a): the parallel stacked graphene layers are randomly oriented without a three-dimensional long-range order. By annealing at high degrees (graphitization), a turbostratic structure can be transformed into a regular layer formation. As a result of its 2D structure, the physical properties of graphite are highly anisotropic. If the crystallites consist only of small twisted graphene layers, they typically reach a size between 1.0 and 2.0 nm and have a turbostratic structure [30]. In that case, the layer distance d_{002} is broadened in comparison to that of graphite (>0.344 nm). The 00*l* peak observed in the X-ray powder diffraction pattern are remarkably broadened [31]. Such a behavior is typical for amorphous carbons produced at lower temperatures (<1300 °C). Members of this group of materials like carbon blacks or activated carbons can have very high specific surfaces (>1000 m^2 g^{-1}) and a high porosity (typical density <1.80 g cm^{-3}) [30].

2.2.2.2 Synthesis of Graphite-like Carbons

Graphite-like carbon materials can be produced from organic polymers (carbon precursors) by subjecting them to a thermal treatment at a temperature of 2000 °C [31]. Depending on the carbon precursor, the temperature, the heating rate, density, and the concentration ratio, some components turn gaseous during that process, while the residue undergoes further changes. Due to the fact that some of the C—C bonds are weaker than C—H bonds, the first molecules that turn gaseous at the beginning of the pyrolysis are aliphatic and aromatic molecules with low molecular weight. Accompanied by the release of these hydrocarbons, cyclization and aromatization within the residue, and the polycondensation of aromatic molecules occur. At approximately 600 °C, mostly

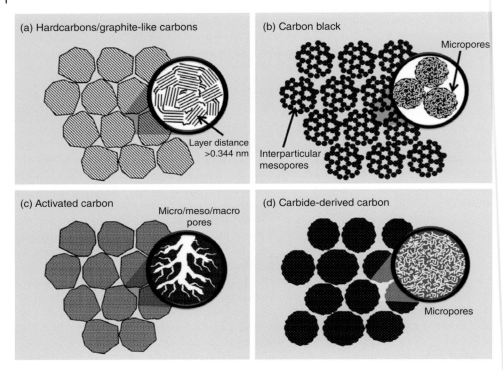

Figure 2.3 Morphologies of different carbons: (a) hard carbons, (b) carbon black, (c) activated carbon, and (d) carbide-derived carbon. Grain size strongly depends on the carbon precursor and post-synthetic treatment of the respective material. For electrode application, grain sizes are adjusted to a size ranging between 1 and 10 μm.

compounds containing carbon, oxygen, and hydrogen atoms are released, usually in the form of CO_2, CO, and CH_4. Depending on the carbon precursor, the residue at that point is either liquid or solid. At temperatures higher than 800 °C, it comes to a clustering of aromatics leading to the separation of H_2 (dehydrogenation). At this temperature, the residue is now a carbon-like solid (carbon material), which usually contains some hydrogen atoms as well as small amounts of heteroatoms (O, N, S, etc.). For the elimination of these impurities, a thermal treatment (carbonation of the pyrolysis product) at temperatures up to 2000 °C is necessary. The carbonation process can be divided into two steps overlapping with the pyrolysis to a certain degree: at first, some hydrocarbon gases emerge (first carbonation) and, subsequently, lighter gases like N_2, H_2 are released (second carbonation).

2.2.2.3 Carbon Black

Carbon blacks are important industrial products generated by the incomplete combustion of gaseous or atomized hydrocarbons. Their structure comprises spherical primary particles that have diameters between 10 and 500 nm. Those primary particles are conjoined to an aggregate by covalent bonds (Figure 2.3b) [32]. Depending on the reaction process and the raw materials used, carbon blacks can be subdivided into furnace black, channel black, lamp black, thermal black, gas black, and acetylene black [33].

The physicochemical properties that result from the production process such as specific surface area, structure (degree and type of aggregation), and impurities (functional groups caused by defects) of the manufactured carbon black determine its capability for electron conduction [30]. Thus, e.g. decreasing the size of the primary particles can cause a number of changes: as the distance between the centers of the particles becomes shortened, a higher surface-to-volume ratio occurs, resulting in a high specific surface area as well as an intense particle aggregation. Due to the higher contact surface between the primary particles, even a very small mass fraction of such a highly branched carbon black structure can be sufficient to allow the electron stream to overcome the critical percolation threshold. Impurities within the carbon black material (especially oxygen) are generally undesired, because they would reduce its electron conductibility. On the other hand, surface functionalities crucially affect dispersion formation. In this context, carbon black properties are highly system dependent, especially with respect to the solvent used for the slurry coating process. Carbon blacks being frequently used as carbon additive in battery electrodes are Super C65, Printex XE-2, and Ketjenblack EC-600JD [34–37].

2.2.2.4 Activated Carbons

In most cases, activated carbons have a granular form and a high quantity of open pores (Figure 2.3c). Both of these factors influence their specific surface as well as the pore size distribution. Especially the pore structure of activated carbons is a crucial parameter for their most important application as adsorbent. Micropores with a diameter <2 nm (IUPAC) are usually achieved by a mild oxidation ("activation") of the carbonized educt. Pores below 0.7 nm are called ultramicropores. Typically, biomass-like plants (wood, coconut shells) are used as a carbon source for this process, but materials like pitch, turf, or soft and hard coal are sometimes used as well. Macropores with a diameter >50 nm (IUPAC) are not important for the adsorption capacity in gas adsorption applications [38, 39], but their presence has a positive influence on the mass transfer, e.g. for adsorption kinetics and during the activation process (generation of microporous wall structures). Mesopores with a diameter of 2–50 nm (IUPAC) form pathways for adsorptive molecules or other species that need to migrate into the micropores. The structure of typical activated carbons is composed of a combination of micro-, meso-, and macropores. While pores with a narrow opening allow the permeation of N_2, O_2, and H_2O, they are, in some cases, not accessible for electrochemical processes (e.g. for the electrolyte molecules and ions) [40–42].

A variety of processes can be used to realize the activation of the carbon material. Physical activation includes oxidation processes by (diluted) gases like oxygen, air, water vapor, or carbon dioxide. The latter process is based on the Boudouard equilibrium, which favors CO formation at a temperature around 400–1000 °C [43].

$$C + CO_2 \rightleftarrows 2CO \tag{2.1}$$

Due to material removal during the oxidation process, closed pores become accessible and existing pores expand, e.g. micropores expand into mesopores. In other words, the oxidation process leads to a shift and widening of the pore size distribution, but also causes a loss in yield. The chemical activation is based on oxidizing agents like phosphoric acid, zinc chloride, or potassium hydroxide induced by a thermal treatment [31]. It can be advantageous to combine the chemical activation with the carbonization process. In particular, the complex KOH activation enables the formation of micropores in a controlled manner [44]. This is realized by the directed intercalation of potassium atoms between individual graphene layers, leading to an irreversible broadening of the layers.

$$6KOH + 2C \rightarrow 2K + 3H_2 + 2K_2CO_3 \tag{2.2}$$

The KOH activation is accompanied by a physical activation that is induced by *in situ* generated water, enabling the production of carbon materials with a specific surface as high as 3000 m^2 g^{-1} [45]. Mass scale production of KOH activation is limited due to corrosion problems under highly caustic conditions at high temperature.

In summary, activated carbons are available with an enormous spectrum of properties and represent the most inexpensive choice for sulfur cathode architectures. However, traditional production pathways are optimized for a high degree of microporosity, and the specific pore volume is often too low to achieve a high sulfur loading. Moreover, their low degree of graphitization may pose limitations in terms of electrical conductivity.

2.2.2.5 Carbide-Derived Carbon

Carbide-derived carbons (CDCs) have narrow pore size distributions and are thus ideal model materials to achieve a fundamental understanding of pore size vs. cathode performance relationships. They are synthesized at high temperatures by a selective extraction of metal- or semi-metal ions from a carbide precursor (ZrC, SiC, Mo$_2$C, TiC, B$_4$C, etc.) [31, 46–48]. For this treatment, supercritical water, halogen gases, such as chlorine (high-temperature chlorination, Eq. ((2.3))), or vacuum decomposition (Eq. ((2.4))) can be used.

$$MC + x2Cl2 \rightarrow MClx + C \tag{2.3}$$
$$MCs \rightarrow Mg + Cs \tag{2.4}$$

Typical CDCs are microporous carbons that have a relatively high specific surface of up to 2800 m^2 g^{-1}. These CDCs are typically microporous (Figure 2.3d, $d < 2$ nm) and the exact size is predetermined by the distribution of carbon atoms within the carbide precursor [49]. By adapting the temperature of the synthesis and/or the post-treatment steps (e.g. activation), the nanotexture of the CDCs can be tailored in order to generate different pore forms and even mesoporous carbons [50]. The synthesis of functional surface groups can be controlled in the same manner [31]. Due to the high adjustability of the CDCs' structural properties, they are attractive active materials for ion transport applications, such as electric double-layer capacitors [48, 49, 51].

2.2.2.6 Hard-Template-Assisted Carbon Synthesis

With the hard template method, it is possible to precisely control the porosity and nanostructure of the carbon material. This synthesis approach relies on the usage of a temperature stable inorganic substance with a nanoscaled architecture that is used as structure directing agent. Nanoparticles (NPs) from SiO$_2$, ZnO, or TiO$_2$ are often used for this kind of synthesis [26, 47, 52, 53]. By nanocasting the carbon precursor with the help of an appropriate infiltration process onto the template, a composite is generated. This method is suitable for various purposes, since there is no need for specific interactions in order to achieve a close precursor/template contact. After the pyrolysis and carbonization, the template has to be removed by a suitable process.

This treatment leaves some voids (pores) within the carbon material. Ideally, those voids show the inverse shape of the used template nanostructure. With the help of this method, various kinds of nanostructures have already been synthesized, for example, CMK-3 (carbon mesostructured by KAIST, CMK) which consists of hexagonally arranged carbon rods and homogeneously distributed 4.5-nm mesopores [54].

2.2.2.7 Carbon Surface Chemistry

The chemical surface composition of carbon materials depends on their production conditions (raw materials, oxidation processes, activation) as well as on their geological origin. Via chemisorption, surface groups with elements like O, S, N, P, and halogens emerge; in most cases, ash with traces of Na, K Ca, Si, Fe, Al, and V can also be detected [38, 55]. Catalytically active metals such as nickel may be present as impurities. They stem either from the precursor or high-temperature alloys used in furnaces for thermal treatment of activated carbons. Depending on the respective (biological) raw material and on the resulting specific surface area of porous carbons, the content of heteroatoms can vary over a broad range. While only small quantities of nitrogen are common, sulfur concentrations (elemental sulfur, carbon–sulfur) of >1% are possible [30]. The oxygen species occur as carbonyl, carboxyl, lactone, quinone, and phenol surface groups or rather structural units. They can have a major influence on the physicochemical properties, e.g. wettability, catalysis/reactivity, and pH-value, as well as on the adsorption affinity for polar impurities, and the residual water content when handled in air.

2.2.3 Carbon/Sulfur Composite Cathodes

The role of the carbon materials in the sulfur cathode is to provide a stable, conductive scaffold for the electrochemical conversion of sulfur species (Figure 2.1). The following structural criteria are expected to impact the electrode performance:

- Accessible specific surface area of the carbon, as this will define the rate of charge transfer;
- Specific pore volume of the carbon (and electrode), as this will define the ability to store sulfur and electrolyte;
- Pore size and shape of the carbon, as this will define the accessibility and mass transport for electrolyte and sulfur species;
- Surface functionality of the carbon, as this will define the wettability and interaction (adsorption/desorption equilibria) with electrolyte and sulfur/sulfide species.

A selection of recently published attractive material approaches is discussed in the following. It should be noted that a fair comparison and objective evaluation of all approaches is difficult due to the complexity of the entire system in terms of the electrode process parameter such as sulfur infiltration conditions, mixing procedures, binders, and additives. Moreover, nonuniform measurement conditions in terms of temperature, charge/discharge rates, and potential ranges essentially hinder an objective comparison of the published systems. The most critical factor is the high variation in electrolyte content, and most studies are carried out with much too high electrolyte amounts, lithium excess, and low sulfur loadings or rather contents in the composites. Capacity values are often calculated on the basis of sulfur mass. For that

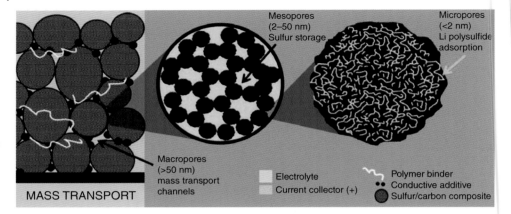

Figure 2.4 Schematic macro- and microscopic structure of a carbon/sulfur composite cathode depicting various types of porosity and their function in the cathode.

reason, the conclusion generated under these conditions often cannot be applied to systems relevant in practice [1, 25, 56–59]. In that context, further efforts are required to establish standardized testing conditions and reporting schemes for sulfur cathodes (Figure 2.4).

2.2.3.1 Microporous Carbons

Purely microporous carbons offer a high specific surface area and a relatively low pore volume. As the pore diameters are in the range of the size of sulfur and polysulfide molecules, a particular strong interaction between the micropore wall and the LiPS is observed [46]. Ultramicroporous carbons ($d < 0.7$ nm) are considered to store sulfur as short S_n species ($n = 2–4$) and thus the formation of S_8^{2-} is suppressed. In order to illustrate the principle, Guo and coworkers used a coaxial core–shell structure consisting of microporous carbon-coated CNTs ∼250 nm in diameter [60]. The very short pore diameter allows the stabilization of short-chained, metastable allotropes S_{2-4} (S_2, S_3, S_4) with a specific surface area of 936 m^2 g^{-1}. However, the unambiguous identification of these species is challenging. The loading of the micropores (0.46 cm^3 g^{-1}) with ∼40 wt% sulfur is accompanied by a remarkable decrease in porosity (82 m^2 g^{-1}, <0.04 cm^3 g^{-1}). These composites can be applied as effective cathode material in carbonate-based electrolytes because polysulfides in solution are suppressed. A capacity retention of 1142 mAh g^{-1} after 200 cycles at C/10 and an excellent rate performance up to 5 C was reported. This result is complementary to the electrochemical properties obtained with 200- to −300-nm carbon spheres with ∼0.7-nm micropores [61]. In this system, the strong adsorption of low-molecular sulfur allotropes and short-chained polysulfides at the carbon surface (844 m^2 g^{-1}) leads to a potential hysteresis, indicating clearly the adsorption energy that needs to be overcome by the electrochemical reaction. A carbon material generated by carbonization of polyacrylonitrile/polymethylmethacrylate with ∼1.3-nm micropores and a specific surface area of 738 m^2 g^{-1} was impregnated by a chemical deposition and thermal post-treatment. With that combination, a high sulfur content of 53.7 wt% can be realized, and in combination with an electrolyte based on LiTFSI in IL/polyethyleneglycoldimethylether (PEGDME) (1 : 1, m:m), a high coulombic efficiency (97.9–98.5%) over 100 cycles at C/20 was achieved [62]. However, the capacity fading from 1080 to 740 mAh g^{-1} after 100 cycles indicates a continuous deactivation or rather incomplete LiPS retention in the relatively large micropores. In order to increase sulfur loading up to ~6.5 mg cm^{-2}, Elazari et al. used a highly porous fabric made of microporous carbon fibers (2000 m^2 g^{-1}, ~1 cm^3 g^{-1}) and ~33.3% sulfur loading [63]. Due to the low sulfur content, a pore volume of 0.57 cm^3 g^{-1} was maintained, leading to good electrolyte wettability and a high capacity of 800–1000 mAh g^{-1} over 80 cycles at ~0.09 C was generated.

On the one hand, microporous carbons are suitable candidates in terms of sulfur utilization and polysulfide retention. On the other hand, the intra-particular entire pore volume with values below 1 cm^3 g^{-1} limits the embedded sulfur content in the composite to values below <55 wt% [64]. Moreover, the electrolyte accessibility in very small micropores is limited or even completely prevented. The associated slow mass transport additionally decreases the sulfur utilization and rate capability.

2.2.3.2 Mesoporous Carbons

In order to create carbon/sulfur composites with high contents (~60–85%) of sulfur being embedded in pores, porous carbons with high pore volumes higher than $\gg$1 cm^3 g^{-1} are required. In 2009, the Nazar group published a CMK-3/sulfur composite that was generated by infiltration of 70 wt% in the mesopores of the CMK-3 carbon scaffold with high values for the specific surface (1976 m^2 g^{-1}) and pore volume (2.1 cm^3 g^{-1}) [21]. Due to the hydrophobic interactions between sulfur and carbon as well as capillary forces, the channel-like mesopores in the range of 3–4 nm could be homogeneously coated with sulfur. To create void space for the electrolyte absorption and volume expansion of the discharge product Li$_2$S, a partial filling of the pores was targeted. In that way, the nanoscaled material morphology of CMK-3 enables the electrical contact of the embedded sulfur, decreases the loss of active material due to polysulfides, and the agglomeration of active material particles. Hence, with this system, stable capacities >800 mAh g^{-1} over 20 cycles at C/10 could be achieved.

Li et al. synthesized carbon particles with varying monomodal pore sizes (7, 12, and 22 nm) and thin (3 nm) pore walls, and evaluated these materials as host material for sulfur [65]. Due to the high pore volume values (1.72, 3.64, and 4.80 cm^3 g^{-1}) and a two-step impregnation process comprising a sulfur/CS$_2$ solution and thermal treatment, sulfur contents as high as 66, 80, and 83 wt% can be realized. A complete sulfur amorphization was only observed for the composite with 66 wt% sulfur content, and the required residual porosity for the discharge product could be seen for all composites. The electrochemical measurements reveal an inverse correlation between sulfur content and starting capacity. The higher the sulfur content, the lower the starting capacity 83 wt%: 1050 mAh g^{-1}, 80 wt%: 1071 mAh g^{-1}, 66 wt%: 1195 mAh g^{-1}. According to Li et al., given that the electrochemical reactions can only take place at the interface sulfur/carbon due to the required electron transfer, large sulfur particles exhibit a lower reactivity. The increased starting capacity of 1250 mAh g^{-1} and a higher reversible capacity of a composite with 50 wt% sulfur loading should underline this thesis.

However, that interpretation appears to be not completely plausible as the discharge profiles of the respective samples differ only regarding the lengths of the lower plateau (Li$_2$S precipitation). For that reason, such phenomena can be explained by a termination due to pore blocking [66] that can be impeded by a higher content of vacant pores or rather a lower sulfur to carbon atom ratio and, hence, lower sulfur loading [57, 67]. Consequently, monomodal mesopores possess not only an important function for the storage of high sulfur amounts but they are also crucial for the mass transport – if not blocked by precipitated particles. In comparison with microporous carbons, the retention and utilization of the active material is lower. One reason for that is the lower specific surface area for the electron transfer and, hence, amount of reaction sites in relation to the high sulfur content [68, 69]. In addition, the larger pore size results in lower adsorption potentials for physisorption.

2.2.3.3 Macroporous Carbons and Nanotube–based Cathode Systems

As purely macroporous carbon scaffolds offer only a relatively low specific surface area and their pore morphology is ineffective for LiPS retention [38, 39], they appear unsuitable as a conductive carbon matrix in sulfur cathodes. However, it could be shown by Watanabe and coworkers that a combination of macroporous carbon and highly viscous electrolyte show stable electrochemical performance [70]. The LiPS mobility is limited by the electrolyte viscosity and the large transport pores maintain the accessibility of the inner particle space even during Li_2S precipitation. Chen et al. used multiwalled CNTs as a core–shell architecture with a thin sulfur coating (57 m.%) for an assembly of a cross-linked secondary structure with macropores between 1 and 5 µm [71]. A functionalization of the CNT surface with carboxylic groups (3 m%) enhanced not only the dispersibility but it also increased the adsorption properties for sulfur species. Given that the electron and ion transport remained intact in the cross-linked porous S/CNT composite, rapid interphase kinetics and high redox activity could be enabled – despite the volume changes. After 100 cycles at ~0.18 C and 200 cycles at ~0.3 C, reversible capacities as high as ~1000 mAh g^{-1} or rather 780 mAh g^{-1} were achieved. The maximum ampacity was at ~0.6 C.

An excellent model system is binder-free vertically aligned CNTs grown by a chemical vapor deposition on metal substrates [72, 73]. The film comprises CNTs with ~100–200 µm length and 7–30 nm diameters, and offers an open intertubular pore structure with high mechanical stability and excellent electric contact to the substrate. Even at high sulfur loadings as high as 70% (estimated thickness of the sulfur layer <50 nm) and without $LiNO_3$ additive, high capacities (1200–1300 mAh g^{-1} and >800 mAh g^{-1} calculated on the composite) in the first 15 cycles at C/13 were possible. The relatively low coulombic efficiency due to the strong LiPS shuttle is attributed to the insufficient LiPS retention in the open pore structure. By adding $LiNO_3$ to the electrolyte, the coulombic efficiency and cycle stability could be increased. Further experiments showed that VA-CNT film (~1 mg cm^{-2}) could be transferred from a nickel foil to a primered aluminum foil by a heat press [25]. Due to the retained open pore structure it was possible to infiltrate sulfur contents between 21.4% and 80.4% (corresponding to 0.23–4.87 mg cm^{-2} sulfur loading). A stable cycling behavior was realized especially by combining low sulfur loadings, high electrolyte amounts, and a high rate. However, these conditions are not suitable for a high-energy cell. Self-discharge tests over 18 and 41 hours as well as the comparison of high (2.5 C) and low (C/5) rates reveal a degradation mechanism that correlates with the cell life. Typically, that mechanism is masked by short measurement durations per cycle (high rates).

2.2.3.4 Hierarchical Mesoporous Carbons

The recent research on micro-, meso-, and macroporous carbon scaffolds indicates that a hierarchical structure with a multimodal pore size distribution is desirable for lithium–sulfur batteries. This approach allows the combination of several advantageous properties in one system, as, for example, a high storage capacity for sulfur, unhindered electrolyte penetration, a high number of accessible redox sites, as well as an improved retention of LiPS. The rates of dissolution and precipitation processes can be enhanced as well by implementing meso- or micropores into the walls of bigger transport pores. Although several research teams followed these premises, the evaluated systems showed many structural diversities. Using soft-template synthesis and subsequent KOH activation, Liang et al. synthesized hierarchical carbons with homogeneous 7.3-nm mesopores and a high micropore content ($1566 \, m^2 \, g^{-1}$) [74]. Utilizing hydrophobic interactions, the team realized the selective filling of the micropore volume. Hence, they increased the sulfur amount stepwise from 11.7% to 37.1%. Using sulfur contents higher than 45%, the mesopores of the composite were filled. Consequently, starting capacities as high as $1585 \, mAh \, g^{-1}$ could only be realized with the lowest sulfur content, accompanied by accelerated cell degradation (56% loss of capacity within 50 cycles).

The implementation of a SiC-CDC with hexagonal mesopore structures (DUT-18) made of microporous carbon rods enabled a system with improved structural and electrochemical properties [49, 75]. Adjusting the chlorination temperature between 700 and 900 °C allowed precise control of size distribution of the micropores (0.7 nm), mesopores (3.0–4.0 nm), and the pores in between (1.3–3.0 nm). After impregnation with sulfur and a thermal removal of excess sulfur, the CDC chlorinated at 900 °C with 57% sulfur loading showed the best performance of all samples. The reduced number of defects within the carbon structure and, hence, a reduced level of impurities like, for example, chlorine residue, were identified to cause the improved performance. Nonetheless, the capacity degraded after 100 cycles at C/5 from ~700 to ~550 $mAh \, g^{-1}$, which made an adaptation of the LiTFSI concentration from 1 to 5 M necessary. Due to short diffusion paths from mesopores (electrolyte reservoir) to micropores, the system achieved an excellent rate capability even at 1 C compensating the increased viscosity. The stable capacity after 100 cycles was increased to ~920 $mAh \, g^{-1}$.

High-temperature chlorination of TiO_2–NP/C composites proved to be a highly efficient way to synthesize porous carbons with a total pore volume of up to $3.1 \, cm^3 \, g^{-1}$ and a micropore content of $0.3 \, cm^3 \, g^{-1}$ [52]. In this hard-template route inspired by the Kroll process, the mesopore size (i.e. 8.5–18 nm) is controlled by the nanoparticle diameter, while the mesopores independently develop ~4.2-nm pore windows. However, *in situ* activation accompanied by gaseous $TiCl_4$ formation yields ~1-nm micropores in the pore walls. Even after infiltration with 66.7–80% sulfur, the Kroll carbon with 18-nm mesopores ($3.11 \, cm^3 \, g^{-1}$) remains completely X-ray amorphous. Consequently, a homogeneous distribution of sulfur in contact with the high specific surface area (up to $1989 \, m^2 \, g^{-1}$) was enabled. Moreover, a high degree of connectivity between micro- and small mesopores or rather unhindered access of the electrolyte through the 18-nm transport pores was achieved. As a consequence, high starting capacities as well as good capacity retention could be realized. Even composites with a high sulfur content showed starting capacities of up to 1046–1115 $mAh \, g^{-1}$ as well as a capacity retention of >70% after 80 cycles at C/1. Additional adjustment of the cathode composition led to an electrode with a relatively high sulfur content of 72% (80% inside the composite) and a sulfur areal loading of 4.92 mg cm^{-2} with a stable capacity of >600 mAh g^{-1}. A bimodal mesopore structure with a pore diameter of 3.1–6 nm was studied for spherical opal-like arranged carbon particles with a diameter of 300 ($\pm$40) nm [76]. Due to the high specific surface area (2445 m^2 g^{-1}) and comparatively high pore volume (up to 2.63 cm^3 g^{-1}), only a minor decrease in the electrochemical performance was observed, when increasing the sulfur content from 50 to 70 m.-%. Given that cathodes with lower sulfur contents of the latter carbon material degrade faster, all composites showed a capacity of 700–730 mAh g^{-1} after 100 cycles at 1 C. It was possible to reduce the irreversible capacity loss of cathodes significantly, starting with a 70 wt% sulfur loading as long as the excess sulfur residue (~14%) was removed from the outer surface of the nanoparticles with CS$_2$ extraction. Wang and coworkers used ~37-μm spherical microparticles with a 75% sulfur content in order to produce cathodes with high sulfur loading as high as ~5 mg cm^{-2} [77]. The carbon material had a specific surface area of 1014 m^2 g^{-1} and was characterized to show broadly distributed (5–25 nm) mesopores and other smaller mesopores located inside the pore walls. The accessible pore volume inside the composites structure decreased during the sulfur infiltration from 2.5 to 0.06 cm^3 g^{-1}. Nonetheless, the C/S composite showed a stable performance of ~800 mAh g^{-1} (>3.5 mAh cm^{-2}) – after a drastic capacity loss during the first five cycles. The retention after 50 cycles at 0.84 mA cm^{-2} was ~91% (740 mAh g^{-1}). An additional implementation of ~10% CNTs into the carbon particles led to the formation of a highly conductive, percolating network that enormously increased the rate capacity.

Li et al. studied a peapod-like carbon structure with interconnected mesopores (22 nm). Furthermore, the 22-nm pores exhibit smaller (2 nm) mesopores being integrated into the thin pore walls (3–4 nm) [78]. The total pore volume of the structure was so high (4.69 cm^3 g^{-1}) that 60–84% amorphous sulfur could be easily infiltrated, preventing an agglomeration. The loading was carried out by melt infiltration and a subsequent thermal treatment at 300 °C. Compared with the previous structures, the peapod-like carbon with the lowest sulfur ratio (60%) showed the highest starting capacity (1106 mAh g^{-1}), as well as a ~100 mAh g^{-1} higher reversible capacity after 50 cycles at C/5. The moderate surface area of 977 m^2 g^{-1} as well as the highly loaded pore system seem to be the most limiting factors. The coulombic efficiency of ≥90% was surprisingly high for an electrochemical investigation without LiNO$_3$ additive. A similar structure had already been synthesized via a chemical vapor deposition (CVD) process.[210] A cathode composite was created by the decomposition of acetylene gas on a nanoporous template made of anodic aluminum oxide (AAO) and the synthesis of sulfur via integrated *in situ* reduction of sulfite [79]. Due to the extremely homogeneous distribution of sulfur in the microporous CNT walls, the ordered S-CNT architecture with 15- to 20-nm mesopores and a specific surface of >600 m^2 g^{-1} had excellent contact between sulfur and carbon. While the composite showed an acceptable LiPS-retention, the free mesopores ensure an accelerated mass transportation. The samples based on binder-free cathode membranes with a sulfur ratio of 23% or 50% showed at ~0.9 C high starting capacities of 960 and 675 mAh g^{-1}. However, the capacity degraded after 100 cycles to 653 mAh g^{-1} (23% S) and 524 mAh g^{-1} (50%). This proved again that C/S composites with low sulfur ratios enhance an accelerated capacity loss.

A structurally analogous system made of cylindrical 3.5- to 8-nm mesopores comprising a porous wall structure (1.7–3.5 nm) was reported [80]. The material had a high specific surface area of 2102 m^2 g^{-1} and a total pore volume of 2.0 cm^3 g^{-1}; consequently, it was possible to load this carbon scaffold with 70% completely amorphous sulfur.

At a sulfur ratio of >60%, the cylindrical transport pores started to be filled up as the limited volume of the wall pores is already overloaded at that point. Therefore, the composite with 70% sulfur showed a reduced electrochemical activity and a reduced starting capacity of 822 mAh g^{-1} at C/10 as well as a considerable overpotential in the lower voltage plateau, respectively. Structures with a reduced sulfur ratio of 60% had a higher starting capacity of 1138 mAh g^{-1} due to their improved contact at the interface. While suffering strong overcharges (shuttle effect), the stable cathode scaffold showed a capacity of ~400 mAh g^{-1} after 400 cycles at C/2. Parallel to these analyses, the Nazar group generated a carbon material for cathodes with high rate capability and exhibiting an identical pore structure with a slightly higher surface area (2300 m^2 g^{-1}) [81]. By varying the sulfur content between 40% and 60%, 50% sulfur content was identified as the best compromise between capacity and cycle stability (~55% of its starting capacity after 100 cycles at 1 C). The unhindered ion transport within the free cylindrical pores and the retention of LiPS inside the pore walls synergistically causes the improved performance of the cathode with 50% sulfur content.

Another porous carbon system comprises microspheres with an adjustable mesoporosity of ~10 nm and a macroporosity of ~20–70 nm and a mixed meso-/macroporosity of ~10 nm and ~20–70 nm, respectively. These spheres are generated using selected SiO$_2$-NP with varying diameter (10-, 40-, and 10/40-nm combination) as templates [82]. These carbon materials exhibit ~1.5-nm micropores and 4- to 6-nm mesopores inside their pore walls. Those pores are selectively filled and blocked between ~61% and 64% sulfur content. Starting capacities of 1278 mAh g^{-1} (10/40 nm NP) and 1158 mAh g^{-1} (40 nm NP) were reached, as long as free meso-/macropores were available as transport paths for lithium ions. Yet, cathodes with solely ~10 nm mesopores (10-nm NP) reached only 940 mAh g^{-1}. The capacity retention after 100 cycles at 1 C clearly correlated with the pore size distribution and proved that small ~10-nm pores (72.3% retention, 680 mAh g^{-1}) perform superior compared to ~20- to 70-nm pores (57.0% retention, 660 mAh g^{-1}). The optimal balance was achieved with a multimodal distribution (7.7% retention, 904 mAh g^{-1}). In accordance with the template size, the specific surface area increased from 1236 m^2 g^{-1} (40 nm NP) to 2485 m^2 g^{-1} (10/40 nm NP) and an impressive 2776 m^2 g^{-1} (10 nm NP).

2.2.3.5 Hierarchical Microporous Carbons

Ding et al. used the self-assembly of colloidal ~300-nm poly(methylmethacrylate-co-butyl acrylate-co-acrylic acid) spheres and ~9-nm SiO$_2$-NP to generate a hierarchic ordered carbon material with meso- and macroporous 3D geometry [83]. The spherical mesopores inside the walls were created in order to offer a defined volume for electrochemical conversion reactions. At the same time, the macropores were connected by ~120-nm-wide pore windows and ordered in an inverse opal structure, enabling the compensation of volume changes. Moreover, they guarantee an unhindered access for lithium ions. Yet, this highly porous composite structure, with a low sulfur content of

50%, achieved only a starting capacity of 1193 mAh g^{-1} and a relatively poor rate capacity (~600 mAh g^{-1} at 1 C and ~450 mAh g^{-1} at 2 C). The open pore structure reduced the retention of LiPS, and, hence, after 50 cycles at C/10 only 74% of the starting capacity could be utilized. Even hierarchical carbon scaffolds consisting of acicular macropores and cross-linked meso- and micropores showed an increased degradation [84]. The described composite with a high sulfur loading of 84% showed after 100 cycles at C/2 a capacity loss of ~68% (406 mAh g^{-1}). The low specific surface area of 903 m^2 g^{-1} is probably one of the factors leading to this poor performance.

2.2.3.6 Hollow Carbon Spheres

Archer and coworkers were the first to try to improve the cathode structure using hollow carbon spheres with a diameter of ~200 nm and a mesoporous shell (~3 nm) [85]. This material was produced via a hard templating method, and loaded with ~70% sulfur. Electrochemical characterization showed an excellent wettability with the tetra(ethylene glycol) dimethyl ether (TEGDME)-based electrolyte as well as a good retention of the LiPS. Furthermore, good electrical bonding of the amorphous sulfur inside the hollowed spheres and inside the porous shell in combination with the mechanical stability helped retain a high reversible capacity as high as 974 mAh g^{-1} (91% starting capacity) after 100 cycles at C/2. The coulombic efficiency was maintained at ≥94% and the rate capacity was relatively high (~450 mAh g^{-1} at 3 C). Due to this success, He et al. produced also ~200-nm porous hollow carbon nanospheres (with 10- to −12-nm-thick shells). Monodisperse SiO$_2$-NP hard templates were used for this purpose (Stöber process) [86]. A modification of the sphere shell via KOH activation and the import of cationic ionomers as a pore-forming agent allow to optimally adjust the structure to meet the requirements of the cathode chemistry. The finally prepared composite with a 70% sulfur content showed a capacity loss of ~35% after 20 cycles at 1 C due to its relatively small specific surface area and small pore volume (~0.4 cm^3 g^{-1} within ~5 nm mesopores). This led to the assumption that the sulfur embedment was incomplete proved by energy dispersive X-ray (EDX) investigations [87]. However, the high specific surface area of the KOH-activated sample of 1800 m^2 g^{-1} considerably improved the homogeneous dispersion of the sulfur [86]. Therefore, the KOH-activated sample showed higher starting capacities than the abovementioned hollow spheres/sulfur composites. An increase of ~30% (up to ~1300 mAh g^{-1} and 70 m% sulfur) starting capacity compared to the latter was achieved. When sulfur concentrations higher than 75% or 80% were used, the starting capacity dropped to 1150 or 940 mAh g^{-1}, respectively. Transmission electron microscopy (TEM) studies indicated a vacant inner part of those spheres. Yet, the capacity retention over 100 cycles at 1 C increased 68% probably caused by the higher micropore content. According to the authors, thin sulfur coatings were mostly observed on the inner particle surfaces. Those coatings presumably blocked the pathways into the sphere and thereby prematurely terminated the filling procedure. Especially large mesopore volumes (~1.1–1.6 cm^3 g^{-1}) were accessible for an ionomer-controlled pore formation. Since the latter allows an unhindered coating of the inner particle surfaces, it was possible to achieve higher cycle stability than using a porous carbon nanosphere (PCNS). Böttger-Hiller et al. used twin polymerization to produce hollow-sphere structures with a microporous shell, a controlled shell thickness and an enclosed ~5.5-nm mesopore. These hollow-sphere structures also showed an excellent capacity retention and reversibility of conversion: After 500 cycles, they still maintained a capacity of $\sim$440 mAh g^{-1} [88]. Especially the high specific surface area of up to 1370 m^2 g^{-1}, the total pore volume of up to 2.33 cm^3 g^{-1}, and the compensation of mechanical strain on the nanolevel were identified as causes for the good retention. Those results were also affirmed in full cell tests with pre-lithiated anodes made from hard carbon (HC). This procedure increased the cycle stability at the same time to nearly 1400 cycles, using only $\sim$10% lithium excess [58]. It should be noted that the small apparent density of this nanoarchitecture influences the energy density.

2.2.3.7 Graphene

In order to find a new and highly conductive scaffold, Li et al. analyzed thermally expanded graphene nanosheets with a specific surface area of 598 m^2 g^{-1}, a relatively high pore volume of 2.61 cm^3 g^{-1}, and a broad pore size distribution below <47.5 nm large mesopores (peak at $\sim$2.5 nm) [89]. The thermally treated graphene/sulfur composite with 67% sulfur content had a characteristic laminar structure made of alternating stacked layers of graphene and finely dispersed amorphous sulfur particles. It was possible to improve the initially insufficient performance of 563 mAh g^{-1} after 35 cycles at a coulombic efficiency of $\sim$90% by implementing an additional coating of reduced graphene oxide (RGO). After 100 cycles at $\sim$0.12 C, the reversible capacity of the RGO-modified TG/S composite still retained a capacity as high as 928 mAh g^{-1} (71.9% retention) and the coulombic efficiency reached nearly 100%, even without the addition of LiNO$_3$ additive.

Even at rates as high as $\sim$3.83 C, stable capacities of $\sim$800 mAh g^{-1} were enabled. Based on a selected group of carbon materials with 80% sulfur loading, Zheng et al. summarized central aspects of the structure–property relationship of sulfur cathodes comparing Ketjenblack (1576 m^2 g^{-1}, 4.86 cm^3 g^{-1}), graphene (890–1120 m^2 g^{-1}, $\sim$6.2 cm^3 g^{-1}), acetylene black (124 m^2 g^{-1}, 0.53 cm^3 g^{-1}), and carbonaceous spheres (76 m^2 g^{-1}, 0.38 cm^3 g^{-1}) [66]. They found that a high specific carbon surface has two advantages: first, it aids the homogeneous dispersion of sulfur and, second, it reduces the real current density at the reaction centers. The latter helps to reduce overcharge effects that can lead to precipitation of Li$_2$S from the highly concentrated LiPS solution. The precipitation step usually takes place during the third discharge plateau. Hence, a constant and complete Li$_2$S precipitation process is crucial. Thus, the formation of an isolating Li$_2$S film is prevented and, eventually, sulfur utilization as well as reversibility are increased. At a higher current density (starting at a rate of 0.5 C), the pore volume becomes the most dominant factor. While adjusted nanostructured carbons allow immense improvements, there are still challenges to be solved, such as the low interactions (physisorption) between hydrophilic LiPS and hydrophobic carbon surfaces, as well as the uncompleted encapsulation of the active material.

2.2.4 Retention of LiPS by Surface Modifications and Coating

By coating the particle surface with a physical barrier, the pores of porous composites become encapsulated, which hinders the release of the active material. Another way of increasing the retention of the LiPS is to modify the composite with amphiphilic and hydrophilic polymers, respectively. In this case, a chemical gradient is formed and an increased bond of the sulfur species to the carbon surface. Functionalized carbon

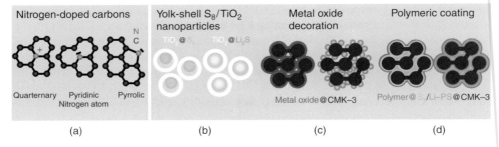

| Nitrogen-doped carbons | Yolk-shell S₈/TiO₂ nanoparticles | Metal oxide decoration | Polymeric coating |

Quarternary   Pyridinic   Pyrrolic
       Nitrogen atom

Metal oxide@CMK–3   Polymer@S/Li–PS@CMK–3

(a)   (b)   (c)   (d)

Figure 2.5 Scheme of different polysulfide retention principles: (a) Nitrogen doping of carbons, (b) TiO₂/sulfur yolk–shell structures, (c) metal oxide decoration of CMK-3, and (d) polymeric coating of CMK-3/sulfur composites.

derivatives like graphene oxide (GO) and nitrogen-doped carbons function analogously (Figure 2.5).

Typically, the binder mechanically stabilizes a cathode film by the adhesion between individual particles and the current collector. Inside the lithium–sulfur battery (LSB), the binder polymer can also function as a part of the electrolyte system as long as it swells in solution and interacts with the LiPS. Thus, a correlation between the cathode performance and the polyethylenoxide (PEO)-CB ratio was observed. While a higher PEO-CB ratio at a constant sulfur concentration of 70% decreases the starting capacity, it also encapsulates the sulfur particles [90]. Consequently, a higher PEO content limits the release of LiPS and, hence, helps sustain the capacity. In addition, Cheon et al. discuss that a high ratio of ion-conductive PEOs will support the stabilization of the carbon matrix and thereby prevent morphology changes caused by dissolution or separation [91].

Given that PEO is an electrical isolator between the contact area of the carbon particles, it is crucial to precisely control the PEO content in the composite to achieve a high electrical conductivity. PEOs often complicate the coating procedure as they mostly show bad adhesion properties [92]. A polyethylene glycol (PEG) modification of the outer, hydrophobic carbon surfaces can also be useful for nanostructured C/S composites due to the fact that the polarity gradient retains polar LiPS inside the carbon structure. CMK-3/S particles with added PEG chains showed an impressive capacity increase of ~300 mAh g⁻¹ after 20 cycles accompanied by a reduced LiPS content and shuttle [21]. In a similar way, even nonporous, highly conductive 50- to 80-nm acetylene black (AB) particles were transformed into a core–shell structure after an oxidative functionalization [57]. The subsequent deposition of 80% sulfur was carried out directly in the PEG shell, causing a slight increase in particle diameter from 80 to 100 nm. Due to the homogenizing and stabilizing effects of PEG, a homogeneous microstructure was realized after the sulfur deposition, suppressing the crystallization of sulfur to agglomerates. While the compound suffered from high capacity losses during the first cycles, it developed a stable capacity of 700–850 mAh g⁻¹ after 100 cycles at ~0.06 C. After 500 cycles, it still showed 577 mAh g⁻¹ capacity. The suppressed diffusion of LiPS in the hydrophilic phase and a mitigated dilatation of the cathode volume with an increased lithium-diffusion coefficient were identified as the main causes for the good composite capacity retention.

Lacey et al. observed that low-molecular weight ether polymers (for example, PEG 20 000) are soluble inside electrolyte systems being typical for lithium sulfur batteries while high-molecular ether polymers (for example, PEO, Mw $\leq$ 4 000 000) tend to swell inside those systems [93]. Several routes to implement PEO derivatives into composites (as coating, binder, or electrolyte additive) were carried out. All variations increased the capacity in the same manner: the capacity was increased about ~150 mAh g^{-1} after 50 cycles. Therefore, the older hypothesis that attributed the increase in capacity to a presumed barrier effect of PEO layers was discarded. It is postulated that the increase in capacity actually must be caused by an improvement in the electron-transfer rate: a PEO-modified electrolyte system was observed to influence the solubility of short-chained LiPS at the cathode/electrolyte interface in such a way that the precipitation of insoluble, passive discharge products is delayed.

The finding that polymers could have a profound influence on the chemistry of sulfur cathodes led to a paradigm change. Thus, Cui and coworkers correlated the capacity loss of C/S composites with interface effects and showed in TEM investigation that polar Li$_x$S clusters as well as Li$_2$S precipitated from the unpolar carbon wall during discharge [94]. This behavior was also identified as a cause of the deposition of Li$_2$S outside of the porous carbon scaffold. The reduction of binding energy between carbon and the sulfur species (S : 0.79 eV; LiS : 0.21 eV; Li$_2$S : 0.29 eV) determined via density functional theory (DFT) simulation on a model system (graphene, Li$_x$S cluster ($x \leq 2$), or S) supports this assumption. Yet, a functionalization of the carbon surface with amphiphilic polymers like Triton X-100 or polyvinylpyrrolidone (PVP) helped overcome the degradation mechanism. In this case, the polymer functions as an agent that establishes intense contact between the nonpolar carbon and the polar sulfur species, e.g. between the dissolved LiPS and the separated Li$_2$S. *Ab initio* simulations were also used to identify the interactions between electron-rich functional groups of macromolecules that contain O/N heteroatoms and Li$_2$S and LiPS, respectively. The coordinative Li—O bonding of esters, ketones, and amides with a double-bond carbonyl-O-atom ($>$C—O) was highlighted as especially effective [95]. PVP as an amide was identified to be a very attractive binder because it has an appropriate concentration of active $>$C—O groups as well as a high thermodynamic impetus for the coating of hydrophobic carbon surfaces [94]. The homogeneous dispersion and effective retention of Li$_2$S/LiPS reduced the loss of sulfur from 27% (polyvinylidene fluoride, PVDF) to 13% (PVP) after 20 cycles. The retention of the starting capacity (~750 mAh g^{-1}) was at 69% after 500 cycles at C/5 [95]. Yushin and coworkers used PVP also as a capping reagent for the *in situ* polymer coating of Li$_2$S-NP within ethanolic solution [96]. That enables steric separation of the highly dispersed particles and, hence, impedes further particle growth after the nanoparticle crystallization.

It was possible to combine the positive effects of PEO and PVP by synthesizing a binary PEO/PVP binder system with nonporous carbon black as carbon matrix. This system surpassed the capacity of a reference cell with a binder made of carboxymethylcellulose (CMC) and styrene-butadiene rubber (SBR) by more than 200 mAh g^{-1} [59]. The binary system also showed, with 800 mAh g^{-1} after 200 cycles at 1 C, an excellent rate capability and capacity retention. Due to the lack of additional porosity, it was necessary to retain the LiPS in the close proximity of the carbon black particle surface during the electrochemical reactions. This was achieved through the formation of PVP–Li$_2$S$_x$ complexes. The importance of those complexes was demonstrated by mixing it with solutions made of Li_2S_6 and PVP (both in DME/ dioxolane [DOL]) which caused a red precipitate. Interestingly, high PVP contents caused an enormous charge hysteresis and consequently decreased the cells' performance.

However, the pores of carbon materials as, for example, highly porous carbon blacks ($>1000 \, m^2 \, g^{-1}$, $>2 \, cm^3 \, g^{-1}$) can be filled or blocked by the infiltration of binder solutions such as PVDF–hexafluoropropylene copolymer, PVDF, and PEO [36]. This is especially revealed for the pore volume of macro- and large mesopores (8–50 nm). These polymers can act as blocking agents, leading to an insufficient electrolyte penetration in the unblocked micro- and small mesopores. As a consequence, the embedded sulfur as well as the specific carbon surface area cannot be utilized for the electrochemical reactions if the binder has no swelling property, as is the case with PVDF, a binder most frequently used [1, 36]. This leads to strong overcharge effects and a low sulfur utilization, two issues that can only be avoided if a swellable binder is used. In order to enable an appropriate adhesion, a gel-like binder state (PEO) after electrolyte contact in contrast to binder dissolution is preferred.

A classical barrier concept was presented by Cui and coworkers applying poly-3,4-ethylendioxythiophene-polystyrenesulfonate (PEDOT:PSS) as ion and electron conductive thin (10–20 nm) amorphous polymer coating on a CMK-3/S composite [97]. The coating process was realized by simple ultrasonic dispersion of the composite particles. Using this coating in comparison to the untreated composite, the polysulfide retention in the pores of the carbon structure was increased. Hence, even without $LiNO_3$ additive, a high coulombic efficiency as high as 96–98% at 0.2 C (without coating: 92–94%) could be achieved. Moreover, barrier coatings prepared by other conductive polymers such as polypyrrole (PPy) [78, 98], or thin RGO sheets were applied [89, 99]. With the described polymers, various porous scaffolds could be encapsulated, leading to increased cycle stability. However, the issue of decreased sulfur content in the entire cathode mass was not discussed.

The modification of the carbon material is another approach for the prevention of active material loss and polysulfide shuttle. It was possible to coat a thin sulfur film ($\sim$66%) via chemical solution deposition from a microemulsion on a quasi-2D scaffold of GO [100]. The combination of high surface area and homogeneously distributed voids was beneficial to realize a close contact with sulfur, to minimize the volume changes as well as to ensure a long-lasting accessibility for ions and electrons. Due to the fact that the surface of the GO has various functional groups with strong adsorption ability for sulfur atoms, LiPS could be effectively retained. X-ray absorption spectroscopy (XAS) analysis and *ab initio* calculations revealed that epoxy and hydroxyl groups are especially involved. The strong chemical interaction hinders sulfur agglomeration and loss of contact enabling a stable electrochemical cycling with $\sim$950 mAh g^{-1} for 50 cycles at C/10. Due to the used IL/PEGDME electrolyte, even without $LiNO_3$ additive, a coulombic efficiency >96% could be observed. The excellent properties of GO as scaffold are also attributed to the carboxylic groups and the associated hopping transport mechanism of the lithium ions. That mechanism as well as the polysulfide retention by electrostatic and steric exclusion principles could be revealed by optical experiments on thin GO membranes (0.12 mg cm^{-2}) [101].

Using a GO/S core–shell composite with 50% sulfur content, highly stable sulfur cathodes were generated by Rong et al. [102]. Due to the slow degradation of 0.02% per cycle even after 1000 cycles at $\sim$0.6 C a capacity as high as $\sim$800 mAh g^{-1} were obtained. Reasons for that performance are not only the improved conductivity but also the minimized charge-transfer overpotential in the GO/S particles with sizes ranging between 500 nm and 10 μm attributes to the improved performance. Using an additional cetyltrimethylammonium bromide (CTAB) modification of the GO/S composite (~82% S) and a novel IL-/LiNO$_3$-based electrolyte system, the number of reversible cycles was increased to 1500 at 1 C and a final capacity of ~740 mAh g^{-1} (0.05 C) [103].

The multifaceted approach showed a slightly increased degradation of 0.039% per cycle. However, the rate capability up to 6 C was excellent and the capacity in the first 40 cycles at C/5 as high as ~1400 mAh g^{-1}. The coulombic efficiency remained at >96.3%. The high concentration of LiNO$_3$ (0.5 M) in combination with the ionic liquid (IL) content in electrolyte was discussed as a reason for the excellent capacity retention. Furthermore, for the highest CTAB content (5 mM) additional C—S bonds were observed.

Analogously, graphene sheets functionalized by nitrogen doping can encapsulate sulfur particles and form composite structures with high sulfur contents (80%) [104]. The integration of N-atoms into the graphene layer increases the electrical conductivity as well as the adsorption of polar species on the carbon surface [105]. Due to their unbound electron pair, negative polarized, pyridinic N-atoms can function as Lewis bases in the edge of aromatic six rings. These polarized pyridinic N-atoms coordinate lithium cations and help immobilize LiPS and Li$_2$S, respectively.

C-atoms in the electron-rich proximity of quaternary N-atoms also form anchor points. Yet, pyrrolic N-atoms are ineffective for this purpose, given that their free electron pair is preferentially integrated into the π-system of the five rings in order to create an aromatic state. Therefore, the donor–acceptor interaction being necessary for the coordination or rather bonding of polar species cannot be as effective. Accordingly, composites with pyridinic N-atoms (12.5%) show after 100 cycles at 0.1 C an increased capacity (compared with pyrrolic N atoms, 11.5%). After more than 500 cycles at 1 C, a stable capacity of ~580–700 mAh g^{-1} can be observed. Both systems were superior to the graphene-based composite. An N-doped, mesoporous carbon with an adjustable nitrogen ratio was successfully produced with the hard template method [106]. While 4–8% nitrogen is sufficient for high conductivity, an acceptable immobilization of the LiPS does not occur below a heteroatom content of 8%. After 100 cycles at C/5 and no LiNO$_3$ additive, the optimized composite with 60% sulfur and a nitrogen content of 8.1% showed a coulombic efficiency of ~94%. With an additional PPv/PEG modification, the reversible capacity of 758 mAh g^{-1} was successfully increased to 891 mAh g^{-1}. Analyzing N-doped carbon materials with a hierarchic mesopore structure by X-ray absorption near edge structures (XANES) and DFT studies, Wang and coworkers found out that oxygenated groups close to pyridinic N-atoms (—COOH) as well as pyrrolic N-atoms (—C—O) interact intensively with S-atoms [107, 108]. At the same time, the coordination structure of the N-atoms changed only insignificantly, indicating an electron density transfer between adjacent carbonyl and carboxyl groups, respectively. This led to a highly negative polarization of the O-atom groups, which effectively retained polar LiPS via chemisorption on the carbon surface. Moreover, the precipitated Li$_2$S particles were more homogeneously distributed. All in all, the general observations are in concert with the effects resulting from amphiphilic surface modification by polymers (e.g. PVP). However, a crucial advantage of the N-doped carbons lies in their electrical conductivity, which guarantees electrochemical access to strongly adsorbed LiPS. The relatively good coulombic efficiency of >90% (for Li–S systems) as well as the evaluation of the *ex situ* adsorption behavior via UV/Vis spectroscopy affirmed these results. Heteroatom doping is a suitable method to improve sulfur cathodes, since it allows controlled modifications of the basic properties of the carbon-material-like electron structure, wettability, adsorption capacity, and surface polarity (alkalinity).

2.2.4.1 Metal Oxides as Adsorbents for Lithium Polysulfides

Metal oxides like SiO_x, AlO_x, VO_x, TiO_x, and $Mg_{0.6}Ni_{0.4}O$ adsorb polar species which enables them to form LiPS reservoirs inside the cathode [109–113]. However, since these isolating species hamper the electron transport, their usage comes hand in hand with a reduced sulfur utilization and a declining rate capability. In other words, a direct electrochemical transformation of the active material is no longer possible if such metal oxides are used. Studies of metalorganic frameworks (MOFs) proved that LiPS were effectively encapsulated within such frameworks and that Lewis acidic metal centers have a positive influence on the adsorption behavior (chemisorption) [114–116]. Yet, the low conductivity of the MOF leads to an insufficient sulfur utilization of merely $\sim$650 mAh g^{-1}. For that reason, thin diffusion retarding layers of metal oxides were mostly used to coat the surface of the conductive C/S composite. For example, using plasma-enhanced atomic layer deposition (PEALD), a <3- to 5-nm-thin Al_2O_3-layer was coated on a microporous carbon fiber which had previously been infiltrated with $\sim$50% sulfur [117]. Samples that had been modified in over 30 (30 Al_2O_3) or rather 50 PEALD cycles (50 Al_2O_3) showed a decreased capacity at C/5 and 70 °C compared to the uncoated reference samples, but cycle stability as well as sulfur retention were significantly increased. Due to the hindered diffusion of unsolvated lithium ions in the barrier, the maximum capacity of the coated samples was observed. However, the electrolyte infiltrates slowly via defects in the cathode matrix and successively increases the reaction kinetics. Lee et al. modified the surface of a CMK-3/S composite with SiO_x and VO_x in a similar way [113]. Via hydrolysis and condensation of the metal oxide precursors, a slow and controlled growth process on an acidized carbon surface was initiated, enabling the adjustment of the film thickness. A thin 2.7% SiO_x coating already led to an increase in coulombic efficiency from 80.7% up to 85.8% without the supplement of a $LiNO_3$ additive. The coulombic efficiency increased even more with a SiO_x content of 26.5%, but such a coating also creates high overvoltage effects. Those overvoltage effects were caused by a higher charge-transfer resistance, e.g. a reduced permeability of the SiO_x layer for lithium-ions and finally led to very low cathode capacity.

Even a C/S composite based on a 12-nm mesoporous carbon framework was completely encapsulated by a SiO_x coating of 7.5%. After 100 cycles at 1 C, it showed a capacity retention of 71%, being nearly two times higher than the reference. Yet, for sulfur cathodes with a very high cycle stability, conductive metal oxides or hybrid structures based on metal oxides and amorphous carbons were used for nanoscale coatings. For example, Seh et al. realized the coating of monodisperse $\sim$800-nm sulfur NPs with a $\sim$15-nm TiO_2 by a wet chemical solution method [109]. The partial dissolution of sulfur from its core–shell structure produced a hollow concept which seems to be an option to compensate the volume expansion during cycling. Since hydrophilic Ti–O- and hydroxyl groups ensured a long-lasting retention of the LiPS within the yolk–shell architecture, the loss of sulfur was reduced to 19%. Compared with these results, pure sulfur and sulfur without a hollow core–shell architecture showed higher sulfur losses. Therefore, a capacity loss of only 0.033% per cycle at 0.5 C and an initial capacity of 1030 mAh g^{-1} could be observed.

A hybrid structure made of ~8- to 20-nm Ti$_4$O$_7$ crystallites (Magneli phase) and carbon was realized by the Nazar group. This hybrid structure combined three positive properties: a surface area of 290 m^2 g^{-1} (micro- and mesopores), a high conductivity, and a chemical bonding of LiPS [118]. The Ti$_4$O$_7$/C-hybrid structure can also act as a redox mediator, since X-ray photon spectroscopy (XPS) analyses detected polarizations between terminal and bridging sulfur atoms caused by an electron density shift to electron-deficient Ti- and/or O-gaps. In over 500 cycles at 2 C, the composite with a 60% sulfur ratio showed a smaller capacity loss of 0.06% per cycle and a higher coulombic efficiency of ~96% (without LiNO$_3$ additive). Other redox mediator concepts with superb cycle stability are based on 3D hybrid structures made of carbon and conductive metal oxides such as MnO$_2$ nanosheets and Sn-doped In$_2$O$_3$-NP [119, 120]. All these approaches combined the effective retention of LiPS with a homogeneous, controlled separation of polar sulfur species at the metal oxide as well as a separation of nonpolar sulfur at the carbon. The MnO$_2$@C/S composite loaded with 75% sulfur showed a starting capacity of 1120 mAh g^{-1} and capacity retention of 92% over 200 cycles at 0.2 C. After over 2000 cycles at 2 C, a marginal change in the cathode structure and only a very small capacity loss of 0.036% per cycle were detected. The efficiency of the LiPS adsorption was confirmed through optical experiments [121]. A correlation between the LiPS-adsorption capacity of the basic framework, the reversible capacity loss of the cell caused by self-discharge and the irreversible capacity reduction caused by the loss of active material (washout of sulfur species) was observed. As expected, polar substances like TiO$_2$, Ti$_4$O$_7$, MnO$_2$, and GO reduced the diffusion ability of the LiPS via chemisorption more than a physisorption by nonpolar carbons like Super P, Vulcan, and others. Furthermore, materials with a high specific surface area showed increased retention effectivity. Since the observed dependency was not linear, it was concluded that the surface determined via nitrogen physisorption measurements is only partially vacant for the adsorption of sulfur species and electrolytes, respectively. As expected, GO interacts stronger with LiPS than it is the case for physisorption by unpolar carbons such as Super P and Vulcan. Moreover, materials with high specific surface area show a more effective retention. However, the observed dependence was not linear. For that reason, it was concluded that the specific surface area calculated from the nitrogen physisorption measurements only partially enable the adsorption of sulfur species or rather electrolyte.

2.3   Cathode Processing

2.3.1   Methods for C/S Composite Preparation

The currently existing sulfur impregnation methods can be classified into two main categories: heat treatment and chemical *in situ* syntheses. Heat treatment can be further divided into two types: sulfur melting diffusion method (under inert atmosphere or in vacuum, with or without additional solution infiltration) [21, 65, 100, 122], and the sulfur vaporizing method [85, 123].

As for the heat treatment, sulfur is typically infiltrated via a melt-diffusion method at 155 °C (minimum viscosity) [21]. In the first step, sulfur and the respective carbon material were ground together to form a homogeneous sulfur to carbon distribution. In some cases, ball milling is used to homogenize the composite before thermal treatment [124]. The subsequent heat treatment at 155 °C leads to sulfur diffusion into the pores of the carbon host material. On the one hand, chemically assisted methods such as dissolving of excess sulfur by, e.g. toluene are applied in order to remove sulfur on the outside of the porous carbon particles [125]. On the other hand, excess sulfur can be removed by physical methods such as evaporation of sulfur [61]. During that procedure, the carbon/sulfur composites are typically sealed in an argon-filled receptacle and the temperature ranges between 300 [61] and 400 °C [123]. Furthermore, combinations of chemically assisted and heat treatment methods are described [74, 126]. Chemical generation of elemental sulfur inside the pores via a redox reaction of a sulfur-containing precursor such as $Na_2S_2O_3$ [109] or Na_2S_x solutions is also an option to generate well-defined model cathodes [127–129].

2.3.2 Wet (Organic, Aqueous) and Dry Coating for Cathode Production

As explained before, the cathode performance is strongly influenced by its microstructure and composition. Besides the choice of components, the actual film formation process will have an important impact as well.

As with Li-ion electrode production, slurry-based processes may be used for sulfur cathode deposition on aluminum foils as current collectors. First, active materials, carbons, and binders are dispersed in a solvent. Subsequently, the dispersion is deposited on the substrate through coating equipment such as slot die, doctor blade, or comma bar systems. In the laboratory scale, this is done sheet by sheet and in pilot or production scale roll-to-roll equipment may be applied. Depending on the equipment configuration, double-sided coating might be applied in one run or in two subsequent runs.

A critical process step is the electrode drying. Temperature profile, gas velocity, and pressure are important process parameters for control over electrode structure as well as speed and completeness of the solvent removal. This step may induce cracks and macroporosity to the electrode film, which may improve the electrolyte wetting and uptake properties, but could also lead to partial delamination. It is important to note that typical drying conditions (elevated temperature and vacuum) may affect the sulfur distribution or even induce sublimation.

The established cathode binder/solvent system in Li-ion technology is PVDF dissolved in N-methyl-2-pyrrolidone (NMP). While this can be adapted for processing carbon/sulfur composites and is widely used in material studies, it has two main disadvantages:

PVDF is soluble in the ether-based electrolyte system, limiting its adhesive and cohesive function and leading to unwanted side effects [36]. In addition, the high boiling point of NMP (203 °C) would negatively affect the sulfur distribution upon drying or even hinder the complete solvent removal from the cathode.

Water is another widely used solvent with PEO, PVP, CMC, and SBR as suitable binders. Water is attractive as a low-cost and low-risk solvent, and the binders were found to be compatible or even beneficial for the sulfur cathode [93, 130, 131]. The aqueous process could potentially be transferred to Li-ion production lines, as most graphite anodes are produced in a similar setup. The main disadvantage of water is again related to the drying step, as the complete removal of water residues from highly porous materials would require elevated temperatures and vacuum conditions.

An alternative and cost-effective process avoids any solvent and drying step and is based on a dry film formation [132, 133]. Upon dry mixing of active material (e.g. carbon/sulfur composites) and a specific binder, the binder starts to form fibrils and binds particle agglomerates. In a subsequent roll-pressing step, freestanding cathode sheets can be produced. Through lamination on aluminum foils, the cathode films can be contacted and used for cell assembly.

2.3.3 Alternative Cathode Support Concepts (Carbon Current Collectors, Binder-free Electrodes)

Due to the high weight fraction of aluminum in Li–S cells, concepts for substitution were proposed in recent literature. Carbon nanomaterials can thereby fulfill two roles simultaneously. They can act as cathode scaffold as described and at the same time work as 3D current collector without metal support [134, 135]. However, the specific resistance of carbon is about 100 times higher than that of aluminum. As a result, the application of this concept would require a redesign of a typical pouch cell setup or is only applicable for low-power requirements.

Based on CNT and graphene, 3D current collectors have been demonstrated in test cells with high utilization and cycle life [79, 136]. Following the concept of reducing inactive material fractions, even binder-free electrodes have been demonstrated on the basis of 3D CNT networks [135, 137].

As these approaches are based on expensive material and/or processes, they might not be competitive yet. However, they open new pathways for a specific Li–S cell development beyond the classical technologies known from Li-ion cells.

2.3.4 Processing Perspective for Carbons, Binders, and Additives

It is important to note that the selection of binders and carbon materials cannot be done independently of the processing conditions. Solvent-based deposition will require certain binders being soluble in the solvent of choice. Dispersability of the carbon/sulfur composites will be an important aspect being influenced by surface functionality and particle morphology. Some additives (such as dispersion aids) might be required just to improve the coating quality (adhesion, homogeneity). Deposition technologies such as slot die or comma bar require a specific slurry rheology to provide consistent results. The rheology might be influenced through solvent content or further additives. The process and equipment of choice and their specific requirements will therefore be an important part to be considered in the optimization and development process for Li–S cathodes.

2.4 Conclusions

The cathode composition and structure takes an important role in the lithium–sulfur cell design and significantly impacts performance. Within the cathode composition, the carbon materials mainly define the rate and completeness of the sulfur conversion reaction. Considering the parameters such as electron transport, LiPS retention, and the compensation of the volume change, nanometer- or micrometer-scaled carbon particles with hierarchical pore structure are favored systems. That can be explained by the fact that a combination of different pore sizes meets the requirements of the dissolution/deposition cathode chemistry better than scaffolds with monomodal pore-size distribution. Micropores effectively suppress the LiPS shuttle mechanism due to high adsorption forces. In addition, they offer a high specific surface area (typically above $2000 \, \mathrm{m^2 \, g^{-1}}$) for close contact and hence, high sulfur utilization. Medium-sized and large mesopores (or small macropores) create a reservoir with high specific pore volume (typically above $2 \, \mathrm{cm^3 \, g^{-1}}$) for sulfur, electrolyte, and LiPS intermediates, and enable high sulfur contents in the composite. The implementation of macropores on material or electrode level will define the free volume of the cathode and therefore the ability for the electrolyte uptake. While a low free volume will result in high energy densities on the cell level, a certain degree of macroporosity will be required in order to accommodate the required electrolyte content.

In order to enhance the charge-transfer kinetics between carbon surface and sulfur species, nitrogen-doped carbon materials with modified electronic properties are suitable candidates. On the one hand, these materials offer a high conductivity for electrons. On the other hand, they interact strongly with polar LiPS species via a free electron pair of pyridinic nitrogen atoms or rather by using the electron-rich proximity of quaternary nitrogen atoms. Nitrogen doping in general does not only lead to a significantly better LiPS retention but it is also beneficial for a more homogeneous Li_2S precipitation. However, little is known about the long-term stability of nitrogen-containing redox sites under the operating conditions in lithium–sulfur batteries, because other degradation effects dominate cell failure. Given that the active material is always in direct contact with the conductive surface of the nitrogen-doped carbon scaffold, active material loss due to irreversible adsorption can be neglected.

The electrostatic interaction, or, rather, the coordinative bond (donor–acceptor interaction), between polar sulfur species and metal oxide crystal faces contributes not only to the immobilization of the sulfur species but also controls the Li_2S precipitation at predefined positions in the cathode. Moreover, a strong interaction is able to weaken S—S bonds in LiPS, and, hence, some metal oxides can act as redox mediators as well as catalysts for the cathode reaction. The implementation of inorganic species in the cathode can be realized by simple addition. Alternatively, a growth of hybrid structures with a close connection between metal oxide and carbon scaffold is possible, leading to a kind of carbon decoration. Both approaches can realize stable capacities and high coulombic efficiencies. However, the conductivity of the metal oxide/carbon scaffold should be ensured as metal oxides are mostly insulators or irreversibly adsorb active material. The deposition of a thin metal oxide film on the cathode is a further way to establish a physical barrier, preventing the washout of sulfur and sulfide species, irreversible reactions with the electrolyte, and a deposition of an insulating Li_2S coating.

Adapted (ion conductive) polymers provide the necessary elasticity of the cathode in order to compensate the repeating volume changes during charging/discharging without blocking the pore structure. Beyond that, an encapsulation of composite particles with a thin polar/amphiphilic polymer film establishes a chemical gradient, preventing the washout of active material. Conductive polymers as a three-dimensional percolation network or as particle coating mainly increase the electron transport within the cathode.

However, for the advancement of the LSB system, a deeper understanding of the cathode chemistry is required. Innovative carbon material approaches completely suppressing PS shuttle are needed. In addition, the less obvious and highly complex cathode processing steps influence the performance of the entire system as well. On the other hand, optimized cathodes can nowadays sustain more than 4000 cycles against hard carbon anodes [8]. Consequently, other components such as the lithium anode and solid–electrolyte interface (SEI) formation in combination with the electrolyte system limit cycling life more significantly than does cathode degradation. In that context, the comparison of cathode degradation against lithiated hard carbons may be recommended as being more meaningful for intrinsic cathode evaluation than using lithium metal anodes as a counter electrode.

References

1 Hagen, M., Hanselmann, D., Ahlbrecht, K. et al. (2015). Lithium–sulfur cells: the gap between the state-of-the-art and the requirements for high energy battery cells. *Advanced Energy Materials* 1401986.

2 Rosenman, A., Markevich, E., Salitra, G. et al. (2015). Review on Li–sulfur battery systems: an integral perspective. *Advanced Energy Materials* 5 (16): 1500212.

3 Fedorková, A., Oriňáková, R., Čech, O., and Sedlaříková, M. (2013). New composite cathode materials for Li/S batteries: a review. *International Journal of Electrochemical Science* 8: 10308–10319.

4 Bresser, D., Passerini, S., and Scrosati, B. (2013). Recent progress and remaining challenges in sulfur-based lithium secondary batteries – a review. *Chemical Communications* 49 (90): 10545–10562.

5 Borchardt, L., Oschatz, M., and Kaskel, S. (2016). Carbon materials for lithium sulfur batteries-ten critical questions. *Chemistry: A European Journal* 22 (22): 7324–7351.

6 Scheers, J., Fantini, S., and Johansson, P. (2014). A review of electrolytes for lithium–sulphur batteries. *Journal of Power Sources* 255: 204–218.

7 Hippauf, F., Nickel, W., Hao, G. et al. (2016). The importance of pore size and surface polarity for polysulfide adsorption in lithium sulfur batteries. *Advanced Materials Interfaces* 3 (18): 1600508.

8 Thieme, S., Brückner, J., Meier, A. et al. (2015). A lithium–sulfur full cell with ultralong cycle life: influence of cathode structure and polysulfide additive. *Journal of Materials Chemistry A* 3 (7): 3808–3820.

9 Agostini, M., Aihara, Y., Yamada, T. et al. (2013). A lithium–sulfur battery using a solid, glass-type P_2S_5–Li_2S electrolyte. *Solid State Ionics* 244: 48–51.

10 Nagao, M., Hayashi, A., and Tatsumisago, M. (2012). Fabrication of favorable interface between sulfide solid electrolyte and Li metal electrode for bulk-type solid-state Li/S battery. *Electrochemistry Communications* 177–180.

11 Dokko, K., Tachikawa, N., Yamauchi, K. et al. (2013). Solvate ionic liquid electrolyte for Li–S batteries. *Journal of the Electrochemical Society* 160 (8): A1304–A1310.

12 Cuisinier, M., Cabelguen, P., Adams, B.D. et al. (2014). Unique behaviour of nonsolvents for polysulphides in lithium–sulphur batteries. *Energy and Environmental Science* 7 (8): 2697.

13 Cheng, L., Curtiss, L.A., Zavadil, K.R. et al. (2016). Sparingly solvating electrolytes for high energy density lithium–sulfur batteries. *ACS Energy Letters* 1 (3): 503–509.

14 Xu, Y., Wen, Y., Zhu, Y. et al. (2015). Confined sulfur in microporous carbon renders superior cycling stability in Li/S batteries. *Advanced Functional Materials* 25 (27): 4312–4320.

15 Fanous, J., Wegner, M., Grimminger, J. et al. (2011). Structure-related electrochemistry of sulfur-poly(acrylonitrile) composite cathode materials for rechargeable lithium batteries. *Chemistry of Materials* 23 (22): 5024–5028.

16 Fanous, J., Wegner, M., Grimminger, J. et al. (2012). Correlation of the electrochemistry of poly(acrylonitrile)–sulfur composite cathodes with their molecular structure. *Journal of Materials Chemistry* 23240–23245.

17 Kolosnitsyn, V., Kuzmina, E., Karaseva, E., and Mochalov, S. (2011). A study of the electrochemical processes in lithium–sulphur cells by impedance spectroscopy. *Journal of Power Sources* 196 (3): 1478–1482.

18 Kolosnitsyn, V., Kuzmina, E., and Mochalov, S. (2014). Determination of lithium sulphur batteries internal resistance by the pulsed method during galvanostatic cycling. *Journal of Power Sources* 252: 28–34.

19 Jeon, B.H., Yeon, J.H., Kim, K.M., and Chung, I.J. (2002). Preparation and electrochemical properties of lithium–sulfur polymer batteries. *Journal of Power Sources* 109 (1): 89–97.

20 Cheon, S., Choi, S., Han, J. et al. (2004). Capacity fading mechanisms on cycling a high-capacity secondary sulfur cathode. *Journal of the Electrochemical Society* 151 (12): A2067–A2073.

21 Ji, X., Lee, K.T., and Nazar, L.F. (2009). A highly ordered nanostructured carbon–sulphur cathode for lithium–sulphur batteries. *Nature Materials* 8 (6): 500–506.

22 Diao, Y., Xie, K., Xiong, S., and Hong, X. (2012). Analysis of polysulfide dissolved in electrolyte in discharge-charge process of Li–S battery. *Journal of the Electrochemical Society* 159 (4): A421–A425.

23 Mikhaylik, Y.V. and Akridge, J.R. (2004). Polysulfide shuttle study in the Li/S battery system. *Journal of the Electrochemical Society* 151 (11): A1969–A1976.

24 Kolosnitsyn, V.S., Karaseva, E.V., and Ivanov, A.L. (2008). Electrochemistry of a lithium electrode in lithium polysulfide solutions. *Russian Journal of Electrochemistry* 44 (5): 564–569.

25 Brückner, J., Thieme, S., Grossmann, H.T. et al. (2014). Lithium–sulfur batteries: influence of C-rate, amount of electrolyte and sulfur loading on cycle performance. *Journal of Power Sources* 268: 82–87.

26 Strubel, P., Thieme, S., Biemelt, T. et al. (2015). ZnO hard templating for synthesis of hierarchical porous carbons with tailored porosity and high performance in lithium-sulfur battery. *Advanced Functional Materials* 25 (2): 287–297.

27 Bartlett, P.D. (1966). Elemental sulfur: chemistry and physics. *Journal of Chemical Education* 43 (12): A1096.

28 Weller, M., Overton, T., Rourke, J., and Armstrong, F.A. (2014). *Inorganic Chemistry*, 6e. Oxford: Oxford University Press.

29 Greenwood, N.N. and Earnshaw, A. (1997). *Chemistry of the Elements*, 2e. Oxford, Boston: Butterworth-Heinemann.

30 Daniel, C. and Besenhard, J.O. (2011). *Handbook of Battery Materials*. Weinheim: Wiley-VCH.

31 Béguin, F. and Frackowiak, E. *Carbon Materials for Electrochemical Energy Storage Systems*. Boca Raton: CRC Press, an imprint of Taylor & Francis; 2009. (Advanced materials and technologies).

32 IARC (2010). Carbon black, titanium dioxide, and talc. In: *Monographs on the Evaluation of Carcinogenic Risks to Humans*, vol. 93. Lyon: World Health Organization; International Agency for Research on Cancer.

33 Donnet, J., Bansal, R.C., and Wang, M. (1993). *Carbon Black: Science and Technology*, 2 , rev. and expandede. New York: Marcel Dekker.

34 Maeno, S. (2006). The structure and characteristics of conductive carbon black "KETJENBLACK EC". *Tanso* 222: 140–146.

35 AkzoNobel (2016). KETJENBLACK highly electro-conductive carbon black: product information. http://www.pcpds.akzonobel.com/PolymerChemicalsPDS/showPDF .aspx?pds_id=260.

36 Lacey, M.J., Jeschull, F., Edström, K., and Brandell, D. (2014). Porosity blocking in highly porous carbon black by PVdF binder and its implications for the Li–S system. *Journal of Physical Chemistry C* 118 (45): 25890–25898.

37 Jozwiuk, A., Sommer, H., Janek, J., and Brezesinski, T. (2015). Fair performance comparison of different carbon blacks in lithium–sulfur batteries with practical mass loadings – Simple design competes with complex cathode architecture. *Journal of Power Sources* 296: 454–461.

38 Cheremisinoff, P.N. and Ellerbusch, F. (1978). *Carbon Adsorption Handbook*. Ann Arbor: Ann Arbor Science Publishers.

39 Bansal, R.C. and Goyal, M. (2005). *Activated Carbon Adsorption*. Boca Raton: Taylor & Francis.

40 Yihong, W., Zexiang, S., and Ting, Y. (2014). *Two-Dimensional Carbon: Fundamental Properties, Synthesis, Characterization, and Applications*, Pan Stanford Series on Carbon-Based Nanomaterials. Boca Raton: CRC Press, an imprint of Taylor & Francis.

41 Xu, F., Tang, Z., Huang, S. et al. (2015). Facile synthesis of ultrahigh-surface-area hollow carbon nanospheres for enhanced adsorption and energy storage. *Nature Communications* 6: 7221.

42 Zhong, H., Xu, F., Li, Z. et al. (2013). High-energy supercapacitors based on hierarchical porous carbon with an ultrahigh ion-accessible surface area in ionic liquid electrolytes. *Nanoscale* 5 (11): 4678–4682.

43 Holleman, A.F. and Wiberg, E. (1995). *Lehrbuch der anorganischen Chemie*. 101., verb. und stark erw. Aufl./von Nils Wiberg. Berlin: Walter de Gruyter.

44 Lozano-Castelló, D., Calo, J.M., Cazorla-Amorós, D., and Linares-Solano, A. (2007). Carbon activation with KOH as explored by temperature programmed techniques, and the effects of hydrogen. *Carbon* 45 (13): 2529–2536.

45 Wang, J. and Kaskel, S. (2012). KOH activation of carbon-based materials for energy storage. *Journal of Materials Chemistry* 22 (45): 23710–23725.

46 Yushin, G., Dash, R., Jagiello, J. et al. (2006). Carbide-derived carbons: effect of pore size on hydrogen uptake and heat of adsorption. *Advanced Functional Materials* 16 (17): 2288–2293.

47 Krawiec, P., Kockrick, E., Borchardt, L. et al. (2009). Ordered mesoporous carbide derived carbons: novel materials for catalysis and adsorption. *Journal of Physical Chemistry C* 113 (18): 7755–7761.

48 Gu, W. and Yushin, G. (2014). Review of nanostructured carbon materials for electrochemical capacitor applications: advantages and limitations of activated carbon, carbide-derived carbon, zeolite-templated carbon, carbon aerogels, carbon nanotubes, onion-like carbon, and graphene. *WIREs Energy and Environment* 3 (5): 424–473.

49 Rose, M., Korenblit, Y., Kockrick, E. et al. (2011). Hierarchical micro- and mesoporous carbide-derived carbon as a high-performance electrode material in supercapacitors. *Small* 7 (8): 1108–1117.

50 Oschatz, M., Borchardt, L., Senkovska, I. et al. (2013). Carbon dioxide activated carbide-derived carbon monoliths as high performance adsorbents. *Carbon* 56: 139–145.

51 Chmiola, J., Yushin, G., Gogotsi, Y. et al. (2006). Anomalous increase in carbon capacitance at pore sizes less than 1 nanometer. *Science* 313 (5794): 1760–1763.

52 Oschatz, M., Thieme, S., Borchardt, L. et al. (2013). A new route for the preparation of mesoporous carbon materials with high performance in lithium–sulphur battery cathodes. *Chemical Communications* 49 (52): 5832–5834.

53 Ryoo, R., Joo, S.H., and Jun, S. (1999). Synthesis of highly ordered carbon molecular sieves via template-mediated structural transformation. *Journal of Physical Chemistry C* 103 (37): 7743–7746.

54 Jun, S., Joo, S.H., Ryoo, R. et al. (2000). Synthesis of new, nanoporous carbon with hexagonally ordered mesostructure. *Journal of the American Chemical Society* 122 (43): 10712–10713.

55 Kinoshita, K. (1988). *Carbon: Electrochemical and Physicochemical Properties*. New York: Wiley-VCH.

56 Wang, M., Wang, W., Wang, A. et al. (2013). A multi-core–shell structured composite cathode material with a conductive polymer network for Li–S batteries. *Chemical Communications* 49 (87): 10263–10265.

57 Miao, L., Wang, W., Wang, A. et al. (2013). A high sulfur content composite with core–shell structure as cathode material for Li–S batteries. *Journal of Materials Chemistry A* 1 (38): 11659–11664.

58 Brückner, J., Thieme, S., Böttger-Hiller, F. et al. (2014). Carbon-based anodes for lithium sulfur full cells with high cycle stability. *Advanced Functional Materials* 24 (9): 1284–1289.

59 Lacey, M.J., Jeschull, F., Edström, K., and Brandell, D. (2014). Functional, water-soluble binders for improved capacity and stability of lithium–sulfur batteries. *Journal of Power Sources* 264: 8–14.

60 Xin, S., Gu, L., Zhao, N. et al. (2012). Smaller sulfur molecules promise better lithium–sulfur batteries. *Journal of the American Chemical Society* 134 (45): 18510–18513.

61 Zhang, B., Qin, X., Li, G.R., and Gao, X.P. (2010). Enhancement of long stability of sulfur cathode by encapsulating sulfur into micropores of carbon spheres. *Energy and Environmental Science* 3 (10): 1531–1537.

62 Rao, M., Li, W., and Cairns, E.J. (2012). Porous carbon–sulfur composite cathode for lithium/sulfur cells. *Electrochemistry Communications* 17: 1–5.

63 Elazari, R., Salitra, G., Garsuch, A. et al. (2011). Sulfur-impregnated activated carbon fiber cloth as a binder-free cathode for rechargeable Li–S batteries. *Advanced Materials* 23 (47): 5641–5644.

64 Hoffmann, C., Thieme, S., Brückner, J. et al. (2014). Nanocasting hierarchical carbide-derived carbons in nanostructured opal assemblies for high-performance cathodes in lithium-sulfur batteries. *ACS Nano* 8 (12): 12130–12140.

65 Li, X., Cao, Y., Qi, W. et al. (2011). Optimization of mesoporous carbon structures for lithium–sulfur battery applications. *Journal of Materials Chemistry* 21 (41): 16603–16610.

66 Zheng, J., Gu, M., Wagner, M.J. et al. (2013). Revisit carbon/sulfur composite for Li–S batteries. *Journal of the Electrochemical Society* 160 (10): A1624–A1628.

67 Mikhaylik, Y.V., Kovalev, I., Schock, R. et al. (2010). High energy rechargeable Li–S cells for EV application. status, remaining problems and solutions. *ECS Transaction* 25: 23–34.

68 Eroglu, D., Zavadil, K.R., and Gallagher, K.G. (2015). Critical link between materials chemistry and cell-level design for high energy density and low cost lithium-sulfur transportation battery. *Journal of the Electrochemical Society* 162 (6): A982–A990.

69 Ji, X. and Nazar, L.F. (2010). Advances in Li–S batteries. *Journal of Materials Chemistry* 20 (44): 9821–9826.

70 Tachikawa, N., Yamauchi, K., Takashima, E. et al. (2011). Reversibility of electrochemical reactions of sulfur supported on inverse opal carbon in glyme-Li salt molten complex electrolytes. *Chemical Communications* 47 (28): 8157–8159.

71 Chen, J., Zhang, Q., Shi, Y. et al. (2012). A hierarchical architecture S/MWCNT nanomicrosphere with large pores for lithium sulfur batteries. *Physical Chemistry Chemical Physics* 14 (16): 5376–5382.

72 Dörfler, S., Hagen, M., Althues, H. et al. (2012). High capacity vertical aligned carbon nanotube/sulfur composite cathodes for lithium–sulfur batteries. *Chemical Communications* 48 (34): 4097–4099.

73 Hagen, M., Dörfler, S., Fanz, P. et al. (2013). Development and costs calculation of lithium–sulfur cells with high sulfur load and binder free electrodes. *Journal of Power Sources* 224: 260–268.

74 Liang, C., Dudney, N.J., and Howe, J.Y. (2009). Hierarchically structured sulfur/carbon nanocomposite material for high-energy lithium battery. *Chemistry of Materials* 21 (19): 4724–4730.

75 Lee, J.T., Zhao, Y., Thieme, S. et al. (2013). Sulfur-infiltrated micro- and mesoporous silicon carbide-derived carbon cathode for high-performance lithium sulfur batteries. *Advanced Materials* 25: 4573–4579.

76 Schuster, J., He, G., Mandlmeier, B. et al. (2012). Spherical ordered mesoporous carbon nanoparticles with high porosity for lithium–sulfur batteries. *Angewandte Chemie International Edition* 51 (15): 3591–3595.

77 Xu, T., Song, J., Gordin, M.L. et al. (2013). Mesoporous carbon-carbon nanotube-sulfur composite microspheres for high-areal-capacity lithium–sulfur battery cathodes. *ACS Applied Materials and Interfaces* 5 (21): 11355–11362.

78 Li, D., Han, F., Wang, S. et al. (2013). High sulfur loading cathodes fabricated using peapodlike, large pore volume mesoporous carbon for lithium–sulfur battery. *ACS Applied Materials and Interfaces* 5 (6): 2208–2213.

79 Zhou, G., Wang, D., Li, F. et al. (2012). A flexible nanostructured sulphur–carbon nanotube cathode with high rate performance for Li–S batteries. *Energy and Environmental Sciences* 5 (10): 8901–8906.

80 Chen, S., Zhai, Y., Xu, G. et al. (2011). Ordered mesoporous carbon/sulfur nanocomposite of high performances as cathode for lithium–sulfur battery. *Electrochimica Acta* 56 (26): 9549–9555.

81 He, G., Ji, X., and Nazar, L. (2011). High "C" rate Li–S cathodes: sulfur imbibed bimodal porous carbons. *Energy and Environmental Science* 4 (8): 2878–2883.

82 Zhao, C., Liu, L., Zhao, H. et al. (2013). Sulfur-infiltrated porous carbon microspheres with controllable multi-modal pore size distribution for high energy lithium–sulfur batteries. *Nanoscale* 6 (2): 882–888.

83 Ding, B., Yuan, C., Shen, L. et al. (2013). Encapsulating sulfur into hierarchically ordered porous carbon as a high-performance cathode for lithium–sulfur batteries. *Chemistry: A European Journal* 19 (3): 1013–1019.

84 Xu, G., Ding, B., Nie, P. et al. (2014). Hierarchically porous carbon encapsulating sulfur as a superior cathode material for high performance lithium–sulfur batteries. *ACS Applied Materials and Interfaces* 6 (1): 194–199.

85 Jayaprakash, N., Shen, J., Moganty, S.S. et al. (2011). Porous hollow carbon@sulfur composites for high-power lithium–sulfur batteries. *Angewandte Chemie International Edition* 50 (26): 5904–5908.

86 He, G., Evers, S., Liang, X. et al. (2013). Tailoring porosity in carbon nanospheres for lithium–sulfur battery cathodes. *ACS Nano* 7 (12): 10920–10930.

87 Juhl, A.C., Ufer, B., Fröba M. (2015) Investigations of the distribution of sulfur in hollow carbon spheres. Poster presented at 4th Workshop Lithium–Sulfur-Batteries, Dresden (10–11 November 2015).

88 Böttger-Hiller, F., Kempe, P., Cox, G. et al. (2013). Twin polymerization at spherical hard templates: an approach to size-adjustable carbon hollow spheres with micro- or mesoporous shells. *Angewandte Chemie International Edition* 52: 6088–6091.

89 Li, N., Zheng, M., Lu, H. et al. (2012). High-rate lithium–sulfur batteries promoted by reduced graphene oxide coating. *Chemical Communications* 48 (34): 4106–4108.

90 Shim, J., Striebel, K.A., and Cairns, E.J. (2002). The lithium/sulfur rechargeable cell. *Journal of the Electrochemical Society* 149 (10): A1321–A1325.

91 Cheon, S., Cho, J., Ko, K. et al. (2002). Structural factors of sulfur cathodes with poly(ethylene oxide) binder for performance of rechargeable lithium sulfur batteries. *Journal of the Electrochemical Society* 149 (11): A1437–A1441.

92 Zhang, Y., Zhao, Y., Sun, K.E., and Chen, P. (2011). Development in lithium/sulfur secondary batteries. *Open Materials Science Journal* 5: 215–221.

93 Lacey, M.J., Jeschull, F., Edström, K., and Brandell, D. (2013). Why PEO as a binder or polymer coating increases capacity in the Li–S system. *Chemical Communications* 49 (76): 8531–8533.

94 Zheng, G., Zhang, Q., Cha, J.J. et al. (2013). Amphiphilic surface modification of hollow carbon nanofibers for improved cycle life of lithium sulfur batteries. *Nano Letters* 1265–1270.

95 Seh, Z.W., Zhang, Q., Li, W. et al. (2013). Stable cycling of lithium sulfide cathodes through strong affinity with a bifunctional binder. *Chemical Science* 4 (9): 3673–3677.

 96 Wu, F., Kim, H., Magasinski, A. et al. (2014). Harnessing steric separation of freshly nucleated Li_2S nanoparticles for bottom-up assembly of high-performance cathodes for lithium–sulfur and lithium–ion batteries. *Advanced Energy Materials* 1400196.

 97 Yang, Y., Yu, G., Cha, J.J. et al. (2011). Improving the performance of lithium–sulfur batteries by conductive polymer coating. *ACS Nano* 5 (11): 9187–9193.

 98 Ma, G., Wen, Z., Jin, J. et al. (2014). Enhanced performance of lithium sulfur battery with polypyrrole warped mesoporous carbon/sulfur composite. *Journal of Power Sources* 254: 353–359.

 99 Wang, H., Yang, Y., Liang, Y. et al. (2011). Graphene-wrapped sulfur particles as a rechargeable lithium–sulfur battery cathode material with high capacity and cycling stability. *Nano Letters* 11 (7): 2644–2647.

 100 Ji, L., Rao, M., Zheng, H. et al. (2011). Graphene oxide as a sulfur immobilizer in high performance lithium/sulfur cells. *Journal of the American Chemical Society* 133 (46): 18522–18525.

 101 Huang, J., Zhuang, T., Zhang, Q. et al. (2015). Permselective graphene oxide membrane for highly stable and anti-self-discharge lithium–sulfur batteries. *ACS Nano* 9 (3): 3002–3011.

 102 Rong, J., Ge, M., Fang, X., and Zhou, C. (2014). Solution ionic strength engineering as a generic strategy to coat graphene oxide (GO) on various functional particles and its application in high-performance lithium–sulfur (Li–S) batteries. *Nano Letters* 14 (2): 473–479.

 103 Song, M., Zhang, Y., and Cairns, E.J. (2013). A long-life, high-rate lithium/sulfur cell: a multifaceted approach to enhancing cell performance. *Nano Letters* 13 (12): 5891–5899.

 104 Wang, X., Zhang, Z., Qu, Y. et al. (2014). Nitrogen-doped graphene/sulfur composite as cathode material for high capacity lithium–sulfur batteries. *Journal of Power Sources* 256: 361–368.

 105 Peng, H., Hou, T., Zhang, Q. et al. (2014). Strongly coupled interfaces between a heterogeneous carbon host and a sulfur-containing guest for highly stable lithium–sulfur batteries: mechanistic insight into capacity degradation. *Advanced Materials Interfaces* 1 (7): 1400227.

 106 Sun, F., Wang, J., Chen, H. et al. (2013). High efficiency immobilization of sulfur on nitrogen-enriched mesoporous carbons for Li–S batteries. *ACS Applied Materials and Interfaces* 5 (12): 5630–5638.

 107 Song, J., Xu, T., Gordin, M.L. et al. (2014). Nitrogen-doped mesoporous carbon promoted chemical adsorption of sulfur and fabrication of high-areal-capacity sulfur cathode with exceptional cycling stability for lithium–sulfur batteries. *Advanced Functional Materials* 24 (9): 1243–1250.

 108 Song, J., Gordin, M.L., Xu, T. et al. (2015). Strong lithium polysulfide chemisorption on electroactive sites of nitrogen-doped carbon composites for high-performance lithium–sulfur battery cathodes. *Angewandte Chemie International Edition* 54 (14): 4325–4329.

 109 Seh, Z.W., Li, W., Cha, J.J. et al. (2013). Sulphur–TiO_2 yolk–shell nanoarchitecture with internal void space for long-cycle lithium–sulphur batteries. *Nature Communications* 4: 1, 1331–9.

 110 Ji, X., Evers, S., Black, R., and Nazar, L.F. (2011). Stabilizing lithium–sulphur cathodes using polysulphide reservoirs. *Nature Communications* 2: 325.

111 Choi, Y.J., Jung, B.S., Lee, D.J. et al. (2007). Electrochemical properties of sulfur electrode containing nano Al_2O_3 for lithium/sulfur cell. *Physica Scripta* T129: 62–65.

112 Song, M., Han, S., Kim, H. et al. (2004). Effects of nanosized adsorbing material on electrochemical properties of sulfur cathodes for Li/S secondary batteries. *Journal of the Electrochemical Society* 151 (6): A791–A795.

113 Lee, K.T., Black, R., Yim, T. et al. (2012). Surface-initiated growth of thin oxide coatings for Li–sulfur battery cathodes. *Advanced Energy Materials* 2 (12): 1490–1496.

114 Demir-Cakan, R., Morcrette, M., Nouar, F. et al. (2011). Cathode composites for Li–S batteries via the use of oxygenated porous architectures. *Journal of the American Chemical Society* 133 (40): 16154–16160.

115 Bao, W., Zhang, Z., Qu, Y. et al. (2014). Confine sulfur in mesoporous metal-organic framework @ reduced graphene oxide for lithium sulfur battery. *Journal of Alloys and Compounds* 582: 334–340.

116 Zheng, J., Tian, J., Wu, D. et al. (2014). Lewis acid-base interactions between polysulfides and metal organic framework in lithium sulfur batteries. *Nano Letters* 14 (5): 2345–2352.

117 Kim, H., Lee, J.T., Lee, D. et al. (2013). Plasma-enhanced atomic layer deposition of ultrathin oxide coatings for stabilized lithium–sulfur batteries. *Advanced Energy Materials* 3 (10): 1308–1315.

118 Pang, Q., Kundu, D., Cuisinier, M., and Nazar, L.F. (2014). Surface-enhanced redox chemistry of polysulphides on a metallic and polar host for lithium–sulphur batteries. *Nature Communications* 5: 4759.

119 Liang, X., Hart, C., Pang, Q. et al. (2015). A highly efficient polysulfide mediator for lithium–sulfur batteries. *Nature Communications* 6: 5682.

120 Yao, H., Zheng, G., Hsu, P. et al. (2014). Improving lithium–sulphur batteries through spatial control of sulphur species deposition on a hybrid electrode surface. *Nature Communications* 5: 1–9.

121 Hart, C.J., Cuisinier, M., Liang, X. et al. (2015). Rational design of sulphur host materials for Li–S batteries: correlating lithium polysulphide adsorptivity and self-discharge capacity loss. *Chemical Communications* 51 (12): 2308–2311.

122 Zheng, G., Yang, Y., Cha, J.J. et al. (2011). Hollow carbon nanofiber-encapsulated sulfur cathodes for high specific capacity rechargeable lithium batteries. *Nano Letters* 11 (10): 4462–4467.

123 Li, N., Yin, Y., and Guo, Y. (2016). Three-dimensional sandwich-type graphene@microporous carbon architecture for lithium–sulfur batteries. *RSC Advances* 6 (1): 617–622.

124 Xu, J., Shui, J., Wang, J. et al. (2014). Sulfur–graphene nanostructured cathodes via ball-milling for high-performance lithium–sulfur batteries. *ACS Nano* 8 (10): 10920–10930.

125 Zhang, F., Huang, G., Wang, X. et al. (2014). Sulfur-impregnated core-shell hierarchical porous carbon for lithium–sulfur batteries. *Chemistry* 20 (52): 17523–17529.

126 Guo, J., Xu, Y., and Wang, C. (2011). Sulfur-impregnated disordered carbon nanotubes cathode for lithium–sulfur batteries. *Nano Letters* 11 (10): 4288–4294.

127 Rao, M., Song, X., and Cairns, E.J. (2012). Nano-carbon/sulfur composite cathode materials with carbon nanofiber as electrical conductor for advanced secondary lithium/sulfur cells. *Journal of Power Sources* 205: 474–478.

128 Yang, X., Zhang, L., Zhang, F. et al. (2014). Sulfur-infiltrated graphene-based layered porous carbon cathodes for high-performance lithium–sulfur batteries. *ACS Nano* 8 (5): 5208–5215.

129 Ji, L., Rao, M., Aloni, S. et al. (2011). Porous carbon nanofiber–sulfur composite electrodes for lithium/sulfur cells. *Energy and Environmental Science* 4 (12): 5053.

130 Jeong, S., Lim, Y., Choi, Y. et al. (2007). Electrochemical properties of lithium sulfur cells using PEO polymer electrolytes prepared under three different mixing conditions. *Journal of Power Sources* 174 (2): 745–750.

131 Peled, E., Goor, M., Schektman, I. et al. (2016). The effect of binders on the performance and degradation of the lithium/sulfur battery assembled in the discharged state. *Journal of the Electrochemical Society* 164 (1): A5001–A5007.

132 Thieme, S. (2012). Dry processing of self-supporting sulfur cathodes. Presentation given at 1st Workshop Lithium–Sulfur-Batteries, Dresden.

133 Thieme, S., Brückner, J., Bauer, I. et al. (2013). High capacity micro-mesoporous carbon–sulfur nanocomposite cathodes with enhanced cycling stability prepared by a solvent-free procedure. *Journal of Materials Chemistry A* 1 (32): 9225–9234.

134 Chung, S. and Manthiram, A. (2014). Low-cost, porous carbon current collector with high sulfur loading for lithium–sulfur batteries. *Electrochemistry Communications* 38 (0): 91–95.

135 Hagen, M., Feisthammel, G., Fanz, P. et al. (2013). Sulfur cathodes with carbon current collector for Li–S cells. *Journal of the Electrochemical Society* 160 (6): A996–A1002.

136 Chen, Y., Lu, S., Wu, X., and Liu, J. (2015). Flexible carbon nanotube–graphene/sulfur composite film: free-standing cathode for high-performance lithium/sulfur batteries. *Journal of Physical Chemistry C* 119 (19): 10288–10294.

137 Kim, H., Lee, J.T., and Yushin, G. (2013). High temperature stabilization of lithium–sulfur cells with carbon nanotube current collector. *Journal of Power Sources* 226: 256–265.

3

Electrolyte for Lithium–Sulfur Batteries

Marzieh Barghamadi[1], Mustafa Musameh[1], Thomas Rüther[2], Anand I. Bhatt[2], Anthony F. Hollenkamp[2] and Adam S. Best[1]

[1] *CSIRO Manufacturing, Research Way, Clayton, 3168, Vic., Australia*
[2] *CSIRO Energy, Research Way, Clayton, 3168, Vic., Australia*

3.1   The Case for Better Batteries

The demand for energy storage systems that are compact, lightweight, and powerful continues to grow, mainly due to the worldwide proliferation of portable electronic devices and improved batteries for electric vehicles (EVs), and related variants (hybrid electric vehicles, HEVs; plug-in hybrid electric vehicles, PHEVs) [1–9]. On an even bigger scale, improved technologies for energy storage will also enable the incorporation of more renewable energy resources into the main (grid-based) energy supply.

Arguably, the greatest challenge for battery technology is meeting the demand for huge increases in specific energy. For EVs to achieve driving ranges exceeding 300 miles (500 km), cell-specific energy of ~500 Wh kg^{-1} is required [10–18]. By contrast, the rate of increase in specific energy for contemporary lithium-ion battery technology is slowing, with the result that reliable battery performance at levels beyond 200 Wh kg^{-1} is still some way off [10–15]. Thus, present-day lithium-ion battery technology is effectively limiting the growth of the EV, and, to some extent, the PHEV market [11, 13]. Quite clearly, an increase in energy density from present-day values (150–200 Wh kg^{-1}) to the targets of 500–700 Wh kg^{-1} requires a major breakthrough in battery materials. In this context, lithium–sulfur batteries attract attention. With elemental sulfur as the positive electrode, the theoretical specific capacity is 1672 mAh g^{-1}. Assuming an equivalent amount of lithium for the negative electrode, complete reaction of Li and S to form Li$_2$S, and an average discharge potential of 2.2 V per cell, the electrode specific energy for Li–S is 2600 Wh kg^{-1} [11, 16–34]. Fully packaged, it is expected that Li–S batteries will operate at close to 500 Wh kg^{-1}. While this level of performance places lithium–sulfur well clear of existing battery systems, the significance of this technology will ultimately depend on whether it can be made into a durable and safe device. Much of the research effort in the past decade has been devoted to the retention of sulfur and its reduction products [17, 18, 20–22]. In terms of safety, most research to date has employed flammable organic electrolyte materials, and lithium metal anodes are widely used, despite the general lack of acceptance by the battery industry. Both of these issues underline the importance of developing new electrolyte media. This chapter discusses the key role played by the electrolyte solution/medium in the Li–S battery.

3.2 Li–S Battery: Origins and Principles

The high specific energy of the Li–S system is the direct result of combining two relatively light elements as the primary active materials: a lithium metal anode and an elemental sulfur cathode. Sulfur, however, has very low conductivity (5×10^{-30} S cm^{-1} at 25 °C), as is also the case with Li$_2$S. With both the reactant and ultimate product being close-to-insulating, establishing effective charge transfer and material utilization requires that the reactant (sulfur) and all reaction products remain integrated with, and confined within, the current collector. This role is now typically filled by one of a range of mesoporous carbon-based composites (see Chapter 2) [20–34]. The lithium anode works reasonably well in the short term, but suffers from well-known forms of degradation (see Chapter 7): (i) loss of lithium due to dendritic growth on charging (ultimately leading to cell short circuits and failure); (ii) corrosion of lithium through direct reaction with polysulfides in solution (also leading to passivation of the anode by deposited lower-order lithium sulfides). The performance of both electrodes is a complex function of the components of the electrolyte solution, where beneficial effects at one electrode are often established at the expense of other characteristics.

The first developments of the lithium–sulfur battery were, not surprisingly, aimed at accessing the maximum available energy dividend through complete utilization (two-electron reduction) of sulfur [35–37]. Difficulties with the incorporation of the insulating cathode material into a suitable host, together with the recognition that the initial reduction products are freely soluble in most organic electrolyte solutions, showed that this new system would require significant development before realizing its promised levels of performance. Recognizing that the early problems with Li–S were largely associated with the dissolution of the initial products of discharging (long-chain polysulfides), Rauh et al. [27] produced a landmark study in which they proposed a positive electrode comprising an inert current collector (carbon) immersed in a solution of higher-order polysulfides (i.e. a catholyte). In solvents like dimethyl sulfoxide or ethers like tetrahydrofuran (THF), the concentration of dissolved polysulfides (as Li$_2$S$_x$) can exceed 10 M [21, 27, 30]. This approach led to the demonstration of an impressive capability for high rates and a wide range of operating temperatures as well as obviating the problems associated with solid sulfur, but was also disadvantaged by the concomitant loss of a portion of the available specific energy, with only around 300 Wh kg^{-1} being the practical achievable value [27]. The significance of this work, and hence its prominence in the literature, is that it unraveled some of the mysteries of polysulfide solution behavior and showed that the obvious complexities did not preclude operation of a rechargeable battery. Since then, most research effort has focused on trying to control the behavior of sulfur species in solution. In the past decade, as a counterpoint to work with liquid electrolyte solutions, several groups have shown that solid electrolyte media can also be used to produce a viable Li–S battery. Before exploring these two approaches in more detail, it is useful to review the basic characteristics of lithium–sulfur electrochemical cells.

During discharging of the lithium–sulfur cell, the anodic reaction (at the negative electrode) is the oxidation of lithium:

$$2Li \rightarrow Li^+ + 2e^- \tag{3.1}$$

while the cathodic reaction (at the positive electrode) is the reduction of sulfur:

$$S + 2e^- \rightarrow S^{2-} \tag{3.2}$$

The overall cell reaction is

$$2Li + S \rightarrow Li_2S \tag{3.3}$$

However, elemental sulfur exists as octasulfur (S_8) rings, which adopt a stable orthorhombic crystal structure, so that Eq. (3.3) is modified to

$$16Li + S_8 \rightarrow 8Li_2S \tag{3.4}$$

The electrochemical reduction of sulfur in lithium–sulfur cells occurs through the formation of a series of intermediate lithium polysulfides with a general formula Li_2S_x ($8 \geq x \geq 2$) followed by the final reduction product, Li_2S [16, 23, 24]. However, the exact number of stable intermediate sulfide ions during the discharge of a lithium–sulfur cell has not yet been identified beyond doubt. During discharge, it is assumed that the elemental sulfur in the solid phase $S_8(s)$ is firstly dissolved in the electrolyte as S_8 (solvated), which is then gradually reduced to lithium polysulfide. Intermediate products of high-order lithium polysulfide (Li_2S_x, $4 \leq x \leq 8$) are soluble in most of the commonly used organic solvents, but the lower-order lithium sulfides (Li_2S_2 and Li_2S) are insoluble [22–26]. During cell discharge, the polysulfide chain length is shortened as the sulfur is further reduced until the final product, lithium sulfide (Li_2S), is formed at the end of discharge. The following chemical species, polysulfide anions or radicals, have been identified during the reduction of sulfur in organic solution: S_8^{2-}, S_7^{2-}, S_6^{2-}, S_5^{2-}, S_4^{2-}, S_3^{2-}, S_2^{2-}, S^{2-}, $S_3^{-\cdot}$, $S_2^{-\cdot}$, $S^{-\cdot}$ [25].

It is well known that the polysulfides undergo a variety of disproportionation and exchange reactions in solution, to form species of varying chain length. Moreover, due to the complexity and solvent dependence of these reactions, the discharge profile of a sulfur electrode can show significant variance with the choice of electrolyte solution [28, 33]. In general though, the reduction of sulfur is mainly defined by the stepwise formation of four intermediate polysulfides, Li_2S_8, Li_2S_6, Li_2S_4, and Li_2S_2, followed by the final reduction process that yields Li_2S [20]. Solubility increases with the length of the sulfide chain – lithium sulfide is essentially insoluble in common organic solvents. The involvement of several chemically distinct species leads to a complicated voltage–time profile during discharging which, for the purposes of description, can be divided into four regions, as shown in Figure 3.1 [38]. The first region (I), where voltage falls only slightly (2.4–2.2 V vs. $Li^+|Li$), is associated with the reduction ($\frac{1}{4}$e per S atom) of S_8 to S_8^{2-}, which is soluble in most electrolyte media. Region II is characterized by a sharp fall in voltage as a further $\frac{1}{4}$e per S atom reduction takes place along with the complex series of reactions and equilibria that result in the polysulfide chain length being reduced to four sulfur atoms (Li_2S_4). From there, region III consists of a relatively stable voltage plateau (~2.1 V vs. Li^+/Li) which is associated with the $\frac{1}{2}$e reduction of the tetrasulfide to the disulfide (Li_2S_2) as well as the deeper (1e) reduction to Li_2S. The final stages of discharging occur in region IV, where the fully reduced sulfide is formed. Both the disulfide and sulfide are insoluble in most electrolyte media [26–28].

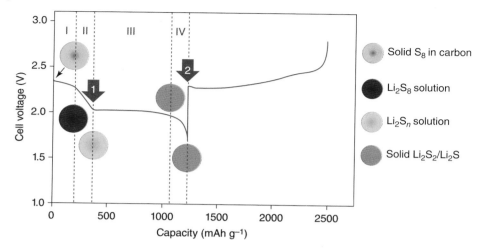

Figure 3.1 Typical discharge–charge curve for a lithium–sulfur cell. Source: Barghamadi et al. 2014 [143]. Reproduced with permission from ACS.

3.3 Solubility of Species and Electrochemistry

As mentioned earlier, and unlike virtually all other batteries, the reduction of the sulfur cathode proceeds through several distinct steps, with each intermediate product having a defined stability and solubility that is a function of the medium (identity of solutes and solvent) in which it is formed. Any soluble sulfur-based species can diffuse out of the electrode, through the separator, to the lithium negative electrode. During charging, or simply from direct chemical reaction with lithium, these soluble sulfur-containing species will be reduced and ultimately deposited as insoluble Li_2S or Li_2S_2 (Eqs. (3.5) and (3.6)). The process by which this occurs is essentially a form of self-discharge; and, more seriously than self-discharge in other systems, it results in a permanent loss of active material. This process accounts for the dramatic loss of discharge capacity with repetitive cycling that was a feature of many early studies on the lithium–sulfur system [30, 31, 35].

$$2Li + Li_2S_x \rightarrow Li_2S_{x-1} + Li_2S \downarrow \tag{3.5}$$

$$2Li + Li_2S_x \rightarrow Li_2S_{x-2} + Li_2S_2 \downarrow \tag{3.6}$$

Migration of sulfide species has further effects as the highly reduced species, Li_2S and Li_2S_2, react with the S_x^{2-} from the electrolyte to form lower-order polysulfides (S_{x-n}^{2-}) (Eqs. (3.7) and (3.8)). On diffusing back to the positive electrode, these species are then reoxidized into S_x^{2-}. The resultant internal "redox shuttle" lowers both the available discharge capacity and the efficiency of recharging. The shuttle mechanism generally becomes active at around 2.4 V and, if the concentrations of participating species are appreciable, it will support a significant parasitic current.

$$Li_2S + S_x \rightarrow Li_2S_{x+1} \tag{3.7}$$

$$Li_2S + Li_2S_x \rightarrow Li_2S_{x-m} + Li_2S_{m+1} \tag{3.8}$$

Unless a larger current is imposed by the charging system, the cell voltage will remain stable at ~2.4 V and the state of charge of the cell will remain well below 100%. While the redox shuttle established by polysulfide interconversion does, in theory, offer a measure of protection against overcharging, it does so by imposing a constant cost on the energetics of the lithium–sulfur system. Unlike in classic overcharge protection mechanisms (e.g. water electrolysis in aqueous battery systems), where the overcharge reaction is largely dormant at normal operating potentials, the dissolution and subsequent reactions of the sulfide species proceed throughout all phases of service. For the Li–S battery, it is a key goal to find ways of limiting the liberation of polysulfides and their subsequent transport through the cell. Hence, the selection and formulation of electrolyte solutions are crucial in deciding how migration of electrode products can influence the electrochemical performance of Li–S cells.

3.4   Liquid Electrolyte Solutions

Generally, lithium battery electrolytes are based on solvents from two groups: (i) organic carbonates, such as ethylene carbonate (EC), propylene carbonate (PC), dimethyl carbonate (DMC), and diethyl carbonate (DEC); (ii) ethers, such as 1,3-dioxolane (DOL), 1,2-dimethoxyethane (DME), and tetra(ethylene glycol) dimethyl ether (TEGDME), tri(ethylene glycol) dimethyl ether (triglyme), diglyme, etc. Electrolyte solutions for battery applications invariably consist of mixtures of these solvents so as to allow optimization against a range of key parameters (volatility, viscosity, conductivity, etc.). Early on, it became clear that carbonate solvents underwent a variety of reactions with reduced sulfur species. These were later characterized by X-ray absorption spectroscopy (XAS) analysis, with the result that this class of solvents is no longer considered for use in Li–S batteries [39]. Chang et al. [40] presented an early study on the optimization of the ratio of TEGDME and DOL on the basis of solvent properties (especially with respect to the polysulfides) and showed that a 1 : 2 volume ratio of TEGDME:DOL gave the best discharge capacity, although lithium triflate ($LiCF_3SO_3$) was the only salt employed. Choi et al. [41] worked at refining the best mixtures of ether solvents by comparing the performance of single- and binary-solvent-based electrolytes with lithium triflate as lithium salt. Their studies on systems with TEGDME, DME/DEGDME (di(ethylene glycol)dimethyl ether) and DOL/TEGDME revealed the beneficial effect of DOL, which is attributed to the lowering of viscosity and improved compatibility with lithium. They also demonstrated that, in addition to the electrolyte composition, the volume of electrolyte solution was also shown to influence the first discharge capacity and cycling stability of the Li–S cells. This idea was developed in greater detail later by Zhang [42]. Ryu et al. [43] investigated discharge characteristics of Li-TEGDME-S cells at low temperatures and found that performance is improved by adding DOL and methyl acetate (MA) to TEGDME solutions. The optimum composition of this mixed electrolyte is MA/DOL/TEGDME (5 : 47.5 : 47.5, v/v). The Li–S cell using the optimum electrolyte has the first discharge capacity of 994 and 1342 mAh g^{-1}[S] at −10 and +20 °C, respectively. However, while the addition of MA improved the initial cell performance, capacity fading was higher after about 20 cycles. Of the other organic additives that were evaluated, toluene added to 1 M lithium triflate in TEGDME electrolyte was found to improve the initial discharge capacity, but only from a very low base, and, as with MA, long-term performance was unclear. Finally, the addition of γ-butyrolactone (GBL) (2.5 vol%), which has a relatively high dielectric constant, was shown to adversely affect the cycle performance [41].

The composition of the electrolyte solution used by most Li–S research groups is up to 1 M of either lithium triflate or lithium bis(trifluoromethanesulfonyl)imide (LiTFSI, $Li(CF_3SO_2)_2N$) in a 1 : 1 mixture of TEGDME and DOL [44]. While as noted earlier, the ether solvents are chosen because they are generally stable in the presence of lithium, it has been shown that dioxolane is susceptible to cleavage and formation of alkoxylithium fragments [45–47]. Similarly, while lithium perchlorate has been used by some groups, the performance is inferior [44, 48], presumably due to the inability to form a stable SEI (solid electrolyte interphase) on the lithium anode. The perfluoroalkylsulfonate electrolyte salts are not completely stable under strongly reducing conditions, but they are known to form a stable interphase layer on a lithium electrode surface [49–52]. It is apparent, however, that this does not create sufficient long-term stability in a Li–S cell, which has led recently to two more changes to the composition of the liquid electrolyte.

At Sion Power, Mikhaylik [53] patented the incorporation of 0.2–0.4 M $LiNO_3$ in the electrolyte solution. The presence of this compound was linked to the limitation of the effects of the polysulfide shuttle. However, it was not until a study by Aurbach et al. [54] detailed the differences in surface chemistry of the lithium anode in the presence and absence of nitrate that some understanding of the mechanism of action became possible. When nitrate was present, the main differences in lithium surface composition were the higher abundance of oxy-sulfur and oxy-nitrogen species. The direct implication of these findings was that their presence made the SEI much less permeable to polysulfide species, and thereby greatly limited the reactions between these compounds and the lithium electrode. Recently, more details of the effects of nitrate on the morphology of the lithium electrode have been provided by Xiong et al. [55] With scanning probe microscopy (SPM) combined with X-ray photon spectroscopy (XPS), they demonstrated the formation of a smooth and compact film on the lithium surface – such film properties (cf. porous, open morphology) would be expected in a surface layer that limited access by bulky (e.g. polysulfide) ionic species. Kim et al. [56] also recently provided new details on the reactivity of nitrate with lithium sulfide, through a study that involved infrared spectroscopy and electrochemical impedance spectroscopy (EIS).

Interestingly, the electrolyte solution promoted by Sion Power also included an appreciable concentration of lithium hexasulfide (Li_2S_6) [53]. In broad chemical terms, this additive will shift the equilibria that involve the higher polysulfides, thereby altering the concentration of a number of other species in solution. It will also increase the viscosity of the electrolyte solution as, for the same concentrations, the viscosity of Li_2S_x solutions (e.g. in sulfolane) is significantly higher than that of corresponding electrolyte solutions made with $LiClO_4$ or $LiCF_3SO_3$ [57]. This is most likely due to the strong self-association of lithium polysulfides in solutions. Whatever the benefits of setting solution concentrations of polysulfide species at high levels, they will ultimately be limited by the point at which high viscosity slows ion movement, particularly in the mesoporous cathodes. Adding another aspect to this discussion, Xiong et al. [58]

very recently provided evidence that links the presence in solution of hexasulfide with the formation of a protective "under layer" of lithium sulfide on the lithium electrode surface. They, in effect, propose that the SEI on lithium in Li–S cells develops a bilayer morphology in which the outer oxy-sulfur layer (promoted by the presence of nitrate) is supported in its protective role by the formation of more reduced forms of sulfur close to the electrode. In a parallel study, this group showed that $LiNO_3$ could also have undesirable effects, such as contributing to the (largely irreversible) oxidation of sulfur to oxy-sulfur species that deposit in the cathode and constitute lost active materials [59].

In recent years, one of the strongest and most prolific research groups advocating the further development of liquid electrolyte solutions for Li–S batteries has been that of Zhang at the US Army Research Laboratories. In a recent review of the liquid electrolyte Li–S battery, they discuss in detail all the issues that are currently hampering commercialization of the technology, with particular attention paid to the problems created by the polysulfide shuttle. After considering a wide variety of strategies for dealing with the effects of this phenomenon, Zhang concludes in favor of promoting the dissolution of polysulfide species and then managing the interaction of these species with cell components, especially the electrodes [60, 61]. In support, they cite earlier work by the group, which has greatly improved our understanding of the mechanism of action of $LiNO_3$ [62, 63]. While the presence of nitrate is clearly beneficial to the extended cycling of the lithium anode, Zhang shows that there are adverse effects at the cathode that can offset the benefit. From extensive voltammetric studies, it is seen that discharging the sulfur cathode to low values (below a critical value of 1.7 V vs. Li|Li$^+$) causes irreversible reduction of lithium nitrate and also renders the ultimate discharge product (Li_2S) even more difficult to recharge [62]. The magnitude of these effects, in terms of electrode impedance, is also quantified by means of impedance spectroscopy. Overall, it is concluded that liquid electrolyte Li–S cells must be provided with what amounts to a reserve of $LiNO_3$ so as to account for the quantities that are inevitably lost to (i) formation of the SEI on the lithium electrode and (ii) trace consumption of each cycle at the cathode due to the difficulty of completely separating sulfur-based reduction from nitrate reduction [63].

A second contribution to our understanding of the influence of electrolyte properties on the Li–S system concerns the role of average concentration of polysulfide species during operation of the cell. Here, the USARL group [42] has shown that by adjusting the ratio of electrolyte solution volume to cathode sulfur loading (E/S), they create a trade-off between good initial capacity performance that rapidly decreases (high E/S makes recharging difficult) and low capacity performance that remains steady with cycling (low E/S leads to highly viscous solutions that impede diffusion but limit loss of polysulfides). This suggests that earlier research on the effects of dissolved higher polysulfides on the properties of lithium anodes should now be qualified, or even reevaluated, by an analysis of the electrolyte composition in terms of E/S. As with the discussion of the role of $LiNO_3$, and also now with regard to dissolved polysulfide species, no aspect of the Li–S battery can ever be treated in simple terms. A summary of the main performance characteristics for solvent-based liquid electrolytes is included in Table 3.1.

Table 3.1 Electrolyte effects of cell cycling.

Electrolyte type and composition	Additive	Indicative conductivity (S cm^{-1})/T (°C)	Initial discharge capacity (mAh g^{-1})/current density/T (°C)	Residual discharge capacity (mAh g^{-1})/cycles	Comments	References
IL: P14-based[a]	P14-TFSI: TEGDME (1 : 1, w/w)	0.2 mol kg^{-1} LiTFSI	6.3 × 10^{-4}/17	887/0.054 mA cm^{-2}/RT	420/20	Adding organic solvent increases ionic conductivity of the electrolyte	[66]
	P14-TFSI: TEGDME (1/2, w/w)	0.2 mol kg^{-1} LiTFSI	8.6 × 10^{-4}/17	—	—		
	P14-TFSI: TEGDME (1 : 1, v/v)	1 mol kg^{-1} LiTFSI	2.3 × 10^{-3}/23	650/0.1 C/RT	450/100	Adding LiNO$_3$ to the electrolyte leads to better capacity retention due to formation of improved SEI layer	[144]
	P14-TFSI: TEGDME (1 : 1, v/v)	1 mol kg^{-1} LiTFSI + 0.1 mol kg^{-1} LiNO$_3$	2.7 × 10^{-3}/23	650/0.1 C/RT	200/100		
	P14-TFSI: PEGDME (1/2, w/w)	0.5 M LiTFSI	4.2 × 10^{-3}/29	450/0.054 mA cm^{-2}/40	250/100	Adding a proper amount of organic modifier to the pure IL leads to higher and more stable capacity due to higher conductivity. Added organic introduces volatility, up to a worst case of ~5% weight loss at 140 °C (TGA). Organic free electrolyte has <1% weight loss at 300 °C	[67]

Electrolyte	Salt	Conductivity	Capacity/rate	Cyclability	Notes	Ref.
P14-TFSI: PEGDME (1:1.5, w/w)	0.5 M LiTFSI	$3.5 \times 10^{-3}/29$	500/0.054 mA cm^{-2}/40	180/100		[67]
P14-TFSI: PEGDME (1:1, w/w)	0.5 M LiTFSI	$3.8 \times 10^{-3}/29$	320/0.054 mA cm^{-2}/40	100/80		[67]
P14-TFSI: PEGDME (1:0.1, w/w)	0.5 M LiTFSI	$2.9 \times 10^{-3}/29$	—	—		[67]
P14-TFSI	0.5 M LiTFSI	$1.7 \times 10^{-3}/30$	120/0.054 mA cm^{-2}/40	20/50		[67]
P14-TFSI: DOL/DME (1:2, v/v)	1 M LiTFSI	—	1073/—/—	648/100	Different percentage of IL is reported. Addition of IL suppresses PSs dissolution and leads to better capacity retention	[145]
DOL:DME (1:1, v/v)	1 M LiTFSI	—	854/—/—	249/100		
P14-TFSI: TEGDME (1:1, v/v)	1 mol kg^{-1} LiTFSI + 0.1 mol kg^{-1} LiNO$_3$	$2.7 \times 10^{-3}/23$	700/0.1 C/RT	500/100	The anion of the IL has direct influence on the polysulfide's solubility and consequent cell performance	[146]
P14-FAP: TEGDME (1:1, v/v)	1 mol kg^{-1} LiTFSI + 0.1 mol kg^{-1} LiNO$_3$	$1.5 \times 10^{-3}/23$	~460/0.1 C/RT	220/100		
P14-OTf: TEGDME (1:1, v/v)	1 mol kg^{-1} LiTFSI + 0.1 mol kg^{-1} LiNO$_3$	$1.9 \times 10^{-3}/23$	~540/0.1 C/RT	200/100		
P14-TCM: TEGDME (1:1, v/v)	1 mol kg^{-1} LiTFSI + 0.1 mol kg^{-1} LiNO$_3$	$3.2 \times 10^{-3}/23$	~580/0.1 C/RT	155/100		

(Continued)

Table 3.1 (Continued)

Electrolyte type and composition	Additive	Indicative conductivity (S cm⁻¹)/T (°C)	Initial discharge capacity (mAh g⁻¹)/current density/T (°C)	Residual discharge capacity (mAh g⁻¹)/cycles	Comments	References
P14-TFSI	0.4 mol kg⁻¹ LiTFSI	1×10^{-3}/25	256/50 mA g⁻¹	220/30	Higher capacity and better capacity retention is achieved after adding organic solvent as a result of improved ion transport due to higher conductivity	[147]
P14-TFSI: DOL (10 : 90, wt%)	0.4 mol kg⁻¹ LiTFSI	4.2×10^{-3}/25	550/50 mA g⁻¹	~600/30		
P14-TFSI	0.5 mol kg⁻¹ LiTFSA	1.4×10^{-3}/30	720/139 mA g⁻¹/30	550/50	Different performances confirm that the anionic structure (subsequently donor ability) of IL directly affects PSs' solubility	[83]
P14-OTf	0.5 mol kg⁻¹ LiTFSA	1.1×10^{-3}/30	650/139 mA g⁻¹/30	190/50		[83]
IL: P13-based[b] P13-TFSI	0.5 mol kg⁻¹ LiTFSA	2×10^{-3}/30	800/139 mA g⁻¹/30	700/50		[83]
P13-BETI	0.5 mol kg⁻¹ LiTFSA	0.47×10^{-3}/30	620/139 mA g⁻¹/30	200/50		[83]
P13-FSI	0.5 mol kg⁻¹ LiTFSA	6.4×10^{-3}/30	1000/139 mA g⁻¹/30	50/50		[83]
P13-TFSI	0.5 mol l⁻¹ LiTFSI		1400/0.1 C/—	500/50	Cell performance is affected by cation of the IL, here P13-based ILs suppress shuttling problems more effectively	[148]

	PMIM^(c)-TFSI	1 mol l⁻¹ LiTFSI		1200/0.1 C/—	100/50		[74]
1b: C₂mim-based	C₂mim-TFSI	1 M LiTFSI	—	1300/50 mA g⁻¹	500/40		[83]
IL: PP13-based^(d)	PP13-TFSI	0.5 mol kg⁻¹ LiTFSA	0.78×10^{-3}/30	600/139 mA g⁻¹/30	520/50	Organic modifier increases capacity by increasing electrolyte conductivity	[83]
	PP13-TFSI/DME (2 : 1, v/v)	1 M LiTFSI		1000/0.2 C	900/50		[75]
	PP13-TFSI	1 M LiTFSI		405/0.1 C	320/10		[75]
IL: DEME-based	DEME-BF₄	0.5 mol kg⁻¹ LiTFSA	1.4×10^{-3}/30	900/139 mA g⁻¹/30	50/50	The capacity of the battery with the BF₄ anion electrolyte declines in the first few cycles due to irreversible reaction between this anion and PSs	[83]
	DEME-TFSA	0.5 mol kg⁻¹ LiTFSA	1.4×10^{-3}/30	780/139 mA g⁻¹/30	650/50		[83]
	DEME-TFSA	0.64 M LiTFSA	1.30×10^{-3}/30	800/139 mA g⁻¹/30	580/100		[82]
	DEME-TFSA	—	3.44×10^{-3}/30	—	—		[82]
IL: P2225-based^(e)	P2225-TFSA	0.5 mol kg⁻¹ LiTFSA	0.92×10^{-3}/30	700/139 mA g⁻¹/30	500/50	Cations of IL can affect the battery performance as a result of difference in viscosity and Li ion transportation	[83]
IL: C₄dmim-based	C₄dmim-TFSA	0.5 mol kg⁻¹ LiTFSA	1.4×10^{-3}/30	750/139 mA g⁻¹/30	580/50		[83]
Solvate IL	Li(G4)-TFSI	—	1×10^{-3}/RT	1100/139 mA g⁻¹/RT	700/50	Solvate IL suppresses PSs' solubility which gives a battery with stable capacity	[68]

(Continued)

Table 3.1 (Continued)

Electrolyte type and composition	Additive	Indicative conductivity (S cm^{-1})/T (°C)	Initial discharge capacity (mAh g^{-1})/current density/T (°C)	Residual discharge capacity (mAh g^{-1})/cycles	Comments	References
Li(G3)$_1$-TFSI		1×10^{-3}/30	1100/0.083 C/30	700/400		[85]
Li(G3)$_4$-TFSI		—	800/0.083 C/30	500/35		[85]
Li(G4)$_1$-TFSI/HFE (1 : 1 : 4)		5.2×10^{-3}/30	1000/0.083 C/30	750/50		[85]
[Li(G4)$_1$][TFSI]		1×10^{-3}/30	900/0.083 C/30	600/50		[85]
Li(G$_3$)-OTf		0.3×10^{-3}/30	280/139 mA g^{-1}/30	150/100		[86]
Li(G$_3$)-NO$_3$		0.31×10^{-3}/30	380/139 mA g^{-1}/30	30/100		[86]
Li(G$_4$)-BETI		0.91×10^{-3}/30	700/139 mA g^{-1}/30	550/100		[86]
Li(G$_4$)-BF$_4$		0.5×10^{-3}/30	400/139 mA g^{-1}/30	100/100		[86]
Solid electrolyte 80Li$_2$S-20P$_2$S$_5$ (mol%)		1×10^{-3}/RT	1200/0.64 mA cm^{-2}/RT	996/200	Battery operates between −20 and 80 °C. The capacity decreases at higher temperature	[101, 109]
80Li$_2$S-20P$_2$S$_5$ (mol%)		$10^{-4} - 1 \times 10^{-3}$/0–80 °C	400/0.05 C/80 °C	400/20		[110]
thio-LISICON$^{f)}$			1600/0.13 mA cm^{-2}	700/10		[115]
EMI-TFSI-LiTFSI-fumed silica (QSE^{g})		4.5×10^{-3}/35	1210/0.05 C/35	~800/45	The type of the cations in QES influences the cell performance. Also, capacities are different with thicker QSE sheets	[149]
DEME-TFSI-LiTFSI-fumed silica(QSE)		1.3×10^{-3}/35	1100/0.05 C/35	~900/45		

Polymer electrolyte							
	PP13-TFSI-LiTFSI-fumed silica(QSE)		1.4×10^{-4}/35	790/0.05 C/35	~300/20		
	PVDF activated in 1 M LiOTf in TEGDME		0.7×10^{-3}/RT	860/0.167 C/RT	311/50	The composition of the organic solvent affects the polymer electrolyte performance; here, the more DOL content leads to the faster capacity decline	[41]
	PVDF activated in 1 M LiOTf in TEGDME:DOL (2:1, v/v)		0.7×10^{-3}/RT	1056/0.167 C/RT	210/50		[41]
	PVDF activated in 1 M LiOTf in TEGDME:DOL (1:1, v/v)		0.7×10^{-3}/RT	1000/0.167 C/RT	220/50		[41]
	PVDF activated in 1 M LiOTf in TEGDME:DOL (1:2, v/v)		0.7×10^{-3}/RT	860/0.167 C/RT	270/50		[41]
	PVDF-TEGDME-LiOTF	M.P. = 100 °C $T_g = -117$ °C	$3:28 \times 10^{-4}$/30	1200/0.14 mA cm^{-2}/RT	400/10	The composition of the organic solvent affects the polymer electrolyte performance; here LiPF$_6$ shows the best result due to higher conductivity	[89]
	PVDF-TEGDME-LiBF$_4$	M.P. = 98 °C $T_g = -99$ °C	6.22×10^{-4}/30	1500/0.14 mA cm^{-2}/RT	400/10		[89]

(Continued)

Table 3.1 (Continued)

Electrolyte type and composition	Additive	Indicative conductivity (S cm^{-1})/T (°C)	Initial discharge capacity (mAh g^{-1})/current density/T (°C)	Residual discharge capacity (mAh g^{-1})/cycles	Comments	References
PVDF-TEGDME-LiPF$_6$	M.P. = 103 °C T_g = −97 °C	1.88×10^{-3}/30	1500/0.14 mA cm^{-2}/RT	500/10		[89]
PEO:LiTFSI (mass ratio 49/1)		4.9×10^{-4}/90	700/0.1 mA cm^{-2}/104°	270/10		[96]
PEMO: LiTFSI (20/1)		4.9×10^{-5}/23, 1.2×10^{-4}/60	220/0.025 mA cm^{-2}/60	30/10		[96]
PEO–LiOTf: 10 mass% ZrO$_2$		$\sim 1 \times 10^{-4}$/$\sim$70	180/0.05 C/70	200/25		[100]
Silica contained polymer electrolyte			1648/0.2 C/	1143/100	Polymer electrolyte facilitates lithium-ion transport and suppresses PS shuttling effect, which leads to better performance	[94]
EC:DEC (1 : 1, v/v)	1 M LiPF$_6$		1610/0.2 C	~350/100		
PVDF/HFP		1.2×10^{-3}/—	440/0.3 mA cm^{-2}/RT	360/25		[88]
Organic electrolyte PEGDME 500	1 M LiTFSI	—	1200/50 mA g^{-1}	400/40		[74]

TEGDME	0.5 M LiTFSI	8.1×10^{-5}/−10	330/10 mA g⁻¹/−10			[43]
						Both conductivity and capacity increase in the present of higher percentage of DOL. Also, using modified amount additives (LiNO$_3$ and MA) improves the battery performance
TEGDME:DOL (1:1, v/v)	0.5 M LiTFSI	9.6×10^{-5}/−10°	760/10 mA g⁻¹/−10			[43]
TEGDME:DOL (1:1, v/v)	1 M LiTFSI	7×10^{-3}/RT	720/0.1 C/—	460/20		[44]
TEGDME/DOL (33:67, v/v)	1 M LiTFSI +0.2 M LiNO$_3$	8×10^{-3}/—	900/0.1 C/—	600/20		[56]
TEGDME:DOL (1:1, v/v) + 5% v MA	0.5 M LiTFSI	1.2×10^{-4}/−10°	900/10 mA g⁻¹/−10			[43]
TEGDME:DOL(1:1, v/v) + 10% v MA	0.5 M LiTFSI	1.4×10^{-4}/−10	600/10 mA g⁻¹/−10			[43]
TEGDME	1 M LiOTf	—	400/0.063 C/RT	390/50		[41]
						The result is a proof of the positive effects of adding DOL to this electrolyte
TEGDME:DOL (1:1, v/v)	1 M LiOTf	—	500/0.063 C/RT	500/50		[41]
TEGDME:DOL (30/70, v/v)	0.5 M LiOTf	15×10^{-3}/—				[40]
TEGDME:DOL (70/30, v/v)	0.5 M LiOTf	1×10^{-3}/—				[40]

(Continued)

Table 3.1 (Continued)

Electrolyte type and composition	Additive	Indicative conductivity (S cm⁻¹)/T (°C)	Initial discharge capacity (mAh g⁻¹)/current density/T (°C)	Residual discharge capacity (mAh g⁻¹)/cycles	Comments	References
TEGDME:DOL (100/00, v/v)	0.5 M LiOTf	0.5×10^{-3}/—				[40]
TEGDME:DOL (00 : 100, v/v)	0.5 M LiOTf	1×10^{-4}/—				[40]
TEGDME:DOL (50/50, v/v)	—	12×10^{-3}/—				[40]
TEGDME:DOL (1 : 1, v/v)	1 M LiClO₄	5×10^{-3}/RT	1000/0.1 C	310/20	Appreciable effect on cell performance from choice of lithium salt anion	[44]
TEGDME:DOL(1:1, v/v)	1 M LiOTf	1.5×10^{-3}/RT	750/0.1 C	450/20		[44]
TEGDME	0.98 M LiTFSA	3.2×10^{-3}/30	950/139 mA g⁻¹/30	500/60		[82]
DOL	0.5 M LiTFSI	9.8×10^{-5}/−10	550/10 mA g⁻¹/−10			[43]
DOL.DME (4/1, v/v)	1 M LiClO₄		900/0.4 mA cm⁻²/RT	320/20	Modified proportion of DOL/DME improves the battery performance. DOL can improve electrode/electrolyte interfacial properties and DME modifies PS solubility	[48]
DOL:DME (2/1, v/v)	1 M LiClO₄		1200/0.4 mA cm⁻²/RT	50/20		[48]

DOL:DME (1 : 1, v/v)	1 M LiClO$_4$	7×10^{-3}/RT	1050/0.4 mA cm^{-2}/RT	510/20		[44, 48]
DOL:DME (1/2, v/v)	1 M LiClO$_4$		1000/0.4 mA cm^{-2}/RT	500/20		[48]
DOL:DME (2/1, v/v)	1 M LiClO$_4$		950/0.4 mA cm^{-2}/RT	260/20		[48]
DOL:DME (1 : 1, v/v)	1 M LiClO$_4$	7×10^{-3}/RT	750/0.1 C	500/20	Battery with LiClO$_4$ shows lower capacity retention	[44]
DOL:DME (1 : 1, v/v)	1 M LiTFSI	11×10^{-3}/RT	1000/0.1 C	520/20		[44]
DOL:DME (1 : 1, v/v)	1 M LiOTf	2×10^{-3}/RT	900/0.1 C	550/20		[44]
DOL:DME (1 : 1, v/v)	1 M LiTFSI	13×10^{-3}/RT	1200/0.1 C/RT	450/120		[77]
DOL:DME (1 : 1, v/v)	0.25 mol kg^{-1} LiTFSI + 0.25 mol kg^{-1} P14TFSI		1000/0.2 mA cm^{-2}	450/200	Adding LiNO$_3$ is effective in forming a stable SEI layer on Li anode	[61]
DOL:DME (1 : 1, v/v)	0.25 mol kg^{-1} LiNO$_3$ + 0.25 mol kg^{-1} P14TFSI		1100/0.2 mA cm^{-2}	500/200		[61]
DOL:DME (1 : 1, v/v)	0.25 mol kg^{-1} LiNO$_3$ + 0.25 mol kg^{-1} TBAOTf		1200/0.2 mA cm^{-2}	500/120		[61]
DOL:DME (1/11 : 1, v/v)	1 mol kg^{-1} LiTFSI	12.1×10^{-3}/23	1150/0.1 C/RT	~900/100	Better capacity retention in LiNO$_3$-containing electrolyte is due to improved SEI layer on the surface of the lithium anode	[144]
DOL:DME (1/11 : 1, v/v)	1 mol kg^{-1} LiTFSI + 0.1 mol kg^{-1} LiNO$_3$	11.9×10^{-3}/23	1100/0.1 C/RT	~300/100		

(Continued)

Table 3.1 (Continued)

Electrolyte type and composition	Additive	Indicative conductivity (S cm^{-1})/T (°C)	Initial discharge capacity (mAh g^{-1})/current density/T (°C)	Residual discharge capacity (mAh g^{-1})/cycles	Comments	References
DOL:DME (1:1, v/v)	1.5 M LiTFSI		850	600/100	Combining LiFSI and LiTFSI leads to better sulfur utilization (due to LiFSI) and improved cycle stability (due to LiTFSI)	[150]
DOL:DME (1:1, v/v)	1.5 M LiFSI		1360	500/100		
DOL:DME (1:1, v/v)	0.5 M LiTFSI +1 M LiFSI		1330	700/100		
DOL:DME (1:1, v/v)	1 M LiTFSI +0.5 M LiFSI		900	650/100		
DOL:DME (1:1, v/v)	1.0 M LiOTf		~1000/	800/10	Other salts are also tested, but the aim of the paper is mainly discussing the key role of the electrolyte solvent on the cell performance	[39]
TEGDME	1.0 M LiOTf		~1050/	~400/10		
PC$^{h)}$:EC$^{i)}$:DEC$^{j)}$ (1:4:5, v/v)	1.0 M LiOTf		350/	~100/10		
EMS$^{k)}$:DEC (8:1, v/v)	1.0 M LiOTf		800/	~100/10		
EMS: DOL:DME (4:1:1, v/v)	1.0 M LiOTf		~1000/	~420/10		

Electrolyte	Salt		Rate	Capacity/cycle	Remarks	Ref
DOL:TEGDME (1:1, v/v)	1 M LiTFSI		800/0.1 C/RT	~400/10		[151]
DOL:DME (1:1, v/v)	1 M LiTFSI		450/0.1 C/RT	~400/10		
DOL:PEGDME (1:1, v/v)	1 M LiTFSI		1100/0.1 C/RT	~550/10		[152]
DOL:DME (1:1, v/v)	1 M LiTFSI +0.2 M LiNO$_3$				Adding TTE reduces self-discharge of Li–S cells. It is also showed that TTE participates in forming a stable SEI on both cathode and anode	
DOL:TTE$^{b)}$ (1:1, v/v)	1 M LiTFSI +0.2 M LiNO$_3$		1400/0.1 C/RT	~1020/50		[153]
DOL:TTE (1:1, v/v)	1 M LiTFSI		1079/0.1 mA cm^{-1}	~320/50	Improved cycling stability is shown after adding LiNO$_3$ due to effective SEI formation on the anode	[154]
DOL:TEGDME (1:1, v/v)	0.5 M LiOTf		1138/0.1 mA cm^{-1}	527/50		
DOL:TEGDME (1:1, v/v)	0.5 M LiOTf +0.4 M LiNO$_3$					
DOL	1 M LiTFSI	0.004/60	~600/0.1 C/RT	~275/10	The organic electrolytes ratio is optimized to achieve the highest conductivity and improve PS solvation ability of the electrolyte	[155]
DOL:TEGDME (75:25, v/v)	1 M LiTFSI	0.2/60	1000/0.1 C/RT	~500/10		
TEGDME	1 M LiTFSI	0.008/60	600/0.1 C/RT	250~10		
DME:DEGDME (1:1, v/v)	1 M LiOTf	0.65×10^{-3}/RT	944/0.063 C/RT	350/50		[41]

(Continued)

Table 3.1 (Continued)

Electrolyte type and composition	Additive	Indicative conductivity (S cm⁻¹)/T (°C)	Initial discharge capacity (mAh g⁻¹)/current density/T (°C)	Residual discharge capacity (mAh g⁻¹)/cycles	Comments	References
1NM3[m]	1 M LiPF$_6$	1.2×10^{-3}/RT	1200	400/50	LiDFOB enables reversible sulfur redox by forming an efficient passivating film on the surface of sulfur cathode	[156]
1NM3	1 M LiPF$_6$ + 2 wt% LiDFOB	—	1300	~500/50		
1NM3	1 M LiTFSI		1300	500/20		
1NM3	1 M LiTFSI+2 wt% LiDFOB		1300	~550/20		
1NM3	1 M LiPF$_6$		~1200	~450/40		
1NM3	1 M LiPF$_6$ + 2 wt% LiNO$_3$		~1200	~450/40		

PS = polysulfide; LiDFOB = lithium difluoro oxalato borate.

a) P14, 1-butyl-1-methyl-pyrrolidinium (also represented as C$_4$mpyr$^+$).
b) P13, 1-propyl-1-methyl-pyrrolidinium (also represented as C$_3$mpyr$^+$).
c) 1-Methyl-3-propyl imidazolium cation.
d) PP13, 1-propyl-1-methyl-piperidinium (also represented as C$_3$mpip$^+$).
e) P2225, triethyl-pentyl-phosphonium.
f) Li$_{3.25}$Ge$_{0.25}$P$_{0.75}$S$_4$.
g) Quasi-solid-state electrolytes.
h) Propylene carbonate.
i) Ethylene carbonate.
j) Diethyl carbonate.
k) Ethyl methyl sulfone.
l) 1,1,2,2-Tetrafluoroethyl-2,2,3,3-tetrafluoropropyl ether.
m) Tri(ethylene glycol)–substituted trimethyl silane.

3.5 Modified Liquid Electrolyte Solutions

3.5.1 Variation in Electrolyte Salt Concentration

First consideration is given to the recent work of Shin et al. [64] who have taken a similar approach to Zhang [42] in looking at solubility effects, but have used lithium-ion concentration as the principal variable. They reason that increasing the concentration of the primary lithium salt (LiTFSI, in their case) will likely suppress the solubility of lithium polysulfides. Having shown that this indeed is the case, they also suggest that the high viscosity of the electrolyte solutions containing >4 M LiTFSI impedes the diffusion of polysulfides to such a degree that the efficiency of charging is greatly improved. The results presented by Shin et al. [64] are certainly impressive, with initial discharge capacities holding at around 1000 mAh g^{-1} for around 50 cycles.

More recently, the idea of pushing salt concentration to even higher values has been presented by Suo et al. [65] on what they term "solvent-in-salt" systems. Here the authors have extended the concentration range of typical battery electrolyte solutions from around 1 mol of solute per liter of solvent up to 7 mol per liter. In the latter region, they show that LiTFSI – (DOL-DME) solutions certainly become more viscous (>60 cP for the 7 : 1 solution), but at the same time some important characteristics appear. Of particular relevance to Li–S batteries, there is a large increase in transference number for the lithium ion (t_{Li+} = 0.7) and an appreciable decrease in the solubility of polysulfide species. XPS data suggest that the SEI on lithium is thicker than usual, and this appears to be consistent with the overall lower voltages recorded during discharge (i.e there is significantly more cell internal resistance for the highest salt concentration.) Nevertheless, the cells with 7 : 1 electrolyte solutions yielded discharge capacity that remained above 800 mAh g^{-1} (based on sulfur) for 100 cycles, at close to 100% coulombic efficiency. These results, together with those of Shin et al. [64], are good grounds for further investigation of these "solutions" with extremely limited liquid component.

3.5.2 Mixed Organic–Ionic Liquid Electrolyte Solutions

The possible effects of high concentrations of bis(trifluoromethanesulfonyl)imide (TFSI) are also critical to a consideration of research conducted on a range of mixed organic–ionic liquid (IL) electrolyte media. The commonly used ILs are the relatively hydrophobic and electrochemically stable salts of the TFSI anion. This anion is typical of the weakly Lewis acidic ionic species that comprise the subgroup of ILs that are used widely in electrochemical applications [49, 51]. Room-temperature ionic liquids (RTILs), also known as organic molten salts, consist only of cations and anions, and exhibit the important properties of good electrochemical stability, low volatility, good thermal and chemical stabilities, and are also environmentally benign. In the context of Li–S battery electrolytes, the use of ILs can decrease the solubility of polysulfides, increase the ionic conductivity of the electrolyte, and, in some cases, lower the viscosity [66–70].

In what appears to be the first reported use of a mixed organic–IL electrolyte solution in a Li–S cell, Shin and Cairns [66, 67] investigated the performance of mixtures of poly(ethylene glycol) dimethyl ether (PEGDME) with *N*-methyl-*N*-butyl-pyrrolidinium

(C$_4$mpyr) TFSI. This appears to be the first reported combination of IL and organic electrolyte, which produced an impressive first cycle capacity of around 1300 mAh g^{-1} for the highest ratio of ether to IL (2 : 1 by mass) [67]. A rapid fall in capacity ensued in all cells though, to around 500 mAh g^{-1}, after approximately 25 cycles. Significantly, the authors recorded voltage–time curves for galvanostatic cycling of Li–Li symmetrical cells and these revealed that impedance steadily grew with cycling, and showed no signs of reaching a plateau. This suggested that further optimization of the proportion of IL in these mixtures was required. As something of a postscript to this work, the Cairns Group [71] more recently employed the same electrolyte solution with an advanced graphene oxide–sulfur nanocomposite and demonstrated a remarkably steady discharge capacity of almost 900 mAh g^{-1} after 50 cycles. Importantly, the same cell in which the electrolyte solution was simply a mixture of 1 M LiTFSI in the same PEGDME showed substantial capacity loss after less than 20 cycles. This clearly emphasizes the role of the IL, with the implication that the likely suppression of solubility of polysulfides on adding the IL is just enough to limit movement of these species to negligible levels.

At around the same time, Kim and coworkers [41, 69, 70] were also investigating mixed IL–organic solvent electrolyte solutions, but were focused on the imidazolium salts of TFSI. They considered the cations 1-ethyl-3-methyl- (emim), 1-butyl-3-methyl- (bmim), and 1-propyl-2,3-dimethyl-imidazolium (dmpim) with the common anion (TFSI) in mixtures with 0.5 M LiSO$_3$CF$_3$ in DME-DOL. The cells which contained emimTFSI and bmimTFSI maintained capacities of 721 and 642 mAh g^{-1} (sulfur), respectively, after 100 cycles. While this compared favorably with the corresponding value of 534 mAh g^{-1} (sulfur), for the IL-free control cells, the latter is unusually low. Arguably of greater interest, in terms of understanding some of the mechanisms that explain capacity loss in Li–S cells, the cell containing dmpimTFSI showed the poorest cycle-life performance. Well before this study, it had been found by several groups that imidazolium TFSI ILs were not suitable for use in lithium-ion batteries because they were generally not stable under strongly reducing conditions [72]. The exception sometimes noted was with cations such as dmpim, where the somewhat acidic proton at C2, is replaced by a methyl group, thereby greatly increasing the difficulty of reduction [73]. In terms of the results of Kim and coworkers [41, 69, 70], perhaps the greater reactivity of the emim and bmim cations keys is with mitigating the effects of the more reduced forms of sulfur (lower polysulfides), which are reasonably strong reductants. It is also interesting to compare these results with those of Wang et al. [74] who evaluated the performance of an electrolyte comprised of 1 M LiTFSI in pure emimTFSI. They recorded discharge capacity that was initially 1200 mAh g^{-1} and gradually fell to around 600 mAh g^{-1} after 40 cycles. This is clearly inferior to that observed by Kim and coworkers [41, 69, 70] and, allowing for slight differences in electrode behavior, suggests that reactivity of the imidazolium cation may become difficult when the neat IL is used as the basis for the electrolyte.

Wang and Byon [75] returned to the first studies of IL electrolytes for Li–S and used a piperidinium TFSI IL, similar to that featured originally by Yuan et al. [76] In Li–S cells with typical mesoporous sulfur–carbon composite cathodes, Wang and Byon [75] evaluated the performance of electrolyte solutions that were 1 M in LiTFSI, and different ratios of *N*-propyl-*N*-methyl-piperidinium bis(trifluoromethanesulfonyl)imide (C$_3$mpipTFSI) and DME. The S–C composite cathode was prepared by precipitation of sulfur, from aqueous sodium thiosulfate solution, on to suspended particles of carbon (Ketchen black). Against lithium metal anodes, these cathodes delivered best performance in a 2 : 1 (v/v) mixture of IL and DME. Initial discharge capacity of ~1000 mAh g^{-1} fell steadily over 50 cycles to around 850 mAh g^{-1}. Given that cell impedance, as measured by EIS, did not change appreciably during this period of service, the authors attributed the bulk of capacity loss to the effects of the polysulfide shuttle. They did not, however, report any analysis of the electrolyte solution at the completion of the charge–discharge service.

In a recent study of Li–S cells with mixed IL–organic electrolyte solutions, Zheng et al. [77] offer some more detailed data that provides an interesting comparison with those of Wang and Byon [75]. Zheng et al. [77] examined the behavior of Li–S cells (typical contemporary configuration) in which the electrolyte solution comprised various proportions of (C$_4$mpyrTFSI) and a 1 : 1 (v/v) mixture of DOL and DME (all 1 M in LiTFSI). By examining the full range of IL:organic ratios, they were able to show that while the rate of capacity loss falls as the ratio of IL increases, the available discharge capacity also decreases. Thus, while Zheng et al. [77] were able to report 120 charge–discharge cycles with <6% capacity loss, the value of specific capacity was ~700 mAh g^{-1}, significantly less than that reported by Wang and Byon [75]. However, in both studies, discharge capacity was trending lower at the end of service, which suggests that a similar mechanism of degradation was acting in both cases.

To obtain information on why their cells were losing capacity, Zheng et al. [77] carried out extensive studies on the lithium metal electrodes removed from both Li–S cells and from Li–Li symmetrical cells, all operated with the mixed IL–organic electrolyte solutions. With greater fractions of IL present, the impedance of the lithium electrode increases significantly when held at open circuit. The magnitude of the increase is, however, greatly attenuated in full Li–S cells, presumably due to the influence of the portion of sulfur that dissolves into the electrolyte and then reacts with the lithium anode. Scanning electron microscopy (SEM) images show that the presence of at least 50% IL ensures a much smoother anode surface, and corresponding XPS analysis indicates that the SEI formed in the IL-containing electrolyte is rich in oxy-sulfur species and other fragments of the IL component ions. As noted by the authors, these findings confirm those from the original surface studies of lithium electrodes in ILs in which the compositional markers of a robust SEI were described [49, 50].

3.5.3 Ionic Liquid Electrolyte Solutions

The solubility of sulfur in various ILs has been studied by Boros et al. [78], and a detailed electrochemical and spectroscopic study of sulfur and polysulfides has been reported by Manan et al. [79]. The latter shows that while the general features of the voltammetry of sulfur and its reduction products are similar in the commonly used ILs, there are some differences with respect to the relative stabilities of the intermediate species, S$_3^{2-}$ and S$_4^{2-}$ [79–81].

Apparently, the first report of operation of a Li–S cell with an IL electrolyte was from Yuan et al. [76], who employed a mixture of 1 M LiTFSI and *N*-methyl-*N*-butyl-piperidinium (C$_4$mpip) TFSI. They indicated that the basis for choosing the IL was to suppress the dissolution of polysulfides, and thereby minimize capacity loss. While they were able to record a reasonable initial discharge capacity (just over 1000 mAh g^{-1} with respect to sulfur), the performance fell away dramatically over 10 cycles. In addition, their control cell used a standard lithium-ion electrolyte solution (LiPF$_6$ in EC – DMC) and thus exhibited relatively poor capacity and cycling behavior. Surprisingly, there was no discussion of the likely protective effects of the IL on the lithium anode.

Yan et al. [38] reported the behavior of a Li–S cell with IL electrolyte and a lithiated Si–C composite anode (replacing metallic lithium). The IL was a somewhat unusual variant of pyrrolidinium (N-methyl-N-allyl-pyrrolidinium) TFSI and was mixed with LiTFSI (0.5 M) to form the electrolyte solution. These cells typically displayed a high initial discharge capacity ($\sim$1450 mAh g^{-1}) followed by rapid loss of capacity, to below 800 mAh g^{-1} within 50 cycles. The cell voltage during discharging was also lower than that of standard Li–S cells. While this was due in most part to the Si-based anode, the shape of the $V–t$ curve during discharging was clearly different from that of the standard curve, which suggested that the changes in electrode and/or electrolyte had altered the stability of the polysulfide species that form during discharge. This issue was not taken up by the authors in discussion.

Watanabe and coworkers [82] have also investigated the characteristics of the Li–S system in an "all-IL" electrolyte, namely, 0.64 M LiTFSI in N,N-dimethyl-N-methyl-N-methoxyethylammonium (DEME) TFSI. This IL is known to be compatible with reversible operation of the lithium electrode as well as providing for good lithium-ion transport when mixed with LiTFSI. By comparison with an electrolyte solution of 1 M LiTFSI in TEGDME, the solubility of all-lithium polysulfides was markedly less in DEMETFSI and, for the higher polysulfides, reduced even further when 0.64 M LiTFSI was mixed with the IL. Electrolyte of the latter composition conferred the best cycling behavior, with Li–S cells exhibiting an initial capacity of 800 mAh g^{-1}, which decreased to around 600 mAh g^{-1} after 100 cycles. The coulombic efficiency of the charge–discharge cycle was more than 98%, which indicates that this research succeeded in its intention of restricting the dissolution of the lithium polysulfides. Presumably, the substantial loss of capacity was attributable to other aspects of cell performance. No data were presented on the evolution of impedance at the lithium electrode, which would have been the likely cause of degradation of cell capacity performance.

The Watanabe Group [83] has also presented a study of the influence of the anions present in a series of pyrrolidinium and tetra-alkylammonium IL electrolytes. They compared the behavior of a series of Li–S cells in which the electrolyte consists of the C$_3$mpyr$^+$ (N-propyl-N-methyl-pyrrolidinium) cation, 0.5 M LiTFSI, and an anion is taken from the homologous series: FSI (bis(fluorosulfonyl)imide – [(FSO$_2$)$_2$N]$^-$), TFSI (bis(trifluoromethanesulfonyl)imide – [(CF$_3$SO$_2$)$_2$N]$^-$), and BETI (bis (pentafluoroethanesulfonyl)imide – [(C$_2$F$_5$SO$_2$)$_2$N]$^-$). The findings showed that while the all-TFSI system did not give the highest initial discharge capacity, it did register by far the strongest charge–discharge cycling performance, with 50 cycles completed at (initially) 800 mAh g^{-1}, down to $\sim$700 mAh g^{-1}. With C$_3$mpyrFSI, the initial capacity was significantly higher (1000 mAh g^{-1}), but the capacity then dropped rapidly to <200 mAh g^{-1} within 10 cycles. The authors suggested that the FSI anion was the least stable of the three related anions and that it was, according to the synthetic chemistry literature, susceptible to nucleophilic attack. They suggested that the most likely nucleophiles in the electrolyte were polysulfides and that this would then explain the presence of lithium sulfate in the cathode after cycling. Clearly, this indicates that researchers wishing to realize the benefit of the lower viscosity of FSI ILs in Li–S electrolytes need to now consider the impact this likely reaction pathway can have on cell performance. Viscosity is certainly an important consideration in these IL electrolytes as the performance of the third anion (BETI), which confers relatively high viscosity on its ILs, was defined by a steady fall in discharge capacity from a low initial value of 600 mAh g^{-1} down to ~200 mAh g^{-1} by cycle 50.

From the Watanabe Group also comes an interesting parallel study on a new class of IL media. Equimolar mixtures of the short-chain oligoethers, known as glymes, and certain lithium salts form complexes which, because of the strength of the ether–lithium bonding, can be regarded as ILs of the general formula [Li(glyme)]$^+$ [84]. Based on the premise that polysulfides are less soluble in such strongly ionic media, Watanabe and coworkers [85] evaluated the charge–discharge behavior of sulfur–carbon composite electrodes in electrolytes comprising either triglyme (G3) or tetraglyme (G4) and LiTFSI. For the composition [Li(G4)]TFSI, Li–S cells maintained discharge capacity of around 800 mAh g^{-1} at a high coulombic efficiency (~97%), while with the shorter-chain glyme, the capacity was initially higher (~1000 mAh g^{-1}) and was still at a useful value of ~700 mAh g^{-1} after 400 cycles, at close to 100% charge efficiency [85]. In an extension of this work, consideration was also given to the influence of other anions on the performance of this type of electrolyte. While substituting TFSI with each of nitrate, triflate, and tetrafluoroborate produced a sharp fall in capacity and capacity retention, the use of BETI, the next longer homolog to TFSI, produced reasonably good performance – initial capacity of 700 mAh g^{-1}, which fell to just under 600 mAh g^{-1} after 100 cycles [86].

From the preceding discussion, it is obvious that the use of ILs as electrolyte (or a component of) is virtually predicated on the presence of TFSI as the major anion. As we have noted briefly for the relevant studies, this is largely because this anion contributes to the formation and maintenance of an effective SEI on the lithium anode. Within the Li–S cell, the operation of the SEI is part of several equilibria involving the key sulfur redox species and, therefore, some further discussion is justified.

Low-oxidation-state forms of sulfur have been detected in the SEI that forms on lithium electrodes that are charge–discharge cycled in TFSI-containing IL electrolytes [50]. Evidence from surface X-ray analytical methods (e.g. XPS) indicates that sulfide is among the reduction products, which also include (predominantly) lithium fluoride as well as a host of fragments from the breakup of the bis(trifluoromethanesulfonyl)imide anion. In the presence of free lithium ions, the sulfide presumably forms the lithium compound, although direct detection of Li$_2$S has so far not been reported.

Importantly, insoluble short-chain lithium sulfides can react with sulfur or long-chain lithium polysulfides to produce soluble polysulfides, as shown in Eqs. (3.7) and (3.8). Thus, unlike in the simple IL electrolyte mixtures employed in the first studies, the presence of sulfur and polysulfides renders the SEI on lithium to some extent labile. With some dissolution of sulfide from the electrode now likely, the thickness, and probably the permeability, of the surface film will be determined by the interplay between the deposition and dissolution processes (Eqs. (3.5)–(3.8)). This in turn makes more difficult the task of establishing and maintaining an effective SEI.

Another aspect of the increased complexity of the behavior of the lithium electrode in Li–S cells is the formation of oxyanions of sulfur and nitrogen, which is promoted by the use of lithium nitrate as an additive to the electrolyte. As noted earlier, a range of oxysulfur species is produced by the reduction and breakdown of TFSI. In general terms, the abundance of these species is enhanced by the addition of TFSI-based IL and/or the presence of $LiNO_3$. Unlike sulfide, the higher oxidation state forms of sulfur do not appear to be involved in any exchange reactions with solution-based species. No information is yet available on whether these compounds continue to accumulate during cell service and hence whether they contribute to loss of discharge capacity.

Overall, at this stage in the development of liquid electrolyte solutions for Li–S batteries, it is clear that none of the combinations of IL and organic solvent, together with variations in the identity of lithium salt (including additives such as nitrate), has proved able to limit significantly the steady decrease in discharge capacity that is always present from the commencement of cycling service. Presumably, even in configurations where combined electrode impedance does not appear to continue growing with cycling, some other process, perhaps as simple as precipitation of lower-order lithium sulfides, progressively drains the electrodes of dischargeable material. Although limited in number, the studies that have considered IL anions other than TFSI suggest that this anion, compared with other fluorosulfonyl species, confers the best characteristics, in terms of both liquid properties and chemical reactivity with the lithium electrode. A summary of the main performance characteristics for modified liquid electrolytes is included in Table 3.1.

3.6   Solid and Solidified Electrolyte Configurations

From the discussion of liquid electrolyte solutions, it seems clear that the causes of capacity loss in Li–S cells are intimately associated with the processes and phenomena that often define behavior in liquids: diffusion, migration, solubility, and precipitation. In battery technology, the development of a solid electrolyte is always attractive as it makes any device inherently safer, in terms of preventing leakage of a potentially dangerous liquid. With respect to the operation of the lithium–sulfur system, some form of solid electrolyte may directly alleviate, or prevent, the processes that appear to be unavoidable in the liquid electrolyte systems investigated to date.

3.6.1   Polymer Electrolytes

3.6.1.1   Absorbed Liquid/Gelled Electrolyte

The notion of allowing a liquid organic electrolyte solution to be absorbed by an appropriate polymer medium is common in lithium-ion technology where the resultant devices are well known as "lithium polymer" batteries. A popular polymer for this purpose is poly(vinylidene difluoride) (PVDF) which can be processed relatively easily into thin, microporous membranes. Choi et al. [41] evaluated the performance of Li–S cells with electrolyte media that were based on ~100 μm thick PVDF membranes that had been soaked with a range of TEGDME/DOL (1 M $LiCF_3SO_3$) solutions. The presence of the polymer caused an approximate doubling of electrolyte resistance and, at 100 μm, the electrolyte thickness was well above what would be optimum in a commercial cell. Nevertheless, the discharge capacities for the best polymer-based cells were comparable with those of liquid-only control cells and, over 50 cycles, registered a similar rate of capacity loss. Overall, though, these results, together with those from

cathode and lithium metal anode they obtained a discharge capacity that peaked at over 1300 mAh g^{-1} and then leveled off at just under 1000 mAh g^{-1}. This performance was much better than that of the control cells (no addition to the electrolyte), the discharge capacity of which fell to less than 400 mAh g^{-1} over the same period of service. The authors attributed the improvement with the addition of phosphorus pentasulfide to in situ formation on the lithium anode surface of ß-Li$_3$PS$_4$ which, as a good lithium-ion conductor, protects the anode from access by reactive species. A second mode of action for the P$_2$S$_5$ additive was that its presence raised the solubility of all sulfide and polysulfide species, thereby largely preventing the accumulation of lower, more reduced, sulfide species in the electrodes. Importantly, the pentasulfide is also stable in the potential window in which the sulfur cathode is typically operated.

As indicated earlier, thio-LISICON (Li$_{3.25}$Ge$_{0.25}$P$_{0.75}$S$_4$) is a well-known superionic conductor and would have been expected to feature prominently in the development of all-solid Li–S cells. However, it has been reported that thio-LISICON is not stable in contact with lithium metal due to reactivity of the germanium centers in the former [114]. Nevertheless, Nagao et al. [115] reported that all-solid-state cells with thio-LISICON as electrolyte and a composite electrode of sulfur and CMK-3 (mesoporous carbon) showed an initial reversible capacity of up to 1300 mAh g^{-1}. However, the cells show significant degradation of charge–discharge capacities after several cycles. The degradation of the charge–discharge characteristics was investigated further in a study that analyzed data from small- and wide-angle X-ray scattering experiments [116]. The results indicated that in this type of cell, two different crystallographic forms of Li$_2$S are formed and that the formation of the high-pressure polymorph is likely linked to the strong interaction between carbon and sulfur in the mesoporous cathode. The same group of researchers also developed a related compound – Li$_{10}$GeP$_2$S$_{12}$ – which shows even higher conductivity, around 10^{-2} S cm^{-1} at room temperature. However, this compound also seems to have limited stability in the presence of elemental lithium [117]. The cells into which it was incorporated by Nagao et al. [116] employed indium anodes, rather than lithium, and the voltammetric behavior reported for reduction processes showed very limited reversibility of the lithium deposition. Unfortunately, at this stage, the great benefits possibly accessible through the use of this highly conductive solid will remain underutilized in Li–S cells.

All-solid-state lithium secondary batteries with a glassy Li$_3$PO$_4$–Li$_2$S–SiS$_2$ electrolyte and lithium sulfide–carbon (Li$_2$S–C) composite positive electrodes, prepared by the spark plasma sintering process, have also been reported [118]. The electrochemical tests demonstrated that the Li$_2$S–C cells showed the initial charge and discharge capacities of ~1010 and 920 mAh g^{-1}-Li$_2$S, respectively, which showed higher discharge capacity and coulombic efficiency ~91% than those for the Li/Li$_2$S–C cells with nonaqueous liquid electrolytes (200–380 mAh g^{-1}-Li$_2$S and ~27%, respectively). The *ex situ* S K-edge X-ray absorption fine structure measurements suggested that the appearance and disappearance of elemental sulfur in the positive electrodes after charging and discharging, respectively, indicating that the ideal electrochemical reaction Li$_2$S $\Longleftrightarrow$ 2Li + S proceeded in the all-solid-state cell. Such ideal electrochemical reactions, due probably to the suppression of the formation of polysulfides in the electrolyte, would result in higher coulombic efficiency and discharge capacity as compared with those of the liquid-electrolyte cells. A summary of the main performance characteristics for the solid electrolytes surveyed is included in Table 3.1.

3.7 Challenges of the Cathode and Solvent for Device Engineering

As discussed in Chapter 2, the challenges of a sulfur-loaded cathode are complex and difficulties in enabling high loading of sulfur onto the cathode and still retaining electrochemical behavior exist. In the context of a battery device, since battery performance is dictated by device weights/volumes, it is essential that high levels of sulfur loading are present. This compensates for the added weight/volume of non-electrochemically active components such as battery management systems, pressure release valves, electrode foils/current collectors, etc. Many manuscripts report satisfactory reports of lithium–sulfur battery performance; however, closer inspection reveals very low loading of sulfur on the cathode and, in many instances, performance figures based on sulfur loading and not device parameters.

In the context of electrolytes, higher levels of sulfur loading provide additional challenges. Firstly, the quantity of electrolyte required has to be optimized for full wetting of the electrodes, including penetration into the sulfur electrode to enable full utilization of sulfur. Secondly, the quantity of electrolyte has to be optimized to be the lowest possible amount to negate detrimental performance penalty effects from non-/underutilized electrolyte. Finally, a higher loading of electrolyte can, under certain circumstances, encourage side product solubility which can have a detrimental effect on battery device performance or lifetime.

3.7.1 The Cathode Loading Challenge

Apart from the challenge of sulfur containment during charge/discharge operation, processes have to be developed for the fabrication of reliable and robust composite cathodes with a sufficiently high loading of active material before the mass production of large-scale LiS batteries can become an alternative to current Li-ion batteries. In a recent article, Abruña et al. [119] evaluated the prospects of specific energy output of LiS technology in comparison to established Li-ion technology. In their study, the authors highlight the fact that in most studies specific capacities are normalized to the mass of sulfur only, whereas a different and more realistic picture emerges when capacities are normalized to the total mass of cathode components. The study found that the relative sulfur loading percentage is the key parameter governing the calculated specific energy; and, notably, the loading has a much more pronounced effect on the volumetric than on the gravimetric specific energy density. In number terms, a 70% sulfur content in the cathode composite is required to exceed the volumetric energy density of typical Li-ion cells which can be a more important parameter than gravimetric energy density in some applications.

Despite the very attractive theoretical specific capacity of sulfur ($1672\ \mathrm{mAh\,g^{-1}}$), the problem of realizing the expected high energy densities is defined by several issues:

- The electrically insulating nature of sulfur and polysulfides;
- Particle size/agglomeration and dispersion of sulfur throughout the cathode composite;
- Pore size of the host materials;

- Contact and wetting of the active material and associated with this the penetration of the electrolyte medium;
- Total material loading and thickness of the cathode coating;
- Large volume expansion upon lithiation and related to this the mechanical strength of the host material and cathode coating;
- Detachment of polar Li_2S_x from the nonpolar conductive carbon host materials;
- Ratio of amount of electrolyte to cathode material;
- Formation of homogeneous crack-free cathode films of sufficient thickness on the current collector.

The central issue, from which many others follow, is the insulating nature of all sulfur species involved in the electrochemical process requiring the blending of sulfur with electrically conducting carbon materials of various natures in amounts sufficient to overcome the insulation issue [120–122]. Clearly, in the initial cycling, this necessity does lower the attainable gravimetric and volumetric capacity significantly since mass contributions are typically in the order of 20–30 wt% to achieve the desired effect on conductivity. Over prolonged periods of cycling, however, the additional carbon material has the positive effect of maintaining capacities at a relatively high level due to its second functionality of mitigating polysulfide loss, as discussed earlier. It is important to note how different weight ratios of sulfur to conducting carbon scale with capacity. Abruña et al. [119] reported a significant decrease in capacity from 1100 below 400 mAh g^{-1} (normalized to the mass of sulfur) when the sulfur content is increased from 50% to 90%, exposing the problem of very low sulfur utilization when the cathode material is deficient in conducting carbon material. Capacities appear slightly different when normalized to the total mass loading (S + C), and somewhat counterintuitive the best performance is achieved with an optimized 70 wt% sulfur loading. Unfortunately, the majority of literature accounts [123] on LiS batteries report capacity data normalized to the sulfur content only and therefore uncertainties remain for a realistic evaluation of novel approaches to the problem.

When considering the areal capacity of typical sulfur composite cathodes, it becomes apparent that values are generally less than 2 mAh cm^{-2} compared to Li-ion technology with 4 mAh cm^{-2}. Given that the operating voltage of LiS batteries is around 2.1 V compared to 3.5 V for Li-ion, the areal loading has to be increased in order to surpass the energy output of current Li-ion technology [122, 124–127].

These considerations limit the range of applications for current LiS technology when comparing gravimetric and volumetric energy densities of the final device. In order to increase areal specific capacity, one can resort to increase the thickness of the electrodes (mass loading cm^{-2}). However, because in the majority of cases the employed carbon materials are of low density, the volumetric energy density is compromised [128, 129]. Volumetric energy density was also highlighted in an earlier review by Cui and coworkers [130]. In number terms, the density of sulfur (2.07 g cm^{-3}) is at best only half of that of transition metal oxides (4–5 g cm^{-3}) typically used in Li-ion technology and the practical volumetric energy density is further reduced by incorporating relatively higher proportions of high-volume electrically conductive carbon materials [131, 132], in particular nanostructured materials of low tap density, to compensate for the insulating nature of sulfur (5×10^{-30} S cm^{-1}) [133] and its discharge products (LiS_2: 10^{-13} S cm^{-1}) [134]. The estimated tap density of the total composite sulfur cathode material (sulfur, carbon, and binder) is only one-third of that of common Li-ion systems (average 3 g cm^{-3})

and therefore the volumetric energy density cannot exceed values of established Li-ion batteries unless significant improvements can be made in cathode design [130].

3.7.2 Cathode Wetting Challenge

The relative voluminous nature of carbon materials is, to a certain degree, essential in order to fulfill their second function, facilitate retention of sulfide species, as well as to ensure sufficient wetting and penetration of the electrolyte. The latter is becoming a serious issue for high coat mass cathodes because it can slow down the redox kinetics in the electrode and lowers utilization of the active sulfur. By investigating the relationship between mass loading and energy density, Xiao and coworkers [122] demonstrated for sulfur-infused nanosized Ketjen carbon interconnected by *in situ* generated carbon black that an optimum performance is reached at an areal sulfur density of approximately $3.5\,mg\,cm^{-2}$, beyond which a further increase in sulfur loading results in a decrease in the areal specific capacity as a result of the trade-off between sulfur loading and sulfur utilization.

Furthermore, this study demonstrated electrolyte wetting issues for thick electrode materials ($5\,mg\,cm^{-2}$) as a function of cycle rates; high capacities of $1200\,mAh(S)\,g^{-1}$ can only be obtained after the first discharge at a very slow discharge rate of $0.05\,C$ and a steep decline in capacity to around $400\,mAh(S)\,g^{-1}$ over 40 cycles was observed at a moderate $0.2\,C$ rate due to insufficient and slow electrolyte uptake which reduced sulfur utilization and ionic conductivity. Realistically, the reported capacities are 40% lower when normalized to the total mass of electrode material ([S/IKB]: carbon: binder $= 80:10:10$).

The performance could be enhanced when the wetting of active material was improved by replacing the traditional carbon with multiwalled carbon nanotube (MWCNT) and graphene. Thus, discharge capacities of $800–900\,mAh\,g^{-1}$ (S) at $0.2\,C$ could be achieved over 90 cycles. In a similar study, Hagen et al. [126] reported in 2013 the achievable capacities of S-CNT cathode composites normalized to sulfur as well as to the full material weight for different areal loadings ($11.6–26.7\,mg\,cm^{-2}$) and ratios of sulfur to CNT ($0.84:1–2.2$). The highest capacities, around $500\,mAh$ (total mass) g^{-1}, were achieved with high sulfur/CNT ratios and the results also showed an optimum mass loading above which capacities decrease. Although no values were reported in the two studies, the volumetric specific energy density may be adjusted, as demonstrated by Xiao and coworkers [122], by applying pressure to the fabricated electrode materials up to a maximum value beyond which capacities decline due to blocking of electrolyte diffusion pathways.

In recent works, several groups adopted the approach of creating 3D interconnected electron and ion conducting channels to overcome the problem of maintaining high sulfur utilization in electrodes with high material loading (up to $17\,mg\,cm^{-2}$). The integration of materials like hollow carbon fibers [124], graphene nanostructures [71, 120, 135], carbon paper [136], and MWCNT/carbon black hybrids [122, 137] has been chosen to create highly conductive scaffolds of interconnected ion channels with abundant void space sufficient to maintain good absorbability of the electrolyte and to accommodate high fractions of well-dispersed sulfur.

In this regard, Zhang and coworkers [71] combined short- and long-range hierarchical CNT conductive networks which can hold high sulfur loadings, ranging from

6.3 to 17.3 mg cm^{-2}. With this cathode architecture holding 54 wt% sulfur, an initial discharge capacity of 995 mAh g^{-1} (6.2 mAh cm^{-2}) and a 60% utilization of sulfur could be achieved, although at a low current density of 0.05 C. Over 20 cycles, the capacity stabilized at around 800 mAh g^{-1} and then slightly decreased to 700 mAh g^{-1} within 150 cycles. The areal capacity of the CNT–S electrode is considerably higher than the value of 4.0 mAh cm^{-2} for commercial Li-ion batteries. The benefit of this paper-like electrode (no Al current collector) is that the areal capacity can be increased almost threefold to 15.1 mAh cm^{-2} by facile stacking of three CNT–S electrodes sheets with an areal sulfur loading of 17.3 mg cm^{-2} as the cathode in a Li–S cell.

Liu and coworkers [137] who investigated high areal loading 3D HCFF/MCNT/CB hybrid cathodes estimated the electrolyte uptake to be 2 orders of magnitude higher than in conventional electrode materials. The hollow carbon fiber foam was used as a 3D current collector and the MWCNT served as a binder substitute. The electrodes studied had sulfur loadings of 6.2 (57), 10.8 (67), and 16.5 (73) mg cm^{-2} (wt% in total mass of electrode material), respectively, and delivered discharge capacities of 1263 (7.83), 1066 (11.5), and 989 mAh g^{-1} (S) (16.3 mAh cm^{-2}) at a 0.1 C rate in the first cycle. Over 50 cycles, the capacities decayed to 1032, 716, and 727 mAh g^{-1} (S). The authors also detailed the sulfur mass loadings normalized to electrode material volume after cycling when the thickness of the electrode coat had significantly shrunk from 2.4 to 0.16 mm as 0.68, 1.0, 1.4 g cm^{-3}; however, no explanation is given whether the compression of the material is simply due to the pressure in the cell or also to effects of cycling and specific volumetric capacities (or energy densities) were not reported [137].

Cheng and coworkers [135] also adopted the approach of interconnected porous current collector by infiltrating a poly(dimethylsiloxane) impregnated graphene foam with a slurry of sulfur, carbon black, and PVDF binder to construct an electrode material with large void spaces to accommodate large amounts of the active sulfur and the penetration of sufficient amounts of electrolyte. Thus, sulfur loadings of 10.1 mg(S) cm^{-2} (approximately 50 wt% in total mass of electrode material) and areal capacities of 13.4 mAh cm^{-2} (1314 mAh g^{-1} (S)) were achieved during the first cycles at a 0.18 C rate. Cycling at variable charge/discharge rates up to 3.6 C for 10 cycles at each rate showed good performance with retention of a relatively high capacity of 450 mAh g^{-1} at 3.6 C. Long-term cycling at a 0.9 C rate showed relatively high capacities around 1000 mAh g^{-1} (S) over approximately 150 cycles, after which more capacity decay down to around 448 mAh g^{-1} (S) after 1000 cycles was observed. Post-cycling autopsy of the cathode material by SEM and X-ray fluorescence (XRF) confirmed the integrity of the material and uniformly embedded sulfur.

The merit of employing porous carbons as materials to facilitate high sulfur loadings and electrolyte wetting was demonstrated by Chung and Manthiram [136] for a simple and low-cost carbon paper typically used in fuel cells which was dip impregnated with a sulfur/carbon black/PVDF blend. The carbon material functioned as a light current collector and hence the tedious procedures of pasting on aluminum could be avoided. Sulfur fractions of 70–80 wt% of total electrode mass were readily achieved by applying simple techniques; however, areal sulfur loadings ranged rather low, 2.1–2.3 mg cm^{-2}, compared to more recent reports. The discharge capacities reported for cells with 70% S content were 1080 (2.3), 961 (2.0), and 767 (1.6) mAh g^{-1} (S) (mA cm^{-2}) at, respectively, 0.1, 0.2, and 0.5 C rates for over 50 cycles. The authors demonstrated the superior performance of their cathode material by comparing it to standard electrodes pasted on aluminum foil with a sulfur content of 80 wt% [136]. The latter delivered 845 mAh g^{-1} (50% utilization) initial discharge capacity followed by 50% capacity loss within the first 10 cycles and further but less rapid decay. The amount of electrolyte absorbed in the porous cathodes is 26.0 µl cm^{-2} (70 wt% S) or 23.0 µl cm^{-2} (80 wt% S), which was much higher than that adsorbed on the conventional flat aluminum supported cathode (9.0 µl cm^{-2}).

Guo and coworkers [138] generated a structural configuration of highly graphitic sp^2 carbon nanocages uniformly covering a graphene backbone. This network not only provides for sufficient electron transport but due to the large pore volume (1.01 cm^3 g^{-1}) can also accommodate a 77 wt% sulfur content. The authors reported very promising performance of this architecture and compared it to a graphene backbone structure coated with a layer of common carbon [138]. The initial discharge capacities were 1375 (1058) mAh g^{-1} (S) (mAh g^{-1} (total material)) and were comparatively high at 943 (726) mAh g^{-1} (S) (mAh g^{-1} (total material)) after 200 cycles at a 0.1 C rate. During variable rate testing, the capacities remained favorable at 1024 (0.5 C), 900 (1 C), and 875 (2 C) mAh g^{-1} (S) (2 C); after 1000 cycles at 1 C, the capacity was still a respectable 706 mAh g^{-1} (S). By comparison, the reference electrode material lacking a graphitized nanocage structure showed a rapid capacity decay from 720 to 350 mAh g^{-1} over 270 cycles. As a rare exception, volumetric specific capacity and energy data are available for this electrode material. With a composite density of 1.60 g cm^{-3}, the initial volumetric capacity was 1692 mAh cm^{-3}.

Cairns and coworkers [120] developed a long-life, high-rate Li–S cell with a high specific energy through a multifaceted approach by uniquely combining cetyltrimethyl ammonium bromide (CTAB)-modified sulfur graphene oxide nanocomposite with an elastomeric styrene-butadiene rubber (SBR)/carboxymethylcellulose (CMC) binder and an IL–TEGDME electrolyte. These cells exhibited a very high initial discharge capacity of 1440 mAh g^{-1} of sulfur at 0.2 C with excellent rate capability of up to 6 C for discharge and 3 C for charge while still maintaining high specific capacity (e.g. ~800 mAh g^{-1} (S)) at 6 C). Very good cycling performance up to 1500 cycles with a final capacity of ~400 mAh g^{-1} (S) was achieved with this cathode/electrolyte configuration. It must further be noted that the authors acknowledge the key parameter of a practical cell, the cell-level specific energy, which is largely determined by the sulfur content (%S), sulfur loading (mg cm^{-2}), and sulfur utilization (mAh g^{-1} (S)). The estimated cell-level specific energy values (including weight of all cell components except the cell housing) was initially ~500, and ~300 Wh kg^{-1} after 1000 cycles, which is much higher than that of currently available Li-ion cells. It is clearly indicated by these data that high cell-level specific energy can be achieved only when the sulfur content is high, and high utilization is obtained.

It must be noted that some of the more sophisticated approaches taken for electrode fabrication [71, 122, 124, 135] to guarantee high active material loading while maintaining good sulfur utilization, electrolyte wetting, and conductivity raise questions about their practicability in large-scale production due to cost of the materials and complexity of the fabrication process. The amount of electrolyte applied in larger cells has to be carefully balanced in order to ensure sufficient penetration and wetting of the active material but at the same time avoid unnecessary excess. While the latter may be neglected in simple coin cell testing, a simple calculation can demonstrate the effect in pouch cell fabrication: the calculated specific gravimetric capacity of a fully packaged pouch cell

Table 3.2 Cell characteristics.

Electrode architecture	Sulfur content (wt%)	Initial capacity (mAh g^{-1} (S)$^{-1}$) (C rate)	Initial capacity (mAh g^{-1} (total material))	Capacity [# cycles] (mAh g^{-1} (S)) (C rate)	Areal loading (mg cm^{-2})	Areal capacity (mAh (S) cm^{-2})	Areal capacity (mAh (total material) cm^{-2})	Volumetric capacity (mAh cm^{-3}) (load per volume)	References
MWCNT/VACNT/ sulfur network	54	995	537	[150] ~150 (0.05)	6.3	6.2	3.3	No data	Zhang
Nano Ketjen black interconnected with conductive carbon black, mw-CNT, graphene	80	1200 (0.05)	720	[90] 800 (0.2)	3.5 (S)	4.7	2.8	No data	Xiao
S infiltrated into CVD-grown CNT on carbon fiber backbone	42–50	800–1100	~450	[50] ~800	2–20 (S) 11.6–26.7 (total material)	Average 7–8	Average 4	No data	[126]
3D HCFF/MCNT/CB hybrid cathodes	57–73	1263–989 (0.1)		[50] 1032–716	6.2–16.5	7.8–16.3	4.4–11.9	No data	[124]
PDMSO-impregnated graphene foam, sulfur, carbon black, PVDF	50	1314 (0.18)	657	[150] 1000 [1000] 448	10.1 (S)	13.4	6.7	No data	Cheng
Carbon paper current collector impregnated with S/CB/PVDF	70–80		756–537 over 50 cycles	[50] 1080–767 (0.1–0.5 C)	2.1–2.3	2.3–1.6	1.6–1.1	No data	[136]
Graphene backbone covered with graphitic sp^2 carbon nanocages, S	77	1375 (0.1)	1058	[200] 943 (0.1) [1000] 706 (1 C)	No data	No data	No data	1692 (1.6)	[71]
CTAB-modified sulfur graphene oxide nanocomposite, SBR/CMC		1440 (0.2)	~800	[1500] ~400 (1 C)	No data	No data	No data	Initial 500 Wh kg^{-1} 300 after 1000 cycles	[120]

assembled with a 7 × 10 cm cathode consisting of 0.46 g electrode composite pasted onto an aluminum substrate (10 × 13 cm, 25 μm) is approximately 70 mAh g^{-1} at a 70% sulfur loading in contact with 1 g of electrolyte. In practical terms, for sufficient wetting the amount of electrolyte may be doubled or tripled, but this will decrease the deliverable specific capacity by 20%.

Conductive polymers such as polypyrrole, polyaniline, polythiophene, and poly-3,4-ethylendioxythiophen (PEDOT) have been widely used in the development of S cathodes for Li–S batteries. Despite their novelty, very few addressed the issue of high S loading. Wang, et al. [139] described the use of core–shell S-aniline composite for making S cathodes with high sulfur loading and utilized capacity. The unique design of the cathode relied on a multilayered core–shell structure of a carbon-polyaniline that is infused with sulfur via the chemical reduction of thiosulfate in aqueous medium. This process enabled high and uniform distribution of sulfur onto this conductive network. By applying this approach, it was possible to prepare composites with high loading of sulfur up to 87%. Due to the conductive nature of this matrix, it was also possible to prepare coatings that can deliver high specific capacity and excellent cycle stability, retaining a reversible discharge capacity of 835 mAh g^{-1} after 100 cycles when the sulfur loading of the cathode was above 6 mg cm^{-2} (Table 3.2).

3.8 Concluding Remarks and Outlook

The standard lithium–sulfur cell, which operates with a liquid electrolyte in which the products of reduction of sulfur are to varying extents soluble, fundamentally embodies a compromise between opposing factors (both chemical and electrochemical). As a result, while this version of the technology typically delivers a high initial capacity, the retention of discharge capacity and charging efficiency are both generally poor, due to the complex behavior of the polysulfides. To some extent, performance can be improved by incorporating nitrate into the electrolyte composition, as this acts to decouple the polysulfide shuttle from one of its driving forces – interaction with the lithium anode. On the timescale of many charge–discharge cycles, though, nitrate is ultimately consumed, by both electrodes, and its influence disappears. The fact that the SEI on lithium remains in dynamic exchange with solution makes protecting this electrode from interaction with redox-active species extremely challenging. Arguably, the best way forward on this front is to consider a shift away from the metallic lithium electrode to one of the other high-specific-energy options, notably silicon. As only a very small number of studies have considered this option [38], it would be unwise to rule it out on the basis of the relatively low levels of performance that have been achieved to date. A more important issue to consider is that the slightly less negative potential of the Li$_{4.4}$Si|Si electrode is likely to significantly decrease the interaction between the anode and TFSI. This in turn offers scope to develop an SEI with a different composition, one that is better suited to the solution-based polysulfide chemistry.

A second approach to improving performance is by enhancing the ability of the cathode to retain sulfur and polysulfides, as shown in a host of recent publications. Even with a specially designed mesoporous electrode, however, cells that are optimized with respect to sulfur:electrolyte ratio still suffer ~50% loss of discharge capacity, albeit after 1000 cycles [140]. In a similar vein, the use of a cathode binder with enhanced affinity for sulfur species has also been shown very recently to provide a benefit that is of the same order, but again it does not address the fundamental limiting mechanisms in the lithium–sulfur system [141]. Ultimately, however, binding sulfur compounds to the cathode can only slow down the effects of diffusion, in much the same way as a chromatographic stationary phase slows the passage of the eluted components, and the degradative aspects of the polysulfide shuttle are eventually observed.

As we have seen, there is now an extensive literature on the modification of the electrolyte for lithium–sulfur batteries, which is largely focused on minimizing the solubility of the polysulfides. The approaches used include blending of electrolyte solvents with ILs, absorption (gelling) of electrolyte solution with a variety of polymers, and use of additives that lower polysulfide solubility. Arguably, the most impressive improvement based solely on electrolyte modification was presented recently by Suo et al. [65], in which very high concentrations of lithium salt are shown to greatly suppress solubility of polysulfides and deliver discharge capacity that falls at a much reduced rate (compared with standard liquid electrolyte cells). Overall, however, there is no indication that any modification to the electrolyte solution, of itself, will produce a lithium–sulfur cell that can maintain good discharge capacity at a constant value for hundreds of charge–discharge cycles.

The ultimate modification that can be made in terms of the solubility of polysulfides is to deploy a true solid electrolyte, based on either a polymer or a ceramic conductor. To date, none of the solid polymer systems has been able to demonstrate performance that is comparable with liquid systems, even when operating temperature is raised above 60 °C (a typical minimum temperature for solid polymer conductivity). The situation with inorganic compounds, such as those based on $Li_2S–P_2S_5$ glasses, is more promising, although even the best results to date are similar to those obtained in standard liquid electrolyte systems, with good initial capacity (around 1000 mAh g^{-1}) falling away steadily within 100–200 cycles [112]. Given the distinctly different discharge voltage–time profiles reported for these true solid electrolyte cells, there is clearly more research effort required to understand the detailed mechanism of the main electrochemical reactions before attempts can be made to improve the performance of the all-solid lithium–sulfur battery.

The best performance recorded to date for laboratory-based lithium–sulfur cells comes, not surprisingly, when materials and configurations are chosen which control the migration of sulfur and polysulfides by more than one means. This review has found that while modifying the electrolyte to reduce the solubility of sulfur species generates an improvement relative to a standard configuration, achieving a discharge capacity of at least half the theoretical value (1675 mAh g^{-1}), and preserving this for hundreds of cycles, requires a combination of some form of active retention of polysulfides in the cathode, as well as suppression of solubility of these species in the electrolyte. Research by Ji et al. [71] provided an excellent example in their studies with a carefully designed cathode of high sulfur affinity (based on reduced graphene oxide) and an electrolyte solution which contained a high proportion of pyrrolidinium TFSI IL. This work, and others with a similar approach, looks to be a promising basis for the development of a durable lithium–sulfur battery technology.

One aspect of the performance of Li–S cells that is rarely discussed in the literature is rate capability. This important issue also limits the application for lithium–sulfur technology. Perusal of Table 3.1 reveals that studies rarely employ discharge current densities greater than 0.1 C. This is no doubt one of the main drawbacks to current approaches that seek to limit the solubility and mobility of the reduction products of sulfur. The fact that the reactions at the positive electrode are solution-based obviously means that the diffusive flux of the reduction products (polysulfides) can limit the rate of discharge. Only in the early work of Rauh et al., where very high concentrations of higher polysulfides (rather than elemental sulfur) were used as the cathode, were much higher rates of discharge achieved. While some increases can be made through the optimization of cell configuration (electrode thickness, etc.), a major improvement in the rate capability of the lithium–sulfur system will require a substantially different approach to the hosting of sulfur–polysulfide electrochemistry. As an indication that such an approach may not require a major departure from the suite of materials currently in use, Kinoshita et al. [142] showed that the performance of Li–S cells featuring the non-polymer solid electrolyte Li_3PS_4 can be dramatically improved by the apparently simple incorporation of around 1 wt% of an imidazolium IL (C_2mimTFSI). Importantly, the improvement persists at 25 °C, at which temperature this and many other solid electrolyte systems barely function. While this composite contains a liquid, there is so little that its properties are more like those of a solid, thereby making the improved behavior remarkable.

Progress over the past 15 years of development of lithium–sulfur cells has seen significant improvements in specific energy and cycle life in configurations that incorporate a range of different electrolyte media. The most recent studies suggest that careful combination of materials that help contain the migration of polysulfides with electrolytes that limit their solubility, without compromising lithium-ion transport, could ultimately lead to devices which realize the goals of specific energy above 500 Wh kg^{-1} for more than 1000 cycles. At this stage, while it seems more than likely that the electrolyte will be a solution (in the liquid-based sense of the word), the progress in solid and solid-like electrolytes, in the hands of a relatively small number of researchers, is quite remarkable. This may well lead to the inherently safer solid-electrolyte version of the lithium–sulfur cell, which is all the more remarkable given that the corresponding lithium-ion cell has so far proved elusive.

Glossary of Terms and Abbreviations

BETI	bis(pentafluoroethanesulfonyl)imide – $[(C_2F_5SO_2)_2N]^-$
FSI	bis(fluorosulfonyl)imide – $[(FSO_2)_2N]^-$
TFSI/TFSA	bis(trifluoromethanesulfonyl)imide – $[(CF_3SO_2)_2N]^-$ (also represented as NTf_2^-)
OTf	trifluoromethanesulfonate (triflate) – $CF_3SO_3^-$
bmim	1-butyl-3-methyl-imidazolium (C_4mim)
P13	1-propyl-1-methyl-pyrrolidinium (also represented as C_3mpyr$^+$)
P14	1-butyl-1-methyl-pyrrolidinium (also represented as C_4mpyr$^+$)
DOL	dioxolane – $(CH_2)_2O_2CH_2$
DME	dimethylether – CH_3OCH_3
TEGDME	tetraethyleneglycol dimethylether – $CH_3O(CH_2CH_2O)_4CH_3$ (also known as "tetraglyme" or "G4")
PEGDME	polyethyleneglycol dimethylether – $CH_3O(CH_2CH_2O)_nCH_3$ (typically a blend of glymes with $n \geq 4$)

References

1 Yang, Z., Zhang, J., Kintner-Meyer, M.C. et al. (2011). Electrochemical energy storage for green grid. *Chemical Reviews* 111 (5): 3577–3613.

2 Rand, D.A.J. (2011). A journey on the electrochemical road to sustainability. *Journal of Solid State Electrochemistry* 15 (7–8): 1579–1622.

3 Tarascon, J.M. and Armand, M. (2001). Issues and challenges facing rechargeable lithium batteries. *Nature* 414 (6861): 359–367.

4 Bruce, P.G., Scrosati, B., and Tarascon, J.M. (2008). Nanomaterials for rechargeable lithium batteries. *Angewandte Chemie* 47 (16): 2930–2946.

5 Mikhaylik, Y.V., Kovalev, I., Schock, R. et al. (2010). High energy rechargeable Li–S cells for EV application: status, remaining problems and solutions. *ECS Transactions* 25 (35): 23–34.

6 Whittingham, M.S. (2008). Materials challenges facing electrical energy storage. *MRS Bulletin* 33: 411–419.

7 Hassoun, J., Reale, P., and Scrosati, B. (2007). Recent advances in liquid and polymer lithium-ion batteries. *Journal of Materials Chemistry* 17 (35): 3668–3677.

8 Goodenough, J.B. and Kim, Y. (2011). Challenges for rechargeable batteries. *Journal of Power Sources* 196 (16): 6688–6694.

9 Burke, A.F. (2007). Batteries and ultracapacitors for electric, hybrid, and fuel cell vehicles. *Proceedings of the IEEE* 95: 806–820.

10 Zu, C.-X. and Li, H. (2011). Thermodynamic analysis on energy densities of batteries. *Energy & Environmental Science* 4 (8): 2614–2624.

11 Bruce, P.G., Freunberger, S.A., Hardwick, L.J., and Tarascon, J.M. (2012). $Li–O_2$ and Li–S batteries with high energy storage. *Nature Materials* 11 (1): 19–29.

12 Thomas, T. (2008). Energy on demand. *The World Electric Vehicle Journal* 2 (2): 1–2.

13 DesignNews 2008. Auto industry working hard to make an electric vehicle battery. http://www.designnews.com/document.asp?doc_id=222703&.

14 Ritchie, A. and Howard, W. (2006). Recent developments and likely advances in lithium-ion batteries. *Journal of Power Sources* 162 (2): 809–812.

15 Broussely, M. and Archdale, G. (2004). Li-ion batteries and portable power source prospects for the next 5–10 years. *Journal of Power Sources* 136 (2): 386–394.

16 Zhang, Y., Zhao, Y., Sun, K.E., and Chen, P. (2011). Developement in lithium/sulfur secondary batteries. *The Open Materials Science Journal* 5: 215–221.

17 Abruña, H.D., Kiya, Y., and Henderson, J.C. (2008). Batteries and electrochemical capacitors. *Physics Today* 61: 43–47.

18 OSTI.GOV (2007). Basic research needs for electrical energy storage. Report of the Basic Energy Sciences workshop on electrical energy storage. http://www.osti.gov/accomplishments/documents/fullText/ACC0330.pdf.

19 Besenhard, J.O. (1998). *Handbook of Battery Materials*. New York: Wiley.

20 Ahn, H.J., Kim, K.W., Ahn, J.H., and Cheruvally, G. (2009). Lithium–sulfur. In: *Encyclopedia of Electrochemical Power Sources* (ed. J. Garche), 155. Elsevier B.V.

21 Zhamu, A., Chen, G., Liu, C. et al. (2012). Reviving rechargeable lithium metal batteries: enabling next-generation high-energy and high-power cells. *Energy & Environmental Science* 5 (2): 5701–5707.

22 Ji, X. and Nazar, L.F. (2010). Advances in Li–S batteries. *Journal of Materials Chemistry* 20 (44): 9821–9826.

23 Kumaresan, K., Mikhaylik, Y., and White, R.E. (2008). A mathematical model for a lithium–sulfur cell. *Journal of the Electrochemical Society* 155 (8): A576.

24 Barchasz, C., Leprêtre, J.-C., Alloin, F., and Patoux, S. (2012). New insights into the limiting parameters of the Li/S rechargeable cell. *Journal of Power Sources* 199: 322–330.

25 Diao, Y., Xie, K., Xiong, S., and Hong, X. (2012). Analysis of polysulfide dissolved in electrolyte in discharge–charge process of Li–S battery. *Journal of the Electrochemical Society* 159 (4): A421.

26 Cheon, S.-E., Ko, K.-S., Cho, J.-H. et al. (2003). Rechargeable lithium sulfur battery: I. Structural change of sulfur cathode during discharge and charge. *Journal of the Electrochemical Society* 150 (6): A796.

27 Rauh, R.D., Abraham, K.M., Pearson, G.F. et al. (1979). A lithium/dissolved sulfur battery with an organic electrolyte. *Journal of the Electrochemical Society* 126: 523–527.

28 Mikhaylik, Y.V. and Akridge, J.R. (2004). Polysulfide shuttle study in the Li/S battery system. *Journal of the Electrochemical Society* 151 (11): A1969.

29 Ellis, B.L., Lee, K.T., and Nazar, L.F. (2010). Positive electrode materials for Li-ion and Li-batteries. *Chemistry of Materials* 22 (3): 691–714.

30 Rauh, R.D., Shuker, F.S., Marston, J.M., and Brummer, S.B. (1977). Formation of lithium polysulfides in aprotic media. *Journal of Inorganic and Nuclear Chemistry* 39: 1761–1766.

31 Peled, E., Sternberg, Y., Gorenshtein, A., and Lavi, Y. (1989). Lithium–sulfur battery: evaluation of dioxolane-based electrolytes. *Journal of the Electrochemical Society* 136: 1621–1625.

32 Chu, M.-Y., De Jonghe, L.C., Visco, S.J., and Katz, B.D. (2000). Liquid electrolyte lithium–sulfur batteries. US patent 6, 030, 720 A.

33 Shim, J., Striebel, K., and Cairns, E. (2002). The lithium/sulfur rechargeable cell effects of electrode composition and solvent on cell performance. *Journal of the Electrochemical Society* 149 (10): A1321–A1325.

34 Ji, X., Lee, K.T., and Nazar, L.F. (2009). A highly ordered nanostructured carbon–sulphur cathode for lithium–sulphur batteries. *Nature Materials* 8: 500–506.

35 Coleman, J.R. and Bates, M.W. (eds.) (1968). Power sources 2. In: *Proceedings of the 6th International Symposium*, Brighton, England (1968).

36 Moss, V. and Nole, D.A. (1970). Battery employing lithium – sulphur electrodes with non-aqueous electrolyte. US patent 3, 532, 543 A.

37 Bhaskara, R.M.L. (1968). Organic electrolyte cells. US patent 3, 413, 154 A.

38 Yan, Y., Yin, Y.-X., Xin, S. et al. (2013). High-safety lithium–sulfur battery with prelithiated Si/C anode and ionic liquid electrolyte. *Electrochimica Acta* 91: 58–61.

39 Gao, J., Lowe, M.A., Kiya, Y., and Abruña, H.D. (2011). Effects of liquid electrolytes on the charge–discharge performance of rechargeable lithium/sulfur batteries: electrochemical and in-situ X-ray absorption spectroscopic studies. *The Journal of Physical Chemistry C* 115 (50): 25132–25137.

40 Chang, D.R., Lee, S.H., Kim, S.W., and Kim, H.T. (2002). Binary electrolyte based on tetra(ethylene glycol) dimethyl ether and 1,3-dioxolane for lithium–sulfur battery. *Journal of Power Sources* 112 (2): 452–460.

41 Choi, J.-W., Kim, J.-K., Cheruvally, G. et al. (2007). Rechargeable lithium/sulfur battery with suitable mixed liquid electrolytes. *Electrochimica Acta* 52 (5): 2075–2082.

42 Zhang, S. (2012). Improved cyclability of liquid electrolyte lithium/sulfur batteries by optimizing electrolyte/sulfur ratio. *Energies* 5 (12): 5190–5197.

43 Ryu, H.-S., Ahn, H.-J., Kim, K.-W. et al. (2006). Discharge behavior of lithium/sulfur cell with TEGDME based electrolyte at low temperature. *Journal of Power Sources* 163 (1): 201–206.

44 Kim, H.-S. and Jeong, C.-S. (2011). Electrochemical properties of binary electrolytes for lithium–sulfur batteries. *Bulletin of the Korean Chemical Society* 32 (10): 3682–3686.

45 Youngman, O., Gofer, Y., Meitav, A., and Aurbach, D. (1990). The electrochemical behavior of 1,3-dioxolane-LiClO$_4$ solutions-I. Uncontaminated solutions. *Elechtrochimica Acta* 35: 625–638.

46 Youngman, O., Dan, P., and Aurbach, D. (1990). The electrochemical behavior of 1,3-dioxolane-LiClO$_4$ solutions-II. Contaminated solution. *Elechtrochimica Acta* 35: 639–655.

47 Gofer, Y., Ein Ely, Y., and Aurbach, D. (1992). Surface chemistry of lithium in 1,3-dioxolane. *Elechtrochimica Acta* 37: 1897–1899.

48 Wang, W., Wang, Y., Huang, Y. et al. (2010). The electrochemical performance of lithium–sulfur batteries with LiClO$_4$ DOL/DME electrolyte. *Journal of Applied Electrochemistry* 40 (2): 321–325.

49 Howlett, P.C., MacFarlane, D.R., and Hollenkamp, A.F. (2004). High lithium metal cycling efficiency in a room-temperature ionic liquid. *Electrochemical and Solid-State Letters* 7 (5): A97–A101.

50 Howlett, P.C., Brack, N., Hollenkamp, A.F. et al. (2006). Characterization of the lithium surface in *N*-methyl-*N*-alkylpyrrolidinium bis(trifluoromethanesulfonyl)amide room-temperature ionic liquid electrolytes. *Journal of the Electrochemical Society* 153 (3): A595–A606.

51 Katayama, Y., Morita, T., Yamagata, M., and Miura, T. (2003). Electrodeposition of metallic lithium on a tungsten electrode in 1-butyl-1-methylpyrrolidinium bis(trifluoromethanesulfone)imide room-temperature molten salt. *Electrochemistry* 71: 1033–1035.

52 Matsumoto, H., Kageyama, H., and Miyazaki, Y. (2003). Effect of ionic additives on the limiting cathodic potential of EMI-based room temperature ionic liquids. *Electrochemistry* 71: 1058–1060.

53 Mikhaylik, Y.V. (2008). Electrolytes for lithium sulfur cells. US patent 7, 354, 680 B2.

54 Aurbach, D., Pollak, E., Elazari, R. et al. (2009). On the surface chemical aspects of very high energy density, rechargeable Li–sulfur batteries. *Journal of the Electrochemical Society* 156 (8): A694–A702.

55 Xiong, S., Xie, K., Diao, Y., and Hong, X. (2012). Properties of surface film on lithium anode with LiNO$_3$ as lithium salt in electrolyte solution for lithium–sulfur batteries. *Electrochimica Acta* 83: 78–86.

56 Kim, H.S., Jeong, T.-G., Choi, N.-S., and Kim, Y.-T. (2013). The cycling performances of lithium–sulfur batteries in TEGDME/DOL containing LiNO$_3$ additive. *Ionics* 19 (12): 1795–1802.

57 Choi, J.-W., Cheruvally, G., Kim, D.-S. et al. (2008). Rechargeable lithium/sulfur battery with liquid electrolytes containing toluene as additive. *Journal of Power Sources* 183 (1): 441–445.

58 Xiong, S., Xie, K., Diao, Y., and Hong, X. (2013). On the role of polysulfides for a stable solid electrolyte interphase on the lithium anode cycled in lithium–sulfur batteries. *Journal of Power Sources* 236: 181–187.

59 Diao, Y., Xie, K., Xiong, S., and Hong, X. (2013). Shuttle phenomenon – the irreversible oxidation mechanism of sulfur active material in Li–S battery. *Journal of Power Sources* 235: 181–186.

60 Zhang, S.S. (2013). Liquid electrolyte lithium/sulfur battery: fundamental chemistry, problems, and solutions. *Journal of Power Sources* 231: 153–162.

61 Zhang, S.S. (2013). New insight into liquid electrolyte of rechargeable lithium/sulfur battery. *Electrochimica Acta* 97: 226–230.

62 Zhang, S.S. (2012). Effect of discharge cutoff voltage on reversibility of lithium/sulfur batteries with $LiNO_3$-contained electrolyte. *Journal of the Electrochemical Society* 159 (7): A920–A923.

63 Zhang, S.S. (2012). Role of $LiNO_3$ in rechargeable lithium/sulfur battery. *Electrochimica Acta* 70: 344–348.

64 Shin, E.S., Kim, K., Oh, S.H., and Cho, W.I. (2013). Polysulfide dissolution control: the common ion effect. *Chemical Communications (Cambridge)* 49 (20): 2004–2006.

65 Suo, L., Hu, Y.S., Li, H. et al. (2013). A new class of solvent-in-salt electrolyte for high-energy rechargeable metallic lithium batteries. *Nature Communications* 4: 1481.

66 Shin, J.H. and Cairns, E.J. (2008). Characterization of *N*-methyl-*N*-butylpyrrolidinium bis(trifluoromethanesulfonyl)imide-LiTFSI-tetra(ethylene glycol) dimethyl ether mixtures as a Li metal cell electrolyte. *Journal of the Electrochemical Society* 155 (5): A368–A373.

67 Shin, J.H. and Cairns, E.J. (2008). *N*-Methyl-(*n*-butyl)pyrrolidinium bis(trifluoromethanesulfonyl)imide-LiTFSI–poly(ethylene glycol) dimethyl ether mixture as a Li/S cell electrolyte. *Journal of Power Sources* 177 (2): 537–545.

68 Tachikawa, N., Yamauchi, K., Takashima, E. et al. (2011). Reversibility of electrochemical reactions of sulfur supported on inverse opal carbon in glyme-Li salt molten complex electrolytes. *Chemical Communications* 47 (28): 8157–8159.

69 Kim, S., Jung, Y., and Park, S.J. (2005). Effects of imidazolium salts on discharge performance of rechargeable lithium–sulfur cells containing organic solvent electrolytes. *Journal of Power Sources* 152: 272–277.

70 Kim, S., Jung, Y., and Park, S.-J. (2007). Effect of imidazolium cation on cycle life characteristics of secondary lithium–sulfur cells using liquid electrolytes. *Electrochimica Acta* 52 (5): 2116–2122.

71 Ji, L., Rao, M., Zheng, H. et al. (2011). Graphene oxide as a sulfur immobilizer in high performance lithium/sulfur cells. *Journal of the American Chemical Society* 133 (46): 18522–18525.

72 Holzapfel, M., Jost, C., and Novak, P. (2004). Stable cycling of graphite in an ionic liquid based electrolyte. *Chemical Communications* 18: 2098–2099.

73 Canal, J.P., Ramnial, T., Dickie, D.A., and Clyburne, J.A.C. (2006). From the reactivity of N-heterocyclic carbenes to new chemistry in ionic liquids. *Chemical Communications* 1809–1818.

74 Wang, J., Chew, S.Y., Zhao, Z.W. et al. (2008). Sulfur–mesoporous carbon composites in conjunction with a novel ionic liquid electrolyte for lithium rechargeable batteries. *Carbon* 46 (2): 229–235.

75 Wang, L. and Byon, H.R. (2013). N-Methyl-N-propylpiperidinium bis(trifluoromethanesulfonyl)imide-based organic electrolyte for high performance lithium–sulfur batteries. *Journal of Power Sources* 236: 207–214.

76 Yuan, L.X., Feng, J.K., Ai, X.P. et al. (2006). Improved dischargeability and reversibility of sulfur cathode in a novel ionic liquid electrolyte. *Electrochemistry Communications* 8 (4): 610–614.

77 Zheng, J., Gu, M., Chen, H. et al. (2013). Ionic liquid-enhanced solid state electrolyte interface (SEI) for lithium–sulfur batteries. *Journal of Materials Chemistry A* 1 (29): 8464–8470.

78 Boros, E., Earle, M.J., Gilea, M.A. et al. (2010). On the dissolution of non-metallic solid elements (sulfur, selenium, tellurium and phosphorus) in ionic liquids. *Chemical Communications (Cambridge)* 46 (5): 716–718.

79 Manan, N.S., Aldous, L., Alias, Y. et al. (2011). Electrochemistry of sulfur and polysulfides in ionic liquids. *The Journal of Physical Chemistry B* 115 (47): 13873–13879.

80 Levillain, E., Demortier, A., and Lelieur, J.P. (2007). Sulfur. In: *Encyclopedia of Electrochemistry* (ed. W.E. Geiger and C.J. Pickett), 253–271.

81 Evans, A., Montenegro, M.I., and Pletcher, D. (2001). The mechanism for the cathodic reduction of sulphur in dimethylformamide: low temperature voltammetry. *Electrochemistry Communications* 3 (9): 514–518.

82 Park, J.-W., Yamauchi, K., Takashima, E. et al. (2013). Solvent effect of room temperature ionic liquids on electrochemical reactions in lithium–sulfur batteries. *The Journal of Physical Chemistry C* 117 (9): 4431–4440.

83 Park, J.-W., Ueno, K., Tachikawa, N. et al. (2013). Ionic liquid electrolytes for lithium–sulfur batteries. *The Journal of Physical Chemistry C* 117 (40): 20531–20541.

84 Tamura, T., Yoshida, K., Hachida, T. et al. (2010). Physicochemical properties of glyme–Li salt complexes as a new family of room-temperature ionic liquids. *Chemistry Letters* 39: 753–755.

85 Dokko, K., Tachikawa, N., Yamauchi, K. et al. (2013). Solvate ionic liquid electrolyte for Li–S batteries. *Journal of the Electrochemical Society* 160: A1304–A1310.

86 Ueno, K., Park, J.-W., Yamazaki, A. et al. (2013). Anionic effects on solvate ionic liquid electrolytes in rechargeable lithium–sulfur batteries. *The Journal of Physical Chemistry C* 117: 20509–20516.

87 Ryu, H.-S., Ahn, H.-J., Kim, K.-W. et al. (2006). Discharge process of Li/PVDF/S cells at room temperature. *Journal of Power Sources* 153 (2): 360–364.

88 Wang, J.L., Yang, J., Xie, J.Y. et al. (2002). Sulfur–carbon nano-composite as cathode for rechargeable lithium battery based on gel electrolyte. *Electrochemistry Communications* 4 (6): 499–502.

89 Shin, J.H., Jung, S.S., Kim, K.W. et al. (2002). Preparation and Characterization of plasticized plymer electrolytes based on the PVDF-HFP copolymer for lithium sulfur battery. *Journal of Materials Science: Materials in Electronics* 13 (12): 727–733.

90 Jeon, B.H., Yeon, J.H., and Chung, I.J. (2003). Preparation and electrical properties of lithium–sulfur-composite polymer batteries. *Journal of Materials Processing
Technology* 143–144: 93–97.

91 Hassoun, J. and Scrosati, B. (2010). A high-performance polymer tin sulfur lithium
ion battery. *Angewandte Chemie International Edition* 49 (13): 2371–2374.

92 Hassoun, J., Sun, Y.-K., and Scrosati, B. (2011). Rechargeable lithium sulfide electrode for a polymer tin/sulfur lithium-ion battery. *Journal of Power Sources* 196 (1):
343–348.

93 Jin, Z., Xie, K., Hong, X., and Hu, Z. (2013). Capacity fading mechanism in lithium
sulfur cells using poly(ethylene glycol)-borate ester as plasticizer for polymer electrolytes. *Journal of Power Sources* 242: 478–485.

94 Jeddi, K., Sarikhani, K., Qazvini, N.T., and Chen, P. (2014). Stabilizing lithium/sulfur
batteries by a composite polymer electrolyte containing mesoporous silica particles.
Journal of Power Sources 245: 656–662.

95 Manthiram, A., Fu, Y., and Su, Y.-S. (2013). Challenges and prospects of
lithium–sulfur batteries. *Accounts of Chemical Research* 46: 1125–1134.

96 Marmorstein, D., Yu, T.H., Striebel, K.A. et al. (2000). Electrochemical performance
of lithium/sulfur cells with three different polymer electrolytes. *Journal of Power
Sources* 89 (2): 219–226.

97 Jeong, S.S., Lim, Y.T., Choi, Y.J. et al. (2007). Electrochemical properties of lithium
sulfur cells using PEO polymer electrolytes prepared under three different mixing
conditions. *Journal of Power Sources* 174 (2): 745–750.

98 Jeong, S.S., Lim, Y.T., Jung, B.S., and Kim, K.W. (2005). *Eco-Materials Processing and Design VI*, Materials Science Forum, 594–597. Brandrain: Trans Tech
Publications.

99 Liang, X., Wen, Z., Liu, Y. et al. (2011). Highly dispersed sulfur in ordered mesoporous carbon sphere as a composite cathode for rechargeable polymer Li/S
battery. *Journal of Power Sources* 196 (7): 3655–3658.

100 Hassoun, J. and Scrosati, B. (2010). Moving to a solid-state configuration: a valid
approach to making lithium–sulfur batteries viable for practical applications.
Advanced Materials 22 (45): 5198–5201.

101 Hayashi, A., Ohtomo, T., Mizuno, F. et al. (2004). Rechargeable lithium batteries,
using sulfur-based cathode materials and $Li_2S–P_2S_5$ glass-ceramic electrolytes.
Electrochimica Acta 50 (2–3): 893–897.

102 Hayashi, A., Hama, S., Morimoto, H. et al. (2001). Preparation of $Li_2S–P_2S_5$ amorphous solid electrolytes by mechanical milling. *Journal of the American Ceramic
Society* 84: 477–479.

103 Hayashi, A., Hama, S., Morimoto, H. et al. (2001). High Li ion conductivity from
mechanically milled glassy powders. *Chemistry Letters* 30: 872–873.

104 Hayashi, A., Hama, S., Minami, T., and Tatsumisago, M. (2003). Formation of
superionic crystals from mechanically milled $Li_2S–P_2S_5$ glasses. *Electrochemistry
Communications* 5 (2): 111–114.

105 Hayashi, A., Ohtomo, T., Mizuno, F. et al. (2003). All-solid-state Li/S batteries with
highly conductive glass–ceramic electrolytes. *Electrochemistry Communications* 5
(8): 701–705.

106 Hayashi, A., Ohtsubo, R., Ohtomo, T. et al. (2008). All-solid-state rechargeable lithium batteries with Li_2S as a positive electrode material. *Journal of Power Sources* 183 (1): 422–426.

107 Nagao, M., Hayashi, A., and Tatsumisago, M. (2012). Fabrication of favorable interface between sulfide solid electrolyte and Li metal electrode for bulk-type solid-state Li/S battery. *Electrochemistry Communications* 22: 177–180.

108 Teragawa, S., Aso, K., Tadanaga, K. et al. (2014). Preparation of $Li_2S–P_2S_5$ solid electrolyte from *N*-methylformamide solution and application for all-solid-state lithium battery. *Journal of Power Sources* 248: 939–942.

109 Nagao, M., Hayashi, A., and Tatsumisago, M. (2011). Sulfur–carbon composite electrode for all-solid-state Li/S battery with $Li_2S–P_2S_5$ solid electrolyte. *Electrochimica Acta* 56 (17): 6055–6059.

110 Agostini, M., Aihara, Y., Yamada, T. et al. (2013). A lithium–sulfur battery using a solid, glass-type $P_2S_5–Li_2S$ electrolyte. *Solid State Ionics* 244: 48–51.

111 Liu, Z., Fu, W., Payzant, E.A. et al. (2013). Anomalous high ionic conductivity of nanoporous beta-Li_3PS_4. *Journal of the American Chemical Society* 135 (3): 975–978.

112 Lin, Z., Liu, Z., Dudney, N.J., and Liang, C. (2013). Lithium superionic sulfide cathode for all-solid lithium–sulfur batteries. *ACS Nano* 7: 2829–2833.

113 Lin, Z., Liu, Z., Fu, W. et al. (2013). Phosphorous pentasulfide as a novel additive for high-performance lithium–sulfur batteries. *Advanced Functional Materials* 23 (8): 1064–1069.

114 Mo, Y., Ong, S.P., and Ceder, G. (2012). First principles study of the $Li_{10}GeP_2S_{12}$ lithium super ionic conductor material. *Chemistry of Materials* 24: 15–17.

115 Nagao, M., Imade, Y., Narisawa, H. et al. (2013). All-solid-state Li–sulfur batteries with mesoporous electrode and thio-LISICON solid electrolyte. *Journal of Power Sources* 222: 237–242.

116 Nagao, M., Imade, Y., Narisawa, H. et al. (2013). Reaction mechanism of all-solid-state lithium–sulfur battery with two-dimensional mesoporous carbon electrodes. *Journal of Power Sources* 243: 60–64.

117 Kamaya, N., Homma, K., Yamakawa, Y. et al. (2011). A lithium superionic conductor. *Nature Materials* 10 (9): 682–686.

118 Takeuchi, T., Kageyama, H., Nakanishi, K. et al. (2010). All-solid-state lithium secondary battery with $Li_2S–C$ composite positive electrode prepared by spark-plasma-sintering process. *Journal of Electrochemical Society* 157 (11): A1196–A1201.

119 Abruna.

120 Song, M.-K., Zhang, Y., and Cairns, E.J. (2013). A long-life, high-rate lithium/sulfur cell: a multifaceted approach to enhancing cell performance. *Nano Letters* 13: 5891–5899.

121 Liang, C., Dudney, N.J., and Howe, J.Y. (2009). Hierarchically structured sulfur/carbon nanocomposite material for high-energy lithium battery. *Chemistry of Materials* 21: 4724–4730.

122 Lv, D., Zheng, J., Li, Q. et al. (2015). High energy density lithium–sulfur batteries: challenges of thick sulfur cathodes. *Advanced Energy Materials* 5.

123 Cairns, E. (2014). *Nano Letters*.

124 Fang, R., Zhao, S., Hou, P. et al. (2016). 3D interconnected electrode materials with ultrahigh areal sulfur loading for Li–S batteries. *Advanced Materials* 28: 3374–3382.

125 Zhou, G., Li, L., Ma, C. et al. (2015). *Nano Energy* 11 (356).

126 Hagen, M., Dörfler, S., Fanza, P. et al. (2013). Development and costs calculation of lithium sulfur cells with high sulfur load and binder free electrodes. *Journal of Power Sources* 224: 260–268.

127 Miao, L.X., Wang, W.K., Yuan, K.G. et al. (2014). A lithium–sulfur cathode with high sulfur loading and high capacity per area: a binder-free carbon fiber cloth–sulfur material. *Chemical Communications* 50: 13231–13234.

128 Evers, S. and Nazar, L.F. (2013). New approaches for high energy density lithium–sulfur battery cathodes. *Accounts of Chemical Research* 46: 1135–1143.

129 Armstrong, A.R., Lyness, C., Panchmatia, P.M. et al. (2011). The lithium intercalation process in the low-voltage lithium battery anode $Li_{1+x}V_{1-x}O_2$. *Nature Materials* 2011 (10): 223–232.

130 Yang, Y., Zheng, G., and Cui, Y. (2013). Nanostructured sulfur cathodes. *Chemical Society Reviews* 42: 3018–3032.

131 Pang, Q., Liang, X., Kwok, C.Y., and Nazar, L.F. (2015). Review – the importance of chemical interactions between sulfur host materials and lithium polysulfides for advanced lithium–sulfur batteries. *Journal of the Electrochemical Society* 162 (14): A2567–A2576.

132 Li, W., Zheng, G., Yang, Y. et al. (2013). High-performance hollow sulfur nanostructured battery cathode through a scalable, room temperature, one-step, bottom-up approach. *Proceedings of the National Academy of Sciences of the United States of America* 110: 7148–7153.

133 Dean, J.A. (ed.) (1985). *Lange's Handbook of Chemistry*, 3e, 3–5. New York: McGraw-Hill.

134 Yin, Y.-X., Xin, S., Guo, Y.-G., and Wan, L.-J. (2013). Lithium–sulfur batteries: electrochemistry, materials, and prospects. *Angewandte Chemie International Edition* 52: 13186–13200.

135 Zhou, G., Li, L., Maa, C. et al. (2015). A graphene foam electrode with high sulfur loading for flexible and high energy Li–S batteries. *Nano Energy* 11: 356–365.

136 Chung, S.-H. and Manthiram, A. (2014). Low-cost, porous carbon current collector with high sulfur loading for lithium–sulfur batteries. *Electrochemistry Communications* 38: 91–95.

137 Yuan, Z., Peng, H.-J., Huang, J.-Q. et al. (2014). *Advanced Functional Materials* 24.

138 Zhang, J., Yang, C.-.P., Yin, Y.-X. et al. (2016). Sulfur encapsulated in graphitic carbon nanocages for high-rate and long-cycle lithium–sulfur batteries. *Advanced Materials* 28: 9539–9544.

139 Wang, M., Wang, W., Wang, A. et al. (2013). A multi-core–shell structured composite cathode material with a conductive polymer network for Li–S batteries. *Chemical Communications* 49: 10263–10265.

140 Cheng, X.-B., Huang, J.-Q., Peng, H.-J. et al. (2014). Polysulfide shuttle control: towards a lithium–sulfur battery with superior capacity performance up to 1000 cycles by matching the sulfur/electrolyte loading. *Journal of Power Sources* 253: 263–268.

141 Seh, Z.W., Zhang, Q., Li, W. et al. (2013). Stable cycling of lithium sulfide cathodes through strong affinity with a bifunctional binder. *Chemical Science* 4 (9): 3673–3677.

142 Kinoshita, S., Okuda, K., Machida, N., and Shigematsu, T. (2014). Additive effect of ionic liquids on the electrochemical property of a sulfur composite electrode for all-solid-state lithiumesulfur battery. *Journal of Power Sources* 269: 727–734.

143 Barghamadi, M., Best, A.S., Bhatt, A.I. et al. (2014). *Energy and Environmental Science* 7: 3902–3920.

144 Barghamadi, M., Best, A.S., Bhatt, A.I. et al. (2015). Effect of LiNO$_3$ additive and pyrrolidinium ionic liquid on the solid electrolyte interphase in the lithium–sulfur battery. *Journal of Power Sources* 295: 212–220.

145 Ma, G., Wen, Z., Jin, J. et al. (2014). The enhanced performance of Li–S battery with P14YRTFSI-modified electrolyte. *Solid State Ionics* 262: 174–178.

146 Barghamadi, M., Best, A.S., Bhatt, A.I. et al. (2015). Effect of anion on behaviour of Li–S battery electrolyte solutions based on *N*-methyl-*N*-butyl-pyrrolidinium ionic liquids. *Electrochimica Acta* 180: 636–644.

147 Xiong, S., Scheers, J., Aguilera, L. et al. (2015). Role of organic solvent addition to ionic liquid electrolytes for lithium–sulphur batteries. *RSC Advances* 5 (3): 2122–2128.

148 Yan, Y., Yin, Y., Guo, Y., and Wan, L.-J. (2014). Effect of cations in ionic liquids on the electrochemical performance of lithium–sulfur batteries. *Science China Chemistry* 57 (11): 1564–1569.

149 Unemoto, A., Ogawa, H., Gambe, Y., and Honma, I. (2014). Development of lithium–sulfur batteries using room temperature ionic liquid-based quasi-solid-state electrolytes. *Electrochimica Acta* 125: 386–394.

150 Hu, J.J., Long, G.K., Liu, S. et al. (2014). A LiFSI–LiTFSI binary-salt electrolyte to achieve high capacity and cycle stability for a Li–S battery. *Chemical Communications (Cambridge)* 50 (93): 14647–14650.

151 Barchasz, C., Leprêtre, J.-C., Patoux, S., and Alloin, F. (2013). Electrochemical properties of ether-based electrolytes for lithium/sulfur rechargeable batteries. *Electrochimica Acta* 89: 737–743.

152 Azimi, N., Xue, Z., Rago, N.D. et al. (2015). Fluorinated electrolytes for Li–S battery: suppressing the self-discharge with an electrolyte containing fluoroether solvent. *Journal of the Electrochemical Society* 162 (1): A64–A68.

153 Azimi, N., Weng, W., Takoudis, C., and Zhang, Z. (2013). Improved performance of lithium–sulfur battery with fluorinated electrolyte. *Electrochemistry Communications* 37: 96–99.

154 Liang, X., Wen, Z., Liu, Y. et al. (2011). Improved cycling performances of lithium sulfur batteries with LiNO$_3$-modified electrolyte. *Journal of Power Sources* 196 (22): 9839–9843.

155 Barchasz, C., Lepretre, J.C., Patoux, S., and Alloin, F. (2013). Revisiting TEGDME/DIOX binary electrolytes for lithium/sulfur batteries: importance of solvation ability and additives. *Journal of the Electrochemical Society* 160 (3): A430–A436.

156 Azimi, N., Xue, Z., Hu, L. et al. (2015). Additive effect on the electrochemical performance of lithium–sulfur battery. *Electrochimica Acta* 154: 205–210.

4

Anode–Electrolyte Interface

Mark Wild

OXIS Energy, E1 Culham Science Centre, Abingdon, Oxfordshire OX14 3DB, UK

4.1 Introduction

In Chapter 1, we discussed the issue of side reactions and solid–electrolyte interface (SEI) formation at the lithium anode in a lithium sulfur cell. In addition to electroplating and electrodissolution of lithium in the core redox chemistry of the cell to release cations into the electrolyte, we note that in a typical construction the anode provides a source of reductant species, whilst excess lithium acts as a lightweight current collector and helps combat poor coulombic efficiency. Thus, the degradation of the anode is a significant contributor to reduced cycle life and limits application. If the energy density of a lithium–sulfur cell is set at 400 Wh kg^{-1}, the thickness of lithium metal is estimated at 25–50 μm (5–10 mAh cm^{-1}). Commercial foils are 70–130 μm in most cases.

Lithium is highly reactive and lightweight, making it an ideal candidate for a battery technology designed for high gravimetric energy density. Unfortunately, lithium reacts with many species it comes into contact with, forming unwanted side products. These unwanted side reactions do not add value and may lead to irreversible loss of lithium and other electrolyte components. Consumption of electrolyte or drying of the cell and or loss of lithium results in accelerated capacity fade.

Lithium has an extremely high theoretical specific capacity of 3860 mAh g^{-1} and a low density of 0.59 g cm^{-3} and the lowest electrochemical potential −3.040 vs. the standard hydrogen electrode.

4.2 SEI Formation

The reactions of lithium and electrolyte components form an SEI on the surface of the lithium that slows reaction of electrolyte components with the anode and can reduce degradation and thus improve cycle life. The SEI covers the surface of the anode, and primary electrochemical reactions occur through the SEI layer. The nature of the SEI layer affects reaction kinetics and can lower cell voltage due to increased internal resistance. Despite this, the SEI layer and its properties are critical to the performance of the anode and the focus of materials research related to the anode in a lithium sulfur cell. While materials research in the electrolyte arena focuses on choosing stable solvent systems or reactive additives that promote a favorable SEI composition, solvents are the main source of organic lithium salts in SEI films [1].

Disordered structure promotes ionic conductivity, while the thickness of the SEI layer increases internal resistance. The film stops growing when electron transfer is blocked, typically tens of angstroms. The compact stratified layer model is commonly used to describe the SEI on a lithium anode. It is considered that the surface film on the anode consists of a porous interphase and a compact interphase consisting of sublayers. The porous outer layer close to the solution is nonuniform because the reduction of solution species cannot take place over the entire film–solution interface, but rather at defects or holes where electrons can tunnel to the surface. The composition of the SEI changes gradually as you move from solution/SEI to SEI/Li. Close to the lithium surface, lower oxidation states are found and the SEI becomes more compact.

The formation of an SEI is a double-edged sword [2] depending on chemical and physical properties. A coarse and inhomogeneous SEI such as the disordered mosaic type from soaking promotes preferential growth through cracks and where the SEI is thinner. An intact and smooth SEI where localized defects are largely eliminated suppresses both intrinsic and induced dendrite growth. Ideally, an SEI should be chemically stable, lithium-ion conductive, compact, uniform, and possess mechanical rigidity and elasticity to accommodate volume change.

4.3 Anode Morphology

In addition to SEI formation during charge and discharge, lithium stripping and plating leads to morphological changes over time. Natural imperfections in the soft metal anodes act as nucleation points, the uneven stripping and plating of lithium over time increases the surface area of the anode and introduces porosity; this is known as 3D mossy growth. While this process increases the anode reactive surface area for electrochemistry, it also promotes continual breaking and reforming of the SEI. This cyclic process depletes reactive SEI forming electrolyte components in the cell over time. While irreversible side reactions nibble away at the active material and perhaps more significantly the anode as current collector. The process accelerates as degradation increases.

Mossy growth [3] is a 3D omnidirectional moss or bushlike growth. 1D growth forms 3D growth by broadening and branching during filament growth. It can be explained by the raisin bread expansion model; there is no preferred direction and the distance between each raisin increases as the loaf expands. It has no growth center, but the movement can be restricted due to its support, where the metal base fixes the moss. Lithium atoms are inserted over the whole structure, and growth does not necessarily occur at tips but at distributed growth points. Growth and dissolution of Li mossy structures is a nonlinear dynamic process, motion appears random and is not dominated by any direction of the electric field in the bulk electrolyte. During dissolution, large parts may become electrically disconnected, which can happen even if the lithium remains attached to its original position at the substrate because the electrical contact sites are substituted by an insulating and passivating SEI layer.

The lithium surface must be smooth to ensure a uniform SEI layer. Stan and coworkers [4] studied the effect of controlling the starting lithium surface roughness and the nature of the native SEI. This was achieved by a simple roll press with a controlled surface finish. This had the effect of reducing overpotentials for plating and deplating in a symmetrical cell.

4.4 Polysulfide Shuttle

Discussed elsewhere, it is worth repeating here that the anode has a role in the polysulfide shuttle mechanism unique to lithium sulfur cells. The anode is able to reduce any high-order polysulfide species, which find their way to the anode surface from the cathode and permeate through the separator dissolved in the electrolyte. Toward the end of charge, high-order polysulfide species accumulate prior to sulfur precipitation at the cathode and diffuse back to the anode. At the anode, these high-order polysulfides are reduced before diffusing back to the cathode to be reoxidized.

The whole process generates an internal parasitic current, in which the polysulfide species transport electrons from the anode to the cathode inside the cell. The effect is to prevent overcharge where the charge current is countered by the shuttle current at the top of charge. This in turn causes internal heating of the cell. The effects must be managed at the level of the battery management system to detect shuttle and manage self-discharge or by materials research to eliminate the parasitic reactions through design of the SEI in the anode case. This chapter focuses on the latter approach.

4.5 Electrolyte Additives for Stable SEI Formation

Electrolyte additives are an effective method to modify electrode/electrolyte interfaces in lithium batteries. Lithium nitrate is a common additive in lithium–sulfur cells to manage the shuttle effect and to form a protective SEI layer. Xiong et al. [5, 6] studied the properties of the surface film on a lithium anode with lithium nitrate salt in the electrolyte. They showed that the effect of adding lithium nitrate is to reduce or eliminate the polysulfide shuttle mechanism until it is depleted, and the effect can be most observed at the end of charge when the cell voltage readily reaches a cutoff voltage (2.4 V) compared to a cell subject to shuttle that will fail to reach its cutoff voltage due to the internal current leveling the voltage. It was proposed that a lithium anode exposed to an electrolyte containing lithium nitrate prevents parasitic reactions between polysulfides and lithium anode, demonstrated by Liang et al. [7].

Lithium was supplied with a native SEI consisting of LiOH, Li_2O, and Li_2CO_3. All are inorganic crystalline salts that provide a rough surface. When immersed in $LiN(CF_3SO_2)_2$/1,3-dioxolane/dimethoxyethane, the lithium reacts to form LiF, $Li_2S_2O_4$, and ROLi creating a new smooth surface. If Li_2S_6 is added, then Li_2S is also detected in the SEI.

Analysis of the SEI formed from an electrolyte composition containing $LiNO_3$/1,3-dioxolane/dimethoxy ethane showed that the top layer consists of both inorganic species such as LiN_xO_y and organic species such as ROLi and $ROCO_2Li$. With increasing depth, the order of the nitrogen in reduction products becomes lower, from $Li_2N_2O_2$ to Li_3N.

Under the top layer, the film mainly consists of $ROCO_2Li$, Li_2O, LiN_xO_y, and Li_3N; these species prevent the continuous electron transfer from lithium metal to polysulfide solutions and electrolyte solutions.

A homogeneous surface film was obtained ensuring a uniform ionic conductivity on the surface of the anode, leading to uniform plating and stripping. Minimizing morphological changes reduces SEI damage and repair, thus reducing electrolyte depletion. Reduction of 1,3-dioxolane synthesizes oligomers of poly-dioxolane *in situ* that are insoluble and adhere to the lithium surface by its −OLi edge groups. These elastomers provide flexibility in the SEI that can better accommodate volume changes compared to many other solvents' SEI, making 1,3-dioxolane a common solvent of choice.

By comparison, the SEI formed from the same electrolyte without lithium nitrate gave a top layer consisting of Li_2CO_3, Li_2O, LiOH, RCH_2Li, and RCH_2OLi with RCH_2OLi and Li_2CO_3 as the main components, suggesting the SEI is composed of lithium metal and mixed solvents. Under the top layer, Li_2CO_3, Li, and Li_2O were the main components similar to the native film found on metallic lithium. The surface of the lithium was very smooth due to ROLi and a polymer film of polymerized 1,3-dioxolane, and the salts formed a regular meshlike pattern on the homogeneous polymer layer, suggesting that the polymer layer formed first.

Impedance spectroscopy determined that the native film on the lithium is disrupted following immersion of the anode in electrolyte. A new surface film forms from reaction of the anode and electrolyte components, a reaction that slows and stops with time as the SEI grows thicker and electron tunneling ceases. If the reaction rate is slow, then the SEI film can form uniformly. However, if the rate of SEI formation is too slow, then when it is damaged during cycling (due to volume changes) inefficient repair may form an uneven surface. The smoother and more compact the SEI, the more low viscosity solvents and lithium polysulfides are separated from reaction at the anode, thus reducing degradation.

This is a valid approach to anode protection in combination with other methodologies; by itself, this technique has been unsuccessful in significantly delaying degradation to meaningfully extend cycle life. Lithium nitrate is ultimately depleted due to constant repair of the SEI as the anode changes volume during cycling. Ultimately, the effect of suppressing the shuttle mechanism is defeated by constant repair of the SEI [8].

Others have used lithium salts in a similar way, such as Li bis(perfluoroethylsulfonyl) imide (LiBETI), which can form stable, thin, uniform, and compact surface films on lithium containing mainly LiF that can also improve cycling efficiency.

Nazar and coworkers [9] used Li_2S_6 and P_2S_5 as electrolyte additives to form a chemically stable thin amorphous layer on the anode consisting of Li_3PS_4 *in situ*. The film reduces the heterogeneity of the SEI, thus allowing a nonimpeding lithium ion flux. The layer has a theoretical lithium transference number of 1, which effectively eliminates ion depletion and strong electric field buildup at the lithium surface, both mechanisms linked to dendrite growth. Stable long-term plating and stripping of symmetrical cells over 2500 hours without short circuit was demonstrated.

Mo and coworkers [10] modeled candidates for stable SEI components from a large materials database from first principles. They found most oxides, sulfides, and halides are reduced by lithium metal due to the reduction of metal cations and that the nitride anion has unique stability against lithium metal either intrinsically or through passivation.

4.6   Barrier Layers on the Anode

An excess of lithium is used as current collector in a lithium sulfur cell and to combat low coulombic efficiency. In lithium ion cells using lithium metal, the formation of lithium dendrites has caused safety concerns due to the potential for internal short circuit. The term dendrite covers a range of structures including needle-like, snowflake-like, tree-like, bushlike, whisker-like, and mosslike. In most lithium sulfur systems, only mossy growth is observed in practice and internal short circuits due to dendritic growth have not been a reported practical issue to date. Barrier layer approaches have been advanced to deal with dendrite growth that has the potential to pierce the separator for lithium ion technology, some of which can be applied to lithium sulfur technology to combat the shuttle effect and other degradation processes.

Most approaches to dendrite prevention in rechargeable lithium cells [1] have focused on the stability and uniformity of the SEI through the use of electrolyte additives. Because lithium metal is thermodynamically unstable in organic solvents, such methods are often short-lived, as discussed earlier. Despite this, their simplicity for scale-up and commercialization make them attractive.

An alternative approach is to form an *ex situ* mechanical barrier on the lithium foil. Examples include polymer coatings or ceramics with a high shear modulus to reduce damage and repair to the protective layer that would otherwise deplete reactive components in the electrolyte. Reel-to-reel coating techniques can be developed for lithium coatings; such techniques are used, for example, in the semiconductor industry.

Barriers rely on forming a strong mechanical layer while attempting to reduce the impact upon the primary electrochemistry taking place. The approach can easily lead to high internal resistance within the cell if the barrier layer blocks electrochemical activity.

Polymer layers can be cast onto lithium and dried; the advantages are that flexibility of the polymer makes them robust to volume changes during cycling. The issue is to find a conductive polymer or to achieve a thin coating that does not significantly increase internal resistance in the cell. Polymer layers are required to be insoluble in the electrolyte and stable in the presence of polysulfides, nucleophiles, and radicals. Wu and coworkers [11] developed a cation-selective polymer blend of Nafion and polyvinylidene difluoride (PVDF), a hierarchically nanostructured composite that provides the flexibility and strength to cope with breathing of the lithium during charge and discharge. The polymer improved rate performance, coulombic efficiency, and cycle life.

Guo and coworkers [12] note that the SEI formed from organic solvents is brittle and cannot withstand mechanical deformation, leading to the formation of cracks. Cracks enhance lithium ion flux and result in dendrite formation and new SEI formation. The recurring breakage and repair of the SEI consumes lithium and electrolyte causing battery failure. Volume change is the main issue that defeats most approaches to forming a stable SEI. The team thus fabricated a smart SEI layer with elasticity using an *in situ* reaction between lithium and polyacrylic acid (PAA). Lithium PAA has good uniform binding properties and is flexible enough to accommodate lithium deformation.

An alternative to flexible polymer layers are ceramic coatings applied via a range of techniques used in the photovoltaic industry, for example, chemical vapor deposition (CVD) or atomic layer deposition (ALD). Ceramic layers may act as thin conductive barrier layers to address internal resistance issues of thicker polymer layers but tend to be brittle and crack during cycling.

Elam and coworkers [13] employed ALD to prepare conformal, ultrathin aluminum oxide coatings on lithium. They found it improved wetting of the surface toward both carbonate and ether electrolytes, leading to dense and uniform SEI formation with reduced electrolyte consumption and a practical 98% coulombic efficiency. During cycling, the surface was smooth and uniform compared to lithium without the coating. The coating allowed lithium ion diffusion and remained intact, resulting in an increase in cycle life with a reduced electrolyte volume in coin cells.

Zhang et al. [14] offer a review of techniques for lithium protection and point to the example of *ex situ* CVD of 2D boron nitride as a flexible, chemically stable, barrier layer with sufficient mechanical strength despite ultrathin thickness, yet noted its contribution to interfacial resistance which must be overcome.

4.7 A Systemic Approach

Hu and coworkers [15] conclude in their review of protected lithium metal anodes that a systemic approach is required to address anode stability. A systemic approach combines modification of the lithium electrode, the organic electrolyte and its additives, the SEI, the separator properties, and the battery configuration and management system itself. The systems approach is a theme that runs throughout this book.

In another review by Zhang and coworkers [3], they conclude that many issues need to be considered for the improvement of the lithium metal anode. Nonaqueous organic electrolyte is the most adopted system for lithium metal anodes, resulting in an unstable interface. The fundamental understanding of the SEI today is inadequate. Solid-state electrolytes offer advantages over liquid in terms of safety and chemical stability, but low ionic conductivity and instability vs. lithium metal reduces their technology readiness level. Control of volume changes could be achieved through a structured matrix in place of a metal foil.

There is a lack of consistency in anode research, making it difficult to directly compare results; the performance of a material depends on the design of the cell and current density. To better understand the underlying mechanism, more smartly designed *in situ* or operando analytical techniques are required. There is a need for thinner commercial foils of 25–50 μm with a tailored SEI layer and lithium plating matrix that can be processed on a roll-to-roll commercial scale. These challenges make anode research a hot topic for next-generation battery research.

References

1 Xu, W., Wang, J., Ding, F. et al. (2014). *Energy & Environmental Science* 7: 513.
2 Gu, Y., Wang, W., Li, Y. et al. (2018). *Nature Communications* 8: 1339.
3 Cheng, X., Zhang, R., Zhao, C., and Zhang, Q. (2017). *Chemical Reviews* 117: 10403–10473.
4 Becking, J., Grobmeyer, A., Kolek, M. et al. (2017). *Advanced Materials Interfaces* 4: 1–9.
5 Xiong, S., Xie, K., Diao, Y., and Hong, X. (2012). *Electrochimica Acta* 83: 78–86.

6 Xiong, S., Xie, K., Diao, Y., and Hong, X. (2013). *Journal of Power Sources* 236: 181–187.

7 Liang, X., Wen, Z., Liu, Y. et al. (2011). *Journal of Power Sources* 196: 9839.

8 Xiong, S., Xie, K., Diao, Y., and Hong, X. (2014). *Journal of Power Sources* 246: 840–845.

9 Pang, Q., Liang, X., Shyamsunder, A., and Nazar, L.F. (2017). *Joule* 1: 871–886.

10 Zhu, Y., He, X., and Mo, Y. (2017). *Advancement of Science* 4: 1–11.

11 Luo, J., Lee, T., Jin, J. et al. (2017). *Chemical Communications* 53: 963–966.

12 Nian-Wu, L., Shi, Y., Yin, Y. et al. (2018). *Angewandte Chemie International Edition* 57: 1505–1509.

13 Chen, L., Connell, J.G., Nie, A. et al. (2017). *Journal of Materials Chemistry A* https://doi.org/10.1039/c7ta03116e.

14 Zhang, X., Cheng, X., and Zhang, Q. (2017). *Advanced Materials Interfaces* 5: 1–19.

15 Yang, C., Fu, K., Zhang, Y. et al. (2017). *Advanced Materials* 29: 1–28.

Part II

Mechanisms

Understanding the mechanism of a lithium–sulfur cell at the molecular level is important on a number of fronts. Firstly, knowledge of the chemistry and physical mechanisms at play within the lithium–sulfur system is the key to focusing on and accelerating materials research efforts (Part I). Secondly, this knowledge can be used to develop a range of physical models at varying levels of detail to further explain and predict the performance of a lithium–sulfur cell (Part III). Such models have practical application forming the basis of battery management systems for applications to realize the potential of lithium–sulfur technology (Part IV).

Unlike well-studied lithium ion intercalation mechanisms, lithium–sulfur technology has a much more complex and intriguing mechanism from dissolution of the active sulfur into the electrolyte. The transition moves from solid sulfur through soluble polysulfides to solid lithium sulfide. An ever-changing equilibrium exists between lithium polysulfides that readily interconvert through association and disassociation reactions via neutral, ionic, and radical species from long chain to the more nucleophilic short chain polysulfides, and the reactivity of polysulfides with cathode, electrolyte, and anode components.

There are many theories around the mechanism of a lithium–sulfur cell and elucidation of that mechanism; reading literature on the subject is perhaps like reading a good murder mystery novel. The mechanism is elusive, changed by cell configuration and choice of materials often designed to influence the cell chemistry, all making direct comparisons from the literature difficult to interpret. Analytical methods often only see part of the puzzle, such as the blind man and the elephant, some techniques only see crystalline species, while others only see dissolved species and yet can miss the radicals or neutral species, for example. Smart *in situ* analysis is growing in popularity and can even be used *in operando* while *ex situ* analysis has the uncertainty of changing the electrolyte composition as an artifact of the analysis. Clearly, there is much left to understand about the mechanism of a lithium–sulfur cell.

A review on the subject by the editors of this book and their coworkers developed a working model for the mechanism of a lithium–sulfur cell, which is useful for the interpretation of electrochemical data and the development of physics-based models of

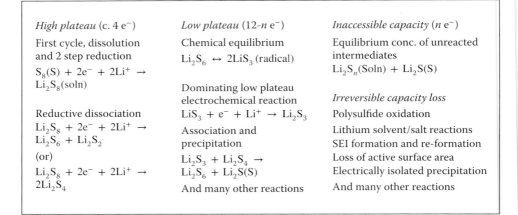

High plateau (c. 4 e⁻)

First cycle, dissolution and 2 step reduction

$$S_8(S) + 2e^- + 2Li^+ \rightarrow Li_2S_8(soln)$$

Reductive dissociation
$$Li_2S_8 + 2e^- + 2Li^+ \rightarrow Li_2S_6 + Li_2S_2$$
(or)
$$Li_2S_8 + 2e^- + 2Li^+ \rightarrow 2Li_2S_4$$

Low plateau (12-n e⁻)

Chemical equilibrium
$$Li_2S_6 \leftrightarrow 2LiS_3 \text{ (radical)}$$

Dominating low plateau electrochemical reaction
$$LiS_3 + e^- + Li^+ \rightarrow Li_2S_3$$

Association and precipitation
$$Li_2S_3 + Li_2S_4 \rightarrow Li_2S_6 + Li_2S(S)$$
And many other reactions

Inaccessible capacity (n e⁻)

Equilibrium conc. of unreacted intermediates
$$Li_2S_n(Soln) + Li_2S(S)$$

Irreversible capacity loss

Polysulfide oxidation
Lithium solvent/salt reactions
SEI formation and re-formation
Loss of active surface area
Electrically isolated precipitation
And many other reactions

Figure 1 A simplified working model of the discharge mechanism of a lithium–sulfur cell that can be used to interpret electrochemical results and cell performance. Source: Reproduced by permission of The Royal Society of Chemistry.

lithium technology. [1] Figure 1 summarizes the mechanism, which is a development from the linear mechanisms proposed by many early authors in the literature.

Figure 1 addresses many of the observed features of lithium–sulfur chemistry. At the cathode/anode surface all available polysulfides (di-anions, anions, radicals, and neutral species) and lithium salts take part in surface-enabled electrochemical and chemical reactions. In solution, chemical association and dissociation reactions maintain a working concentration of electrochemically active intermediates until polysulfide phase transitions take place. These precipitation, crystallization, and dissolution processes give rise to an increase in internal resistance, which hinders further activity. Intermediate species are "trapped" in solution at their equilibrium concentrations until subsequent cycles release that inaccessible capacity, contributing to variable and reversible capacity from cycle to cycle and with changing C-rate.

Depicted as high and low plateau regions electrochemically, the chemical transitions are considered as a continuous multi-species system. The system is modified by controlling morphology, porous structure, and the choice of solvents and additives to modify the SEI, ion mobility, and rate of chemical reaction. The charge mechanism is expected to be similar but not an exact reversal of the discharge process and is likely to be more direct.

The main features of the system include charge transfer at the solid/electrolyte interface, dissolution, and precipitation at the end of charge and discharge, lithium sulfide precipitation during discharge seeding at the start of the low plateau, multiple mechanistic pathways coexisting and/or dominated depending on operating conditions and design of the cell, discharge proceeds through a predominantly dissociative mechanism with a dominant electrochemical cycle in the low plateau based at least around the most stable $S_6^{2-}/2S_3$ radical equilibrium, asymmetry between charge and discharge, inaccessible capacity from trapped polysulfide species in solution, rates of irreversible capacity loss due to degradation, and rates of polysulfide shuttle most evident on charge and on storage self-discharge to the low plateau.

Chapter 5 takes a theoretical look at the stability of common solvents to the reactive polysulfides found in solution. Chapter 6 takes an illuminating look at the mechanism of the lithium–sulfur cell from the perspective of the discharge product lithium sulfide whose properties limit the theoretical potential of the technology considerably. Chapter 7 takes a detailed look at degradation mechanisms including the main causes, methods of measuring degradation, models to analyze the effects of degradation, and approaches to limit degradation.

Reference

1 Wild, M., O'Neill, L., Zhang, T. et al. (2015). *Energy. Environ. Sci.* 8 (12): 3477–3494.

5

Molecular Level Understanding of the Interactions Between Reaction Intermediates of Li–S Energy Storage Systems and Ether Solvents

Rajeev S. Assary[1,2] and Larry A. Curtiss[1,2]

[1] *Argonne National Laboratory, Materials Science Division, Argonne, IL 60439, USA*
[2] *Argonne National Laboratory, Joint Center for Energy Storage Research, Argonne, IL 60439, USA*

5.1 Introduction

Breakthroughs in secondary and grid storage battery technologies are essential for a sustainable future [1–4]. An energy storage system that utilizes lithium–sulfur redox chemistry is promising in this context [5, 6]. After more than three decades of active fundamental and applied research and developmental efforts, the Li–S energy storage systems still offer significant challenges such as cycling efficiency, formation of insulating Li–S layers on the anode, anode reactivity with electrolyte, poor understanding of solid–electrolyte interface, stability of nonaqueous electrolyte, solubility of polysulfide and conductivity, and other safety issues [7–9]. In spite of these challenges, this is a very appealing battery reaction system due to the high theoretical capacity (10 times that of the present Li ion) and the fact that sulfur is relatively abundant and less toxic compared to the cathodes in the conventional lithium ion batteries. To address the aforementioned challenges and to convert the Li–S electrical energy storage system to a game-changing battery technology, further molecular scale understanding of the processes at the cathode, anode, and electrolyte is essential. Our focus in this chapter is on the electrolytes; in particular, we discuss the polysulfide properties and reactivity toward various ether electrolytes using predictive simulations.

Experimental investigation of electrochemical performance and the mechanistic understanding using analytical techniques are the dominant areas of fundamental research associated with the Li–S cell. In recent years, exceptional progress has been witnessed in terms of Li–S cell performances, designing of new cathode architectures, and optimizing electrolytes [7, 8]. However, limited progress has been achieved in terms of modeling the exact mechanisms of speciation and reactivity of lithium polysulfide (LiPS) during discharging and charging processes [10]. Recently, atomistic models for the chemical reactions associated with the Li–S cell have been investigated using the predictive first principle models [10–16]. Selected computational modeling efforts focused on the electrolytes are discussed here. Using the G4MP2 [17] level of theory, Assary et al. have predicted the fragmentation of S_n ($n = 2–8$) and LiPS (Li_xS_y, $x = 1–2$, $y = 1–8$) and simulated the redox windows of probable intermediates in the bulk solution [11]. Siegel and coworkers [18] have investigated the nature of the redox end

Lithium–Sulfur Batteries, First Edition. Edited by Mark Wild and Gregory J. Offer.
© 2019 John Wiley & Sons Ltd. Published 2019 by John Wiley & Sons Ltd.

members in the lithium–sulfur battery from periodic density functional methods, and Balbuena and coworkers [19] have investigated the reactivity at the lithium metal anode surface.

One of the challenges of modeling Li–S chemistry is the intrinsic complexity of the properties of LiPS and their reactivity, as shown in Figure 5.1. In principle, LiPSs may exist as $[Li_xS_y]^{m-}$ (where $x = 0-2$, $y = 0-8$, and $m = 0, -1, -2$) in the electrolyte and depend on the state of charge of the system. The properties of LiPS such as partial charges, unpaired spins, and chelating abilities with organic electrolytes are largely unknown. Broadly speaking, the possible properties of interest involving polysulfides include association, dissociation, charge transfer at the solid electrolyte interphase, solubility, and chemical reactivity toward the anode and organic electrolytes. Species-selective concentration and lifetime of these intermediates are also important parameters of the LiPS during the battery operating conditions. All of these are largely unknown. Thus, there is a clear need for fundamental understanding of the polysulfides in the cell, in particular their reactivity toward electrolyte solvents.

In understanding the chemical reactivity of polysulfide with the electrolytes and subsequently for designing better electrolytes, computational modeling can play an integral part. In particular, simulations based on density functional theory can be used to predict the thermodynamic and kinetic feasibilities of chemical reactions occurring in the condensed phase or at the electrode surface. In this chapter, we focus on the reactivity of the dominant polysulfides with ether solvent molecules as shown in Figure 5.1. The ether class of solvents including linear (monoglyme, diglyme (DGDME), triglyme (TGDME), and tetraglyme) (TEGDME) and cyclic (tetrahydrofuran (THF), dioxolane (DOL)) ethers are considered for use as the organic electrolyte. Additionally, novel fluorinated dimethoxy ethane (DME) (monoglyme) is also considered due to the interest in fluorinated solvents to limit the polysulfide shuttling effect and to maintain the capacity retention and coulombic efficiency.

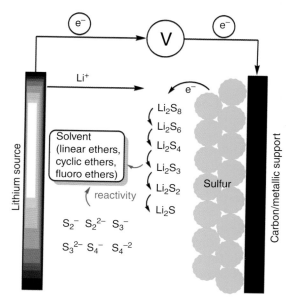

Figure 5.1 Schematic of probable (electro)chemical reactions of polysulfide with the solvent molecule in a Li–S cell.

5.2 Computational Details

All computations presented in the chapter were performed using the Gaussian 09 software [20]. The B3LYP/6–31+G* level of theory was used to compute the structures, electronic energies, vibrational frequencies, and free energy corrections of all species. Solvation free energies were computed employing the solvation model based on density (SMD) [21] by performing single point energy evaluations using an acetone dielectric medium ($\varepsilon = 20.49$) with gas phase optimized structures. The redox potentials of LiPS are adopted from our previous publication [11]. The solvation energy contributions are added to the gas phase enthalpies and free energies to approximate enthalpies ($H_{soln} = H_{gas\,phase} + \Delta G_{solv}$) and free energies ($G_{soln} = G_{gas\,phase} + \Delta G_{solv}$) in solution. The reaction barriers presented in this chapter are apparent enthalpy barriers ($\Delta H^{\dagger}$), computed as the difference between the enthalpy of the transition state structure ($H^{\dagger}$) and the sum of enthalpies of reactants in solution at 298 K. Here, we have used the solution phase enthalpy approximation because the main free energy contributions (G_{CDS}: cavitation, dispersion, and solvent structure terms) cancel out (or are negligible) when computing the apparent barriers ($H_{(TS)} - H_{(reactants)}$). Therefore, the dominant solvation contributions are from the electronic, nuclear, and polarization terms (G_{ENP}), which are included in the computation of solution enthalpies. A similar approach was successfully employed elsewhere [11, 22]. Thus, the enthalpy barriers ($\Delta H^{\dagger}$) are the sum of the gas phase enthalpy barrier and the solvation energy (E^{SMD}):

$$\Delta H^{\dagger}_{Soln} = (H^{\dagger}_{gas} - H^{reactants}_{gas}) + (E^{\dagger}_{SMD} - E^{reactants}_{SMD})$$

The rate constant (k) is computed using the Eyring equation:

$$(k) = \frac{k_b T}{hc^0} \exp^{-\frac{\Delta G^{\dagger}}{RT}}$$

Here, instead of free energy of activation ($\Delta G^{\dagger}_{soln}$), the enthalpy of activation in solution ($\Delta H^{\dagger}_{soln}$) is used due to the uncertainties in computing the entropic contributions in solution from gas phase approximations. Using the computed enthalpy of solution, the rate constants are approximated and subsequently the half-life of the reactions is computed assuming a first-order chemical reaction. For instance, the barrier heights ($\Delta H^{\dagger}_{soln}$) of 25 and 27 kcal mol^{-1} can be translated to first-order chemical reactions with half-lives of 66 hours and 80 days, respectively. Also, chemical reactions that require barriers of 30 kcal mol^{-1} or more are kinetically very slow and are unlikely to occur at room temperature.

5.3 Results and Discussions

The ether solvent molecules have three distinct types of chemical bonds such as C—H, C—O, and C—C bonds. Among these bonds, the "C—O" bond is the most *polar*. Our previous computational study [11] suggests that the C—O bond cleavage of ether solvent molecule (tetraglyme) by lithium–sulfur intermediates is preferred compared to the C—H or C—C bonds. We note that the C—H and the C—C bond breaking of tetraglyme via chemical reactions is neither thermodynamically nor kinetically feasible during the Li–S battery-operating conditions. Therefore, the C—O bond stability of ether solvents is the critical parameter describing chemical stability and the discussion related to the chemical stability of a commonly used DME solvent molecule is presented in Section 5.3.1. From the lessons learned based on the DME deep-dive study, we have performed computations to assess the chemical stability of nine more ether solvent (linear and cyclic) molecules, the details of which are discussed in Section 5.3.2. In the case of fluoroether molecules, the presence of polar "C—F" group introduces another critical parameter in addition to the C—O bond stability. Therefore, the chemical stability of C—O and C—F bonds of fluoroether molecules is also discussed in Section 5.3.3.

5.3.1 Reactivity of Li–S Intermediates with Dimethoxy Ethane (DME)

Reactions of LiPS and polysulfide anions are nucleophilic in nature. These intermediates do not attack the C—H bonds, but prefer carbon(C)-hetero(X)-atom/group, where X can be any heteroatom (or group such as OCH_3) capable of polarizing the carbon center. In fluorinated ethers (described in Section 5.3.3), the heteroatoms are "—O" and "F" respectively. The nucleophilic reaction sites of the DME and the fluorinated DME solvent molecule are schematically shown in Figure 5.2. In Figure 5.2, the subplots (a) and (b) represent the C—O bond cleavage in DME (or any ether molecules) initiated by the nucleophilic attack at the primary and secondary carbon positions, respectively. The subplots (c) and (d) represent the C—F bond cleavage of the fluorinated DME initiated by the nucleophilic attack at the primary and secondary carbon positions, respectively.

To understand the kinetic feasibility of possible C—O bond breaking reactions of DME molecules by possible reaction intermediates in the lithium–sulfur systems, the reaction enthalpies and activation enthalpies in solution are computed. The reaction intermediates considered here are $S^{-(rad)}$, $S_2^{-1,(rad)}$, $S_3^{-1,(rad)}$, S_2^{2-}, S_3^{2-}, S_4^{2-}, $LiS^{(rad)}$, LiS^{-1}, Li_2S, Li_4S_3, LiS_2^{-}, Li_2S_2, Li_2S_3, and Li_2S_4. The computed enthalpies of reactions (ΔH_{rxn}) and activation enthalpies $(\Delta H^{\ddagger})$ are also given in Table 5.1. The computed activation enthalpies $(\Delta H^{\ddagger})$ vs. Li–S cell reaction intermediates for the C—O bond breaking chemical reaction are shown in Figure 5.3. The optimized structures of selected reaction products formed as part of the reactions (entries 1–11 of Table 5.1) are shown Figure 5.4. The C—O bond cleavage of DME by the intermediates occurs via a nucleophilic SN_2 mechanism, where the negatively charged (anionic sulfur site) or radical nucleophile (species with unpaired spin) attacks the carbon atoms of the C—O bond. Based on the computed solution phase activation barriers $(\Delta H^{\ddagger})$, the C—O bond

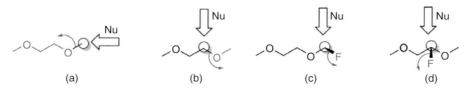

Figure 5.2 Schematic of the preferred nucleophilic reaction sites of ether molecules. The labels "a" and "b" denote C—O bond cleavage of dimethoxy ethane via nucleophilic attack by lithium polysulfides or sulfides. Similarly the labels "c" and "d" denote the C—F bond cleavage of fluorinated dimethoxy ethane via nucleophilic attack of lithium polysulfides/sulfides. The leaving groups are shown in dark gray.

Table 5.1 Computed solution phase enthalpies of reaction (ΔH_{rxn}) and activation ($\Delta H^{\ddagger}$) for the C—O bond cleavage of dimethoxy ethane in the presence of reactive intermediates.

Entry	Reactive Intermediate	Reaction	ΔH_{rxn} (298 K)	$\Delta H^{\ddagger}$ (298 K)
1	Li_2S_3	$DME + Li_2S_3 \rightarrow$ Complex (**1P**)	−23.2	27.1
2	S_2^{2-}	$DME + S_2^{2-} \rightarrow MeEtO^- + CH_3S_2^{1-}$	−0.1	30.4
3	Li_2S_4	$DME + Li_2S_4 \rightarrow$ Complex (**3P**)	4.30	31.1
4	Li_2S_2	$DME + Li_2S_2 \rightarrow$ Complex (**4P**)	−9.2	31.5
5a	LiS_2^{-1}	$DME + LiS_2^{-1} \rightarrow$ Complex (**5aP**)	11.7	31.7
5b	LiS_2^{-1}	$DME + LiS_2^{-1} \rightarrow MeEt\text{-}S_2Li + CH_3O^-$	10.4	31.7
6	S_3^{2-}	$DME + S_3^{2-} \rightarrow MeEtO^- + CH_3S_3^{1-}$	11.0	35.1
7	Li_2S	$DME + Li_2S \rightarrow$ Complex (**7P**)	−40.7	36.4
8	S_4^{2-}	$DME + S_4^{2-} \rightarrow MeEtO^- + CH_3S_4^{1-}$	16.9	37.9
9	LiS^{-1}	$DME + LiS^- \rightarrow MeEtO^- + CH_3SLi$	7.5	38.3
10	Li_4S_3	$DME + Li_4S_3 \rightarrow$ Complex (**10P**)	−45.6	39.7
11	$S^{1-,rad}$	$DME + S^{1-,rad} \rightarrow MeEtO^- + CH_3S^{rad}$	20.0	41.1
12	$S_2^{1-,rad}$	$DME + S_2^{1-,rad} \rightarrow MeEtO^- + CH_3S_2^{rad}$	28.7	43.0
13	$S_3^{1-,rad}$	$DME + S_3^{1-,rad} \rightarrow MeEtO^- + CH_3S_3^{rad}$	40.6	49.2
14	LiS^{rad}	$DME + LiS^{rad} \rightarrow MeEtO^{rad} + CH_3SLi$	13.2	49.8

All energetics are reported in kcal mol^{-1}. A B3LYP/6–31+G(d) level of theory is used. The solvent contributions are modeled using the SMD model using acetone dielectric model.
DME, $CH_3OCH_2CH_2OCH_3$; MeEtO, $CH_3OCH_2CH_2O$; MeEt, $CH_3OCH_2CH_2$; rad, radical. The three-dimensional structures of selected reaction products (**1P, 3P, 4P, 5aP, 7P, 10P**) are shown in Figure 5.4.

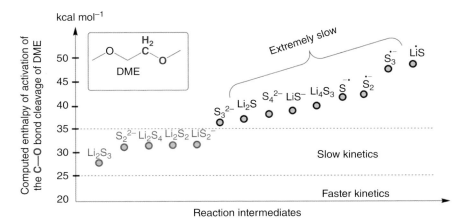

Figure 5.3 Computed activation enthalpies ($\Delta H^{\ddagger}$, kcal mol^{-1}) required for the C—O bond cleavage (shown in dark gray color) of DME by various polysulfide intermediates in acetone dielectric medium ($\varepsilon = 20$). The 25 kcal mol^{-1} activation barrier can be approximated to a first-order chemical reaction with a half-life of 66 hours at 298 K. The 27 kcal mol^{-1} activation barrier can be approximated to a first-order chemical reaction with half-life of 1900 hours at 298 K.

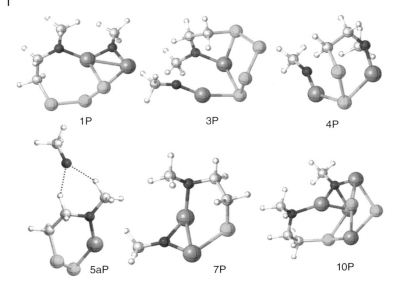

Figure 5.4 Computed geometries of selected complexes formed from the reaction of selected polysulfides with DME. The reaction energies associated with the formation of the complexes are presented in Table 5.1.

breaking by various intermediates can be classified into three categories based on the half-life of the reactions: (i) faster kinetics ($\Delta H^\dagger < 25\,\text{kcal mol}^{-1}$), (ii) slower kinetics ($\Delta H^\dagger = 25\text{–}35\,\text{kcal mol}^{-1}$), and (iii) extremely slow kinetics ($\Delta H^\dagger = 35\text{–}50\,\text{kcal mol}^{-1}$) as shown in Figure 5.3.

Most importantly, from Figure 5.3, it is clear that none of the possible reaction intermediates in a lithium–sulfur cell force *fast kinetics* for the C—O bond cleavage of DME. This information suggests that the DME molecule does not undergo spontaneous reaction(s) with reactive intermediates in the Li–S cell at room temperature. However, the C—O bond cleavage of DME by intermediates such as Li_2S_3, S_2^{2-}, Li_2S_4, Li_2S_2, and LiS_2^- belongs to a slow kinetic regime (Figure 5.3) suggesting that the half-lives of the reactions are on the order of months to years at room temperature. Intermediates shown in the "extremely slow" regime do not pose any threat to the stability of the solvent in terms of reactions with electrolytes. We note that intermediates such as S_3^{2-}, Li_2S, LiS^-, Li_4S_3, $S^{-1,(\text{rad})}$, $S_3^{-1,(\text{rad})}$, $S_2^{-1,(\text{rad})}$, $LiS^{(\text{rad})}$, and S_4^{2-} belong to this category.

In general, short chain LiPSs were found to be relatively more reactive compared to longer chain polysulfide or sulfide anions. Based on the computed transition state structures for selected reactions shown Figure 5.5, it is evident that the lithium ions are chelated with the ether oxygen atoms of the solvent molecule, thereby increasing the sulfur nucleophilicity. Additionally, upon lithium chelation with ether oxygen atoms, the sulfur atoms from the LiPS would form increased interactions with the carbon atoms of the DME molecule to facilitate the cleavage of the C—O bond.

5.3.2 Kinetic Stability of Ethers in the Presence of Lithium Polysulfide

From the reactivity pattern of shorter LiPS with DME, we have carried out a similar analysis of other solvent molecules including linear (glymes) and cyclic ethers (DOL

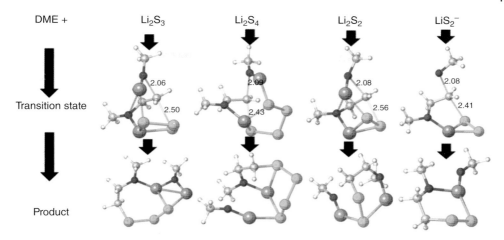

Figure 5.5 Three-dimensional structures of selected transition state structures and products associated with the C—O bond cleavage of dimethoxy ethane molecule computed using density functional theory. Selected bond lengths (Å) are shown.

and THF). In addition to the DME molecule, nine more ether molecules are considered for computing the reaction barriers of the C—O bond cleavage by Li_2S_2, Li_2S_3, Li_2S_4, and Li_2S species. The linear ethers considered here are monoglyme (DME), DGDME, TGDME, and TEGDME. The cyclic ethers considered are DOL, methyl dioxolane (MDOL), dimethyl dioxolane (DMDOL), THF, methyl tetrahydrofuran (MTHF), and dimethyl tetrahydrofuran (DMTHF). The computed activation barriers ($\Delta H^{\ddagger}$) and the schematic of the ether molecules are shown in Figure 5.6.

Computations suggest that C—O bond breaking in ethers requires reaction barriers in the range of 20–70 kcal mol^{-1}. The Li_2S_2 and Li_2S_3 molecular species are found to be the most reactive (neutral) intermediates in the solution toward the C—O bond cleavage. For DGDME, TGDME, and TEGDME solvent molecules, the reaction with Li_2S_2 requires the lowest activation barriers. The computed barriers are 23.8, 21.4, and 21.8 kcal mol^{-1} respectively for the DGDME, TGDME, and TEGDME solvent molecules. It should be noted that the C—O bond breaking barrier of DME by Li_2S_2 is 31.5 kcal mol^{-1}. The relatively lower C—O bond breaking activation barrier ($\Delta H^{\ddagger} < 25$ kcal mol^{-1}) for the longer chain ethers (with three or more oxygen atoms) is likely due to their chelating ability with Li_2S_2. Three or more oxygen atoms in the solvent molecule may effectively chelate with the lithium ions of the Li_2S_2, which in turn weakens the Li–S bonds and enables sulfide ions to act as better nucleophiles to cleave the C—O bond. For shorter chain (DME) or cyclic ethers, the activation barrier for the Li_2S_3 mediated C—O bond cleavage is kinetically preferred compared to the other polysulfides. It should be noted that the Li_2S_3 mediated C—O bond cleavage of ethers requires a minimum of ~27 kcal mol^{-1} barrier. It is worth pointing out that a barrier height of 27 kcal mol^{-1} can be translated to a first-order chemical reaction with a half-life of ~80 days, suggesting the slow nature of the reaction.

The computed activation enthalpies for the C—O bond cleavage in cyclic ethers such as DOL and THF are similar to that of DME (>27 kcal mol^{-1}) suggesting the chemical stability of these molecules in the presence of polysulfide. The lack of chelating ability

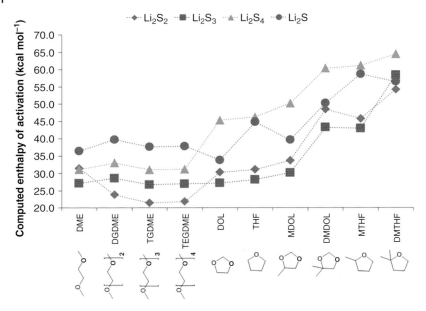

Figure 5.6 Computed enthalpies of activation (kcal mol⁻¹) required for the C—O bond cleavage of linear and cyclic ethers by Li_2S_2, Li_2S_3, Li_2S_4, and Li_2S molecular species in the solution phase. Abbreviations: DME, monoglyme methane; DGDME, diglyme; TGDME, triglyme; TEGDME, tetraglyme; DOL, dioxolane; THF, tetrahydrofuran; MDOL, methyl dioxolane; MTHF, methyl tetrahydrofuran; DMDOL, dimethyl dioxolane.

of individual DOL or THF molecules with lithium ion is the reason for the high reaction barrier for the C—O bond cleavage by the lithium polysulfides. Cyclic ethers are more likely to react with sulfide dianion ($S_2{}^{2-}$) than LiPS; however, the extent of reactivity depends on the concentration of dianion species in the solution. Other cyclic solvents such as MDOL, DMDOL, MTHF, and DMTHF are significantly more stable in the presence of LiPS, on the basis of the computed C—O bond breaking reaction barriers (>30 kcal mol⁻¹). The higher barriers for these methylated cyclic ethers are likely due to the added steric hindrance provided by the methyl groups in the C—O bond breaking transition state.

5.3.3 Linear Fluorinated Ethers

Recently, fluorinated ethers have been introduced as cosolvents to minimize the solubility of longer chain polysulfides to limit polysulfide shuttling [23–26]. These fluorinated ethers have shown positive impacts in sulfur utilization and kinetics during the cycling of the Li–S cell. However, the nature of the interaction of fluoroether molecules with the metal anode and the longer term chemical stability in the presence of LiPS are largely unknown. Having understood the rate controlling chemical reactivity of polysulfide with the ether molecules (unsubstituted), in this subsection we now focus on the reactivity of polysulfide toward fluorinated DME molecules. A schematic of 16 ethers (labeled "A"

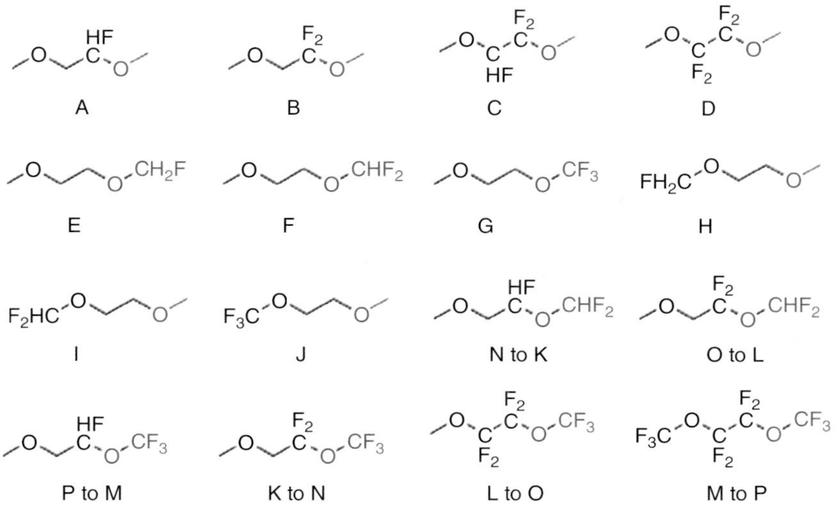

Figure 5.7 Schematic of the fluorine-substituted dimethoxy ethane (DME) species investigated for their chemical reactivity.

to "P") considered in the computations is shown in Figure 5.7 ranging from mono fluorinated ether (A) to a fully fluorinated ether (P). The central difference in reactivity between the ether (Section 5.3.1) and the fluorinated ethers (Figure 5.7) is the presence of polar "C—F" bonds and their influence on the chemical stability of adjacent "C—O" bond, especially in the presence of polysulfide nucleophiles (Figure 5.2). The computed reaction barriers for cleaving the "C—O" ($\Delta H^{\dagger}_{C-O}$) and "C—F" ($\Delta H^{\dagger}_{C-F}$) bonds of the 16 fluoroether molecules by S_2^{2-}, Li_2S_2, and Li_2S_3 are tabulated in Table 5.2. Based on the computed reaction barriers of the fluoroether molecules (entries 2–17 of Table 5.2), it is now possible to compare their relative kinetic stability to that of the DME molecule (entry 1) in the presence of most reactive intermediates. It is important to note that the lowest activation barrier (27.1 kcal mol^{-1}) for the DME molecule (kinetically most likely reaction) corresponds to the C—O bond cleavage by the Li_2S_3 species in the solution. The computed reaction barriers for the C—O and C—F bonds are discussed below. Selected transition state structures for C—O and C—F bond cleavage of fluoroether molecules A, G, P, and L are shown in Figure 5.8.

Replacing one β-hydrogen atom by a fluorine atom results in the formation of species "A," as shown in Figure 5.7. The introduction of a fluorine atom has a negative influence in terms of the C—O and the C—F bond stability (entry 2). In species A, the activation barrier required to cleave the C—O bond by S_2^{2-} is lower than that in the DME molecule. This is due to the presence of the highly electronegative "—F" substituent, where the "O—CHF" bond of species "**A**" becomes more polar (partial positive charge on carbon increases), and subsequently more active toward nucleophilic attack (especially from the sulfide dianion). Note that the "—OCHF—CH$_2$" group of species formed upon the C—O bond cleavage of species "A" is a better leaving group than the "—OCH$_2$—CH$_2$" group during the nucleophilic attack.

Table 5.2 Computed reaction barriers (kcal mol^{-1}) for cleaving the "C—O" ($\Delta H^{\ddagger}_{C-O}$) and "C—F" ($\Delta H^{\ddagger}_{C-F}$) bonds of 16 novel fluoroether molecules (entries 2–16) by S_2^{2-}, Li_2S_2, and Li_2S_3 computed at 298 K using the B3LYP/6–31+G(d) level of theory. Entry 1 corresponds to the computed reaction barriers of a DME molecule.

Entry		$\Delta H^{\ddagger}_{C-O}$				$\Delta H^{\ddagger}_{C-F}$		$\Delta H^{\ddagger} > 25$	[d]$\Delta H^{\ddagger} > 25$
	Species	S_2^{2-}	Li_2S_2	Li_2S_3	[c]S_2^{2-}	Li_2S_2	Li_2S_3	S_2^{2-}/Li_2S_n	Li_2S_n
1	DME	[a]30.4	31.5	27.1	NA	NA	NA	Yes	Yes
2	A	[a]19.6	42.8	39.5	18.1	10.8	9.7	No	No
3	B	[a]10.7	70.7	63.2	29.1	28.2	31.4	No	Yes
4	C	[a]9.4	71.0	71.9	28.2	32.5	32.8	No	Yes
5	D	[a]7.6	79.1	72.6	37.7	44.3	43.5	No	Yes
6	E	[b]28.1	16.6	20.7	15.8	10.3	14.3	No	No
7	F	[b]14.6	13.8	14.8	26.5	19.8	23.8	No	No
8	G	[b]9.6	14.7	11.6	39.8	34.4	38.4	No	No
9	H	27.4	16.6	20.7	15.8	10.3	14.3	No	No
10	I	25.4	13.8	14.8	26.5	19.8	23.8	No	No
11	J	26.1	14.7	11.6	39.8	34.4	38.4	No	No
12	K	33.5	23.7	21.7	19.7	18.0	19.8	No	No
13	L	18.3	45.7	37.4	30.3	30.6	33.7	No	Yes
14	M	46.2	17.9	16.0	22.8	21.5	25.9	No	No
15	N	20.8	27.2	28.9	34.8	37.0	42.0	No	Yes
16	O	2.9	39.6	42.0	33.2	48.5	52.5	No	Yes
17	P	25.2	42.4	45.0	29.9	45.6	51.1	Yes	Yes

a) SN_2 attack on primary carbon, $CH_3-S_2^{1-}$ is the leaving group.
b) SN_2 attack on secondary carbon atom, "fluoromethoxy ($-CHF-OCH_3$)" species is the leaving group.
c) SN_2 attack and the $-F$ is leaving group.
d) The S_2^{2-} ion is excluded.

This reactive behavior is slightly increased when two, three, and four β-hydrogen atoms are replaced by fluorine atoms; these species are labeled as B, C, and D, respectively in Figure 5.7. The computed C—O bond breaking enthalpy barriers by sulfide dianion for species A, B, C, and D are 19.6, 10.7, 9.4, and 7.6 kcal mol^{-1}. This clearly indicates the kinetic feasibility of the C—O bond cleavage in species A, B, C, and D in the presence of nucleophiles such as S_2^{2-}. This trend does not exist for the reaction of a species such as Li_2S_2 and Li_2S_3 with fluoroether molecules A–D for the C—O bond cleavage. The computed reaction barriers for the C—O bond cleavage are in the range of 40–70 kcal mol^{-1} (entries 2–5 of Table 5.2). These high reaction barriers are likely due to the following reasons: (i) the steric interactions from the fluorine atoms in the transition state, (ii) lack of ability of the oxygen atoms of the ether to chelate the lithium ion of the LiPS due to the presence of the electronegative fluorine atoms.

The DME molecule has six α-hydrogen atoms located in the terminal CH_3 groups. Replacing these α-hydrogen atoms by fluorine atoms sequentially at one of the CH_3 sites

Figure 5.8 The computed "C—F" and "C—O" bond breaking transition state structures of selected fluorinated DME derivatives (A, G, P, L) by the Li_2S_3 reaction intermediate. The values shown in parenthesis are the computed activation enthalpies ($\Delta H^\dagger$) in kcal mol^{-1}. Selected bond lengths shown in the figure are in angstrom units. The labels (A, P, L, and G) are consistent with Figure 5.7 and Table 5.2.

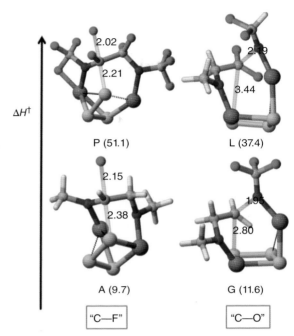

affords three fluoroether molecule species denoted as "E," "F," and "G" in Figure 5.7. The presence of two and three fluorine atoms at the terminal methoxy group enhances its characteristics as a good leaving group when nucleophilic attack occurs at the secondary carbon (CH_2) center (Figure 5.2b). For fluoroether molecules "E," "F," and "G," the C—O bond breaking process is more likely via a nucleophilic attack at the secondary carbon position ($\Delta H^\dagger < 30$ kcal mol^{-1}) than at the primary carbon position (($\Delta H^\dagger > 40$ kcal mol^{-1}). It should be noted that the activation barrier for the C—O bond cleavage of these molecules is less than 20 kcal mol^{-1} when Li_2S_2 or Li_2S_3 species acts as the nucleophile. Fluoroethers H, I, and J show a behavior similar to E, F, and G due to the similarity in the structure and bonds. Fluoroether molecules with more than one fluorine atom in the alpha and beta positions, such as species K, L, M, N, O, and P of Figure 5.7, require relatively high reaction barriers ($\Delta H^\dagger > 30$ kcal mol^{-1}) for C—O or C—F bond cleavage in the presence of Li_2S_2 or Li_2S_3 reaction intermediates. These species, except the fully substituted fluoroether (P) molecule, are reactive toward the sulfide dianion ($\Delta H^\dagger < 23$ kcal mol^{-1}).

Based on the computed reaction barriers that assess the kinetic feasibility of the C—O and the C—F bonds of fluoroether molecules (Table 5.2), the following important points can be noted:

In the absence of sulfide dianion, seven novel fluoroether molecules (B, C, D, L, N, O, P; see Figure 5.7 and Table 5.2) have the "C—O" and "C—F" bonds with adequate kinetic stability based on the computed activation barriers ($\Delta H^\dagger > 25$ kcal mol^{-1}).

In the presence of sulfide dianion in the bulk, the DME and the fully fluorinated DME (P of Figure 5.7 and entry 17 of Table 5.2) are stable against any possible nucleophilic attack that initiates irreversible decomposition of the solvent molecule.

5.4 Summary and Conclusions

Molecular level understanding of the chemical reactivity of the solvent molecule with sulfur-containing reactive intermediates is essential to predict the longer term stability of the Li–S energy storage system. In this chapter, first principle simulations are reported for the a priori prediction of the kinetic feasibilities of LiPS reactions with ether solvents and fluorinated ether solvents. Based on the extensive simulations, the following questions have been addressed:

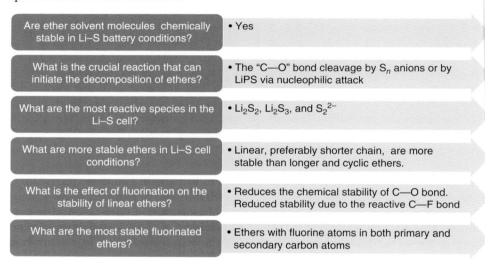

Are ether solvent molecules chemically stable in Li–S battery conditions?	• Yes
What is the crucial reaction that can initiate the decomposition of ethers?	• The "C—O" bond cleavage by S_n anions or by LiPS via nucleophilic attack
What are the most reactive species in the Li–S cell?	• Li_2S_2, Li_2S_3, and S_2^{2-}
What are more stable ethers in Li–S cell conditions?	• Linear, preferably shorter chain, are more stable than longer and cyclic ethers.
What is the effect of fluorination on the stability of linear ethers?	• Reduces the chemical stability of C—O bond. Reduced stability due to the reactive C—F bond
What are the most stable fluorinated ethers?	• Ethers with fluorine atoms in both primary and secondary carbon atoms

We expect that the data presented in this chapter provide a library of fundamental knowledge of the reactivity and stability of ether solvent molecules and fluorinated ether molecules in the presence of any sulfur intermediates likely present in the Li–S stationary/transportation battery settings. The guidelines obtained regarding the reactivity of the fluorinated ether candidates can be used to engineer stable cosolvents for the Li–S energy storage systems.

Acknowledgments

This work was supported as part of the Joint Center for Energy Storage Research, an Energy Innovation Hub funded by the U.S. Department of Energy, Office of Science, and Basic Energy Sciences. We gratefully acknowledge the computing resources provided on "Blues," a 320-node computing cluster operated by the Laboratory Computing Resource Center at Argonne National Laboratory.

References

1 Armand, M. and Tarascon, J.M. (2008). Building better batteries. *Nature* 451 (7179): 652–657.

2 Dunn, B., Kamath, H., and Tarascon, J.-M. (2011). Electrical energy storage for the grid: a battery of choices. *Science* 334 (6058): 928–935.

3 Tarascon, J.M. and Armand, M. (2001). Issues and challenges facing rechargeable lithium batteries. *Nature* 414 (6861): 359–367.

4 Thackeray, M.M., Wolverton, C., and Isaacs, E.D. (2012). Electrical energy storage for transportation-approaching the limits of, and going beyond, lithium-ion batteries. *Energy & Environmental Science* 5 (7): 7854–7863.

5 Evers, S. and Nazar, L.F. (2013). New approaches for high energy density lithium–sulfur battery cathodes. *Accounts of Chemical Research* 46 (5): 1135–1143.

6 Bruce, P.G., Freunberger, S.A., Hardwick, L.J., and Tarascon, J.-M. (2012). Li-O$_2$ and Li-S batteries with high energy storage. *Nature Materials* 11 (1): 19.

7 Manthiram, A., Fu, Y., Chung, S.-H. et al. (2014). Rechargeable lithium–sulfur batteries. *Chemical Reviews* 114 (23): 11751–11787.

8 Yin, Y.X., Xin, S., Guo, Y.G., and Wan, L.J. (2013). Lithium–sulfur batteries: electrochemistry, materials, and prospects. *Angewandte Chemie International Edition* 52 (50): 13186–13200.

9 Zhang, S.S. (2013). Liquid electrolyte lithium/sulfur battery: fundamental chemistry, problems, and solutions. *Journal of Power Sources* 231: 153–162.

10 Wild, M., O'Neill, L., Zhang, T. et al. (2015). Lithium sulfur batteries, a mechanistic review. *Energy & Environmental Science* 8 (12): 3477–3494.

11 Assary, R.S., Curtiss, L.A., and Moore, J.S. (2014). Toward a molecular understanding of energetics in Li–S batteries using nonaqueous electrolytes: a high-level quantum chemical study. *The Journal of Physical Chemistry C* 118 (22): 11545–11558.

12 Vijayakumar, M., Govind, N., Walter, E. et al. (2014). Molecular structure and stability of dissolved lithium polysulfide species. *Physical Chemistry Chemical Physics* 16 (22): 10923–10932.

13 Wang, L., Zhang, T., Yang, S. et al. (2013). A quantum-chemical study on the discharge reaction mechanism of lithium–sulfur batteries. *Journal of Energy Chemistry* 22 (1): 72–77.

14 Ward, A.L., Doris, S.E., Li, L. et al. (2017). Materials genomics screens for adaptive ion transport behavior by redox-switchable microporous polymer membranes in lithium–sulfur batteries. *ACS Central Science* 3 (5): 399–406.

15 Rajput, N.N., Murugesan, V., Shin, Y. et al. (2017). Elucidating the solvation structure and dynamics of lithium polysulfides resulting from competitive salt and solvent interactions. *Chemistry of Materials* 29 (8): 3375–3379.

16 Fan, F.Y., Pan, M.S., Lau, K.C. et al. (2016). Solvent effects on polysulfide redox kinetics and ionic conductivity in lithium–sulfur batteries. *Journal of the Electrochemical Society* 163 (14): A3111–A3116.

17 Curtiss, L.A., Redfern, P.C., and Raghavachari, K. (2007). Gaussian-4 theory. *The Journal of Chemical Physics* 126 (8): 084108.

18 Park, H., Koh, H.S., and Siegel, D.J. (2015). First-principles study of redox end members in lithium–sulfur batteries. *The Journal of Physical Chemistry C* 119 (9): 4675–4683.

19 Camacho-Forero, L.E., Smith, T.W., Bertolini, S., and Balbuena, P.B. (2015). Reactivity at the lithium–metal anode surface of lithium–sulfur batteries. *The Journal of Physical Chemistry C* 119 (48): 26828–26839.

20 Frisch, M.J., Trucks, G.W., Schlegel, H.B. et al. (2009). *Gaussian 09*. Wallingford, CT: Gaussian, Inc.

21 Marenich, A.V., Cramer, C.J., and Truhlar, D.G. (2009). Universal solvation model based on solute electron density and on a continuum model of the solvent defined by the bulk dielectric constant and atomic surface tensions. *The Journal of Physical Chemistry B* 113 (18): 6378–6396.

22 Assary, R.S., Zhang, L., Huang, J., and Curtiss, L.A. (2016). Molecular level understanding of the factors affecting the stability of dimethoxy benzene catholyte candidates from first-principles investigations. *The Journal of Physical Chemistry C* 120 (27): 14531–14538.

23 Azimi, N., Weng, W., Takoudis, C., and Zhang, Z. (2013). Improved performance of lithium–sulfur battery with fluorinated electrolyte. *Electrochemistry Communications* 37: 96–99.

24 Nazar, L.F., Cuisinier, M., and Pang, Q. (2014). Lithium–sulfur batteries. *MRS Bulletin* 39 (05): 436–442.

25 Scheers, J., Fantini, S., and Johansson, P. (2014). A review of electrolytes for lithium–sulphur batteries. *Journal of Power Sources* 255: 204–218.

26 Xu, K. (2004). Nonaqueous liquid electrolytes for lithium-based rechargeable batteries. *Chemical Reviews* 104 (10): 4303–4418.

6

Lithium Sulfide

Sylwia Waluś

OXIS Energy, Culham Science Centre, Abingdon, Oxfordshire OX14 3DB, UK

6.1 Introduction

Lithium sulfide (Li_2S) plays multiple roles in lithium–sulfur (Li–S) cells, with the important one being the ultimate solid discharge product, formed during the reduction of elemental sulfur (S_8) according to the general reaction [1, 2] $S_8 + 16\,Li^+ + 16\,e^- \rightarrow 8\,Li_2S$. It is an insulating, both electronically and ionically [3] ($\sigma \sim 10^{-13}\,S\,cm^{-1}$ at $25\,^{\circ}C$) solid compound, poorly or non-soluble in most of the organic solvents, with Li^+ diffusivity [4] being as low as $10^{-15}\,cm^2\,s^{-1}$. Li_2S is considered by many as a Li–S cell enemy, being the main cause of capacity fade during cycling [5, 6] and limitation in reaching full discharge capacity [7–9]. However, its formation is indispensable to extend the discharge capacity, as transformation of S_8 to the final reduced form of Li_2S involves $16e^-$. On the other hand, the advantage of Li_2S being a lithiated counterpart of sulfur is that it can be used as an active material during cathode formulation. Li_2S-based cathodes give the opportunity of being paired with metallic Li-free anodes (such as silicon, tin, hard carbons), eliminating the problems faced when using pure Li metal as anode, such as poor coulombic efficiency and mossy or dendritic growth leading to safety issues [10, 11]. The predicted specific energy of a Li_2S/Si system ($1550\,Wh\,kg^{-1}$ [12]) exceeds that with existing Li ion technology by even four times. Finally, the application of Li_2S compound as a solid electrolyte $Li_2S–P_2S_5$ was demonstrated in all-solid-state Li–S batteries [3, 13, 14], although this concept has not been widely explored in the literature yet.

Tremendous work has been done so far, including application of advanced characterization methods, to increase the knowledge on mechanistic understanding of the Li–S cells from a Li_2S formation and dissolution perspective, its onset time and morphology being dependent on cycling conditions and other cell components. Similarly, development of Li_2S cathodes to improve their cycling performances as well as to understand the initial charge mechanism is also impressive. This chapter gives a summary overview on the most relevant findings with respect to lithium sulfide, both as a discharge product in a standard Li–S configuration and as an active material for Li_2S-based cathodes.

Lithium–Sulfur Batteries, First Edition. Edited by Mark Wild and Gregory J. Offer.
© 2019 John Wiley & Sons Ltd. Published 2019 by John Wiley & Sons Ltd.

6.2 Li$_2$S as the End Discharge Product

6.2.1 General

During discharge of a Li–S cell, the reduction process of elemental sulfur S$_8$, which leads to the formation of Li$_2$S, is accompanied by formation of intermediate species, the so-called lithium polysulfides (Li$_2$S$_n$, $2 < n \leq 8$), with different chain lengths and physical properties, such as solubility in the electrolyte. In parallel to the electrochemical reactions, chemical interactions, i.e. disproportionation and dissociation, occur [15]. It is generally well accepted that a typical discharge voltage profile of a Li–S cell can be divided into three regions, each being governed by different reactions thus resulting in different capacity. A simplified schematic is shown in Figure 6.1.

The mechanistic details of all these interactions are very complex and despite intensive research during the last years, they are not fully understood. It is important to keep in mind that these reactions are very sensitive to many factors, such as cell design, structure of the cathode, the composition of the electrolyte, and its amount [2]. Therefore, the precise pathways of the reactions may be significantly different between the systems studied. Nevertheless, the lower discharge plateau contributes undoubtedly to the major portion of the total discharge capacity (1256 mAh g$_S^{-1}$ from the total of 1675 mAh g$_S^{-1}$). In this region, reduction of soluble medium-order polysulfides S$_4^{2-}$ to insoluble and insulating short chain polysulfides, i.e. Li$_2$S or Li$_2$S$_2$/Li$_2$S, occurs. The formation of the latter is still not completely clear and it will be further discussed in Section 6.2.2. A general reaction of Li$_2$S formation during the lower discharge plateau can be written as in Eq. (6.1).

$$2S_4{}^{2-} + 12e^- \rightarrow 8S^{2-} \tag{6.1}$$

However, the formation of solid species is often split into two compounds [6], Li$_2$S and Li$_2$S$_2$, where in the initial part of the lower plateau soluble S$_4^{2-}$ species are reduced

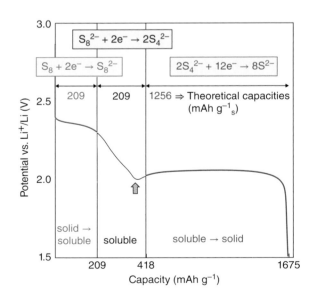

Figure 6.1 Discharge voltage profile of a Li–S cell with three characteristic regions of a voltage curve marked, with corresponding (simplified) reduction reactions and resulting capacities. Source: From Ref. [16].

to Li_2S or Li_2S_2, as shown in Eqs. (6.2) and (6.3). During the final part of the discharge curve a solid-to-solid reduction of Li_2S_2 to Li_2S occurs (Eq. (6.4)).

$$S_4{}^{2-} + 8Li^+ + 6e^- \rightarrow 4Li_2S \tag{6.2}$$

$$S_4{}^{2-} + 4Li^+ + 2e^- \rightarrow 2Li_2S_2 \tag{6.3}$$

$$Li_2S_2 + 2Li^+ + 2e^- \rightarrow 2Li_2S \tag{6.4}$$

The presence of the characteristic voltage dip at the transition between both plateaus is indicative of the beginning of the solid-phase crystallization [17]. It is explained that the concentration of S^{2-} species at this point is high enough so that the solid form starts to precipitate. Some reports assign it to the electrolyte viscosity, being the highest at the transition point due to the highest concentration of S_4^{2-} soluble species [18, 19]. Therefore, Li ion transport encounters difficulties and the voltage dip reflects concentration polarization [6].

Li_2S (or Li_2S/Li_2S_2) precipitation during discharge occurs on the conductive area of the cathode, typically the carbon-type additives such as Ketjenblack, carbon nanotube (CNT), etc. As a result, the pores of the cathode are clogged, and the active surface necessary for redox reactions to occur is diminished and finally blocked. Postmortem scanning electron microscopic (SEM) studies of the morphology of discharged cathodes clearly show the formation of a thick layer fully covering the surface of the cathode [8]. The nonconductive layer of Li_2S (or Li_2S/Li_2S_2) significantly slows down the diffusion of Li^+ and hinders further redox reactions. Once the surface is completely covered, further lithiation is impeded and rapid voltage drop at the very end of discharge occurs. Consequently, complete conversion of sulfur to Li_2S is rarely achieved, even if a large amount of conductive agent is added [4]. Formation of an insulating layer causes the internal resistance of the cell to increase. Electrochemical impedance spectroscopy (EIS) has been widely used [8, 16, 20] to observe the resistance evolution during cycling. Evidence of the formation of an insulating layer on the cathode surface can be seen in the medium to low frequency range, where a significant increase in the resistance (size of the semicircle) is observed. A direct consequence of the increased resistance in the cell is an increase in the cell temperature toward the end of discharge. This effect is more pronounced in large capacity cells as well as when higher currents are passed. Such internal increase of the temperature in the cell, especially when not evenly distributed between different layers of the cell, may cause accelerated degradation of the cell performance. An example of the temperature behavior of a 20 Ah Li–S cell (OXIS Energy Ltd.) at different discharge currents is shown in Figure 6.2.

Conversion of sulfur to Li_2S during discharge is accompanied by a volumetric expansion of the cathode, theoretically as large as 80% [4, 19]. Experimental studies reported on a 22% cathode thickness increase at the end of discharge [21]. The reason for this lies in different densities of both solid products, i.e. 2.03 and 1.66 g cm^{-3} for sulfur and Li_2S, respectively [19]. Volume expansion implies stress on the porous carbon-binder structure, which is believed to be detrimental to the electronic percolation within a cathode and eventually leads to the cathode structure collapsing upon aging [22]. This effect may be strongly enhanced in cathodes that are thick and highly loaded with sulfur, which are required to meet the demands of high specific energy for commercial applications [23].

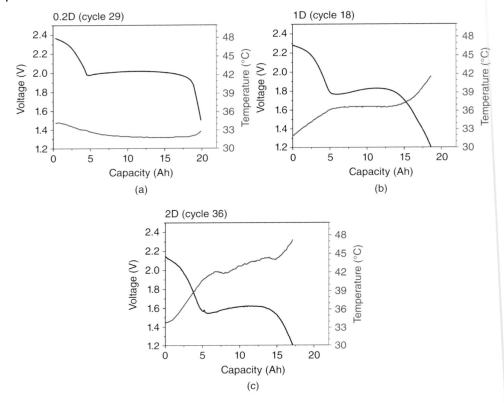

Figure 6.2 Voltage profiles (black) and surface temperature (blue) of a 20 Ah Li–S cell (OXIS Energy) discharged in a convection oven at 30 °C and at different currents: 4 A (a), 20 A (b), 40 A (c).

During the charge process of Li–S cells, solid precipitates (Li_2S or Li_2S/Li_2S_2) are oxidized to soluble polysulfide species (Li_2S_n, $2 < n < 8$), which are further oxidized to elemental sulfur. In the literature, the charge process is far less studied than the discharge one; nevertheless, it is equally complex. One of the main difficulties of charge is the oxidation of solid Li_2S, which is the opposite of the easiness of sulfur reduction. Both compounds, S_8 and Li_2S, are insulating, with the conductivity of sulfur ($\sigma = 5 \times 10^{-30}$ S cm^{-2}) [2, 24] being even lower than that of Li_2S. However, the kinetics of sulfur reduction is much faster than that of the oxidation of Li_2S because of the solubility aspect. In the voltage profile, characteristic kick (overpotential) at the beginning of charge is usually attributed to the initial step of Li_2S oxidation. The higher the overpotential, the more difficult the process. The general opinion about Li_2S oxidation is contradictive, with some studies providing evidence of complete Li_2S reoxidation and clearing out the cathode surface [7, 25], while other studies show that Li_2S is not fully oxidized during the charge and it builds up as a more resistive and thicker layer as the cycling progresses, causing significant increase in cell resistance and capacity fade [5, 7].

Li_2S or Li_2S/Li_2S_2 can be formed not only on the cathode surface during the discharge process but also on the anode side during the charge, when the shuttle phenomenon [26–28] occurs in the Li–S cell. During the shuttle, long chain polysulfides can diffuse from the cathode to the anode side, where they can react with lithium to form solid precipitates, according to the following reactions: $2Li + Li_2S_n \rightarrow Li_2S + Li_2S_{n-1}$ or/and $2Li + Li_2S_n \rightarrow Li_2S_2 + Li_2S_{n-2}$. This leads to irreversible losses of an active material, buildup of an insulating layer on the anode surface, which blocks access to fresh lithium, and increase in the resistance of the cell.

6.2.2 Discharge Product: Li_2S or Li_2S_2/Li_2S?

It is generally well accepted that Li_2S_2 is one of the intermediate discharge products of a Li–S cell. As previously mentioned, lower discharge plateau is assigned to the formation of solid and insoluble product(s), due to the reduction of soluble S_4^{2-} species, according to the reactions shown in Eqs. (6.2) and (6.3). However, there are still many questions remaining unanswered, with the main one being the actual presence of Li_2S_2. To account for this uncertainty, in the literature the end discharge product is often described as a Li_2S/Li_2S_2. If Li_2S_2 is formed during the discharge, mechanistically it is still unclear whether the reduction of S_4^{2-} species occurs sequentially (i.e. Li_2S_2 is formed before the Li_2S) or concurrently (both precipitates are formed at the same time). Contradicting opinions regarding the role and existence of the Li_2S_2 compound in a cell persist. It is believed that Li_2S_2 is only slightly soluble in most of the organic solvents and therefore should exist in a solid form during battery cycling. However, it has not been isolated from a Li–S cell yet. An early work [29] suggested that Li_2S_2 is stable in solid phase; however, its presence in the discharged sulfur cathode has not been fully confirmed. There has also been an early claim that Li_2S_2 is the main product of the lower discharge plateau by the output capacity from the cell and not necessarily by the material characterization [30]. Through the decades, the equilibrium phase diagram of a Li–S system was studied, without Li_2S_2 being conclusively observed [31].

Tremendous work has been done so far to detect different polysulfide species with different experimental techniques, applied via *ex situ*, *in situ*, or even *operando* mode. Despite that, not much has been reported on the detection of Li_2S_2. X-ray diffraction (XRD) seems to be an appropriate technique for solid species detection as long as they appear in a crystalline form. From these studies, applied both *ex situ* [24, 32, 33] and *in situ* [25, 34–38], there is no convincing proof showing the existence of Li_2S_2 and only crystalline Li_2S phase was detected. Only the additional quantitative analysis of the *in situ* XRD data reported by Waluś et al. [36] proposed that the formation of Li_2S_2 may start in the second part of the lower discharge plateau, in parallel to the formation of Li_2S, which commences right at the beginning of the lower plateau. This hypothesis suggests that the reduction reactions involved during the lower plateau are not consecutive but simultaneous. To explain why experimental detection of Li_2S_2 by XRD is difficult (not to say impossible), theoretical first principle calculations were employed [39–41]. Feng et al. [41] predicted the crystalline structure of Li_2S_2, which suggests that multiple crystal structures with similar total energies coexist (Figure 6.3). Therefore, when Li_2S_2 appears, it is more likely to appear as a mixture, and is difficult to be detected by XRD. Moreover, Li_2S_2 is thermodynamically unstable in a Li–S cell; thus it is subjected to spontaneous disproportionation and exists only in the nonequilibrium regime upon cell operation.

Although the XRD technique seems to be appropriate to analyze solid species, its limitation lies in the inability to detect the amorphous phases or the slightly soluble species (the assumption of Li_2S_2 being slightly soluble may apply here). Other techniques, often

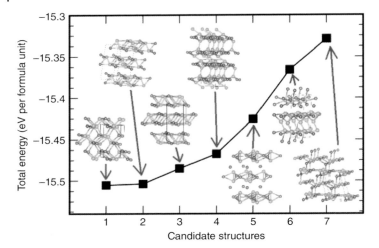

Figure 6.3 Proposed crystalline structures of the Li_2S_2 compound and their total energies, obtained from first principle calculations. S and Li atoms are represented by yellow and blue balls. Source: Feng et al. 2014 [41]. Reproduced with permission from Elsevier.

coupled together, gave more insight into the existence of S_2^{2-} ions. Barchasz et al. [15] reported on the presence of S_2^{2-} ions at the potential $\sim$1.95 V through *ex situ* UV–Vis absorption studies coupled with chronoamperometry techniques. Later *in situ* UV–Vis work by Patel et al. [42] also confirmed the presence of S_2^{2-} species. In the work of Kawase et al. [43], where polysulfide species were analyzed by coupling UV–Vis, LC/MS with NMR and applying the organic conversion technique, formation of S_2^{2-} species along the lower discharge plateau was also confirmed. Moreover, an original insight into the reduction mechanism was given, which will be described in more detail in Section 6.2.3. *In situ* Raman spectroscopy studies [44] supported by calculated Raman peak location of the S_2^{2-} anion did not provide sufficient evidence of the S_2^{2-} species formation. *Operando* X-ray absorption spectroscopy by Cuisinier et al. [45] and Lowe et al. [38] also cast further doubts on the formation and existence of the Li_2S_2 compound. However, Lang et al. [46] have recently observed the formation of Li_2S_2 and distinguished the morphology and growth process of both Li_2S_2 and Li_2S insoluble products, with the use of *in situ* atomic force microscopy (AFM) combined with *ex situ* spectroscopic methods (Raman, X-ray photoelectron spectroscopy [XPS]).

It is important to underline again that each individual system described above may differ with respect to the electrolyte composition, the structure of the cathode, cell design, etc. These (and many more) factors, as previously mentioned, can significantly change the reaction pathways and potentially the nature of the species formed. As an example, recent work of Zaghib's group [47] (which previously explained the reason for not detecting Li_2S_2 by XRD [41]) presented XRD experimental results on Li_2S_2 being formed in the so-called "solvent-in-salt" concept cell [48], i.e. in a highly concentrated electrolyte solution. According to the authors, Li_2S_2 is one of the major active species involved in the Li–S cell operation. However, it is counterintuitive that Li_2S_2, previously

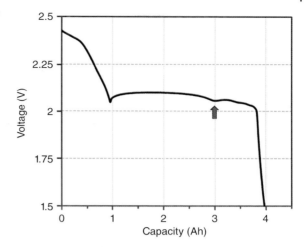

Figure 6.4 Discharge voltage profile of a 3.4 Ah cell (OXIS Energy), cycled at C/20 slow regime and at 30 °C. Red arrow indicates the appearance of another low discharge plateau.

assumed to be one of the end discharge products, turned out to appear at the end of charge or at the beginning of discharge.

Taking a step back from the analytical and spectroscopic characterization techniques, sometimes a discharge voltage profile displays the feature of a lower discharge plateau being split into two sub-plateaus appearing at a similar voltage, hypothetically meaning that two different reduction steps are involved (Figure 6.4). The appearance of this feature has not yet been assigned to cycling conditions or the cell design, such as current regimes or sulfur/electrolyte ratio. This is however an observation rather than a statement and it clearly points out the importance of applying different characterization techniques to decipher the reality occurring inside the cell.

6.2.3 A Survey of Experimental and Theoretical Findings Involving Li_2S and Li_2S_2 Formation and Proposed Reduction Pathways

It is clear that, despite all the effort made toward understanding the working mechanism of Li–S cells, research is still at the beginning of exploring the full picture of the Li–S operational map. Moreover, since the reaction pathways are extremely sensitive to many factors, there are often significant discrepancies between the proposed mechanisms, as the research is conducted on sometimes very different systems and cell configurations. In this section, a global picture of the recent mechanistic understanding about the precipitation of Li_2S/Li_2S_2 product(s) will be given, with the main focus on the onset time of its precipitation and reactions leading to its formation.[1]

Although it is widely accepted that the formation of solid precipitates usually occurs along the lower discharge plateau, the onset time of Li_2S formation (and re-oxidation during charge) observed by different *in situ*[2] techniques is sometimes very different.

---

1 For a full picture of the redox reactions occurring from the beginning to the end of discharge, the reader is directed to other sources [4].

2 The advantage of *in situ* and operando techniques over the *ex situ* ones is that they allow to probe the dynamic changes occurring under the cell operation and to spot precisely the exact moments of these transitions. In addition, *in situ* tests eliminate the artifacts of postmortem treatments required for the sample preparation for the *ex situ* tests.

As an example, the results obtained by See et al. [49] when using *in situ* ^{7}Li NMR revealed the formation of Li^{+} containing solid species near the beginning of the upper discharge plateau. These results are in complete contradiction to the work of Huff et al. [50] who applied ^{6}Li and ^{33}S MAS NMR and demonstrated the formation of Li$_2$S only at the very end of discharge, close to 1.5 V.

XRD has been widely used in Li–S systems to investigate the formation of solid precipitation products. *In situ* XRD studies done by Nelson et al. [35] did not detect any traces of the crystalline Li$_2$S; however, a suggestion about amorphous phase formation has been made. Cañas et al. [34] detected the appearance of Li$_2$S peaks at ~60% DoD, whereas Lowe et al. [38] observed rapid increase of the crystalline Li$_2$S phase only at the very end of discharge. The latter system, however, was a gel polymer electrolyte, contrary to the most studied ones, i.e. the liquid-ether-based type. Demir-Cakan et al. [5, 51] reported on the detection of crystalline Li$_2$S through the end of discharge. In contrast to the previous work and in agreement with the presence of a little dip in the voltage profile suggesting precipitation of a solid product, Waluś et al. [25, 36] reported on Li$_2$S precipitation starting exactly at the beginning of the lower discharge plateau, irrespective of the C-rate (slow or moderate) or cycle number. Further work of Kulisch et al. [37] also detected the crystalline Li$_2$S signal; nevertheless, the precise moment of its appearance was inconclusive.

The limitation in deciphering the reaction pathways solely from XRD data is due to the lack of information about soluble species also being present in the cell as their reduction leads to the formation of solid precipitates. Therefore, an assumption of Li$_2$S$_4$ being the reduced soluble species along the lower discharge plateau is often made [25, 34, 36]. More profound analysis of the hypothetic reactions can be obtained if XRD analysis is coupled with another technique, ideally being sensitive to the soluble species, which allows for speciation of polysulfide anions and radicals. Several advanced characterization techniques have been used to date, such as NMR [43, 49, 50, 52], UV–Vis [15, 42], XAS [38, 45], Raman [22, 44, 53, 54] and many more, to successfully analyze Li$_2$S$_x$ species of both natures.

Cuisinier et al. [45] in their pioneering work on lithium polysulfide speciation by *in situ* and *operando* XANES demonstrated that the beginning of Li$_2$S formation started with delay, approximately in the middle of the lower discharge plateau. This delay was explained by the fact that using carbon composite cathode helps retain polysulfides, thus delaying their diffusion from the cathode structure. The authors also made a statement that their results do not support the hypothesis about the Li$_2$S precipitate being the main reason for the limited capacity, but instead, unreacted sulfur. Recently, a new finding concerning the role of Li$_2$S in preventing the complete sulfur utilization during discharge was proposed by Amine's group [55]. *In situ* TEM was used to observe the morphological and structural evolution of a sulfur particle under real battery operation conditions. A bespoke designed microcell was used with the absence of liquid electrolyte. It was found that a crust of Li$_2$S was formed on the bulk of sulfur particles due to preferential Li^{+} diffusion at the surface. The insulating nature of such formed Li$_2$S crust made the radial diffusion of Li^{+} ions through the interface to the bulk more difficult as compared to that along the contour of the particle.

The work of Wang et al. [56] with the *in situ* EPR investigation on the role of S$_3^{\cdot-}$ radical in the electrochemical processes in a Li–S cell highlighted new facts, which

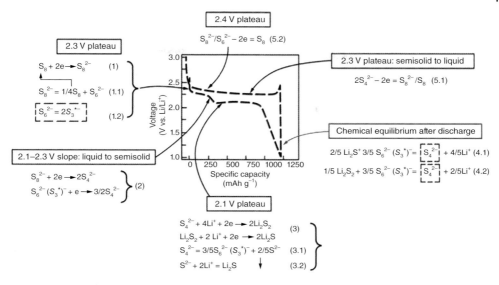

Figure 6.5 Proposed reaction mechanism of a Li–S cell cycled at C/5 rate in a voltage range 1–3 V. Source: Wang et al. 2015 [56]. Reproduced with permission from Electrochemical Society.

may be crucial for further understanding the redox mechanisms as well as developmental improvement. As shown in Figure 6.5, it is proposed that during the lower discharge plateau, electrochemical reduction of S_4^{2-} species to Li_2S_2 and further to Li_2S occurs concurrently with chemical disproportionation of S_4^{2-} into S_6^{2-} (the source of S_3^{-} radical) and S^{2-}. According to the authors, the electrochemical formation of Li_2S is sensitive to the current density and the cutoff voltage, while chemically derived Li_2S depends on the cell design and electrolyte composition. These results also show that the end of discharge product is a radical-containing mixture rather than pure Li_2S/Li_2S_2, which may have a significant impact on the charge process and oxidation reactions (see Section 6.2.6).

Lowe et al. [38] in their combined studies by *in situ* XRD and XAS suggested the following reaction: $S_3^{2-} + Li_2S_n \leftrightarrow Li_2S + S_{n+2}^{\ 2-}$ ($n = 2, 3$) to dominate in the lower discharge plateau along which shorter chain polysulfides are being reduced through electrochemical mediation of S_3^{-} radical anions and equilibration with more reduced species (S_3^{2-}). After the concentration of S_6^{2-} species (being the source of S_3^{-} radical) becomes too low, the cell voltage drops, lower polysulfides disproportionate, and the formation of crystalline Li_2S is detected. It is worth noting that similar conclusions concerning the dominant reduction reaction along the lower discharge plateau ($S_3^{-} + e^{-} \rightarrow S_3^{2-}$) were proposed by Assary et al. [40] in their theoretical predictions.

In the earlier work of Barchasz et al. [15] three techniques were coupled, i.e. UV–Vis, HPLC, and EDR, all applied via *ex situ* mode. The authors proposed that the lower discharge plateau is associated with the formation of S_3^{2-} (due to the reduction of S_4^{2-} according to the following reaction: $3Li_2S_4 + 2e^{-} + 2Li^{+} \rightarrow 4Li_2S_3$) and its further reduction to the insoluble species (according to $2Li_2S_3 + 2e^{-} + 2Li^{+} \rightarrow 3Li_2S_2$ and $Li_2S_2 + 2e^{-} + 2Li^{+} \rightarrow 2Li_2S$).

Recently, an original approach of Kawase et al. [43] was presented where different polysulfide species formed in a Li–S cell were analyzed with a combination of ^{33}S NMR and LC/MS following the application of the organic conversion technique (synthesis and analysis of derivatives of the polysulfides). It was found that the dominant reaction occurring along the lower discharge plateau is a simple reduction of Li_2S_3 according to $S_3^{2-} + 2e^- \rightarrow S_2^{2-} + S^{2-}$. Starting from the second half of the plateau toward the end of the discharge, Li_2S_2 is reduced to Li_2S ($S_2^{2-} + 2e^- \rightarrow 2S^{2-}$) but only partially, since most of the Li_2S_2 was detected to be intact at the very end of discharge.

A completely different approach, where a single-step transformation from S_8 directly to Li_2S/Li_2S_2 was achieved without formation of soluble and intermediate lithium polysulfide species, was proposed by Helen et al. [57]. In this work an ultramicroporous carbon–sulfur composite cathode and conventional carbonate-based electrolyte were used. It is worth mentioning that carbonate solvents cannot be used in conventional Li–S cells where the formation of the soluble polysulfides occurs, as they are vulnerable to the nucleophilic attack from the polysulfides and degrade rapidly [58]. However, carbonate solvents can be used in the systems with the nano-size sulfur confined in the nano-pores of the carbon [59], where the molecules of solvents do not enter. As a consequence, the reduction pathway of S_8 is different from that widely described. *Ex situ* XPS applied to analyze the surface of the cathodes at various charge and discharge stages showed the formation of only one intermediate species, namely Li_2S_2, which directly converted to Li_2S at the end of discharge. During charge a complete disappearance of Li_2S was observed, meaning its full conversion to Li_2S_2. However, unreacted Li_2S_2 was still detectable at the very end of charge and the authors assigned this to be the main reason for irreversible capacity loss in the first cycle.

It is also important to notice that although the final discharge products Li_2S_2 and Li_2S are considered as poorly soluble (or even insoluble) in most of the aprotic solvents, the choice of the electrolyte may affect the interactions with soluble polysulfides in a different way, thus leading to different discharge products. As an example, Gorlin et al. [60] used the electrolyte with either high (dimethylacetamide, DMAC) or low (1,3-dioxolane–1,2-dimethoxyethane, DOL–DME) dielectric constant and studied the conversion of soluble polysulfide species to the solid end of discharge products by the use of *operando* XANES techniques. In both cases, they found a similar onset of Li_2S formation, starting already at 25–30% DoD (4–5 e^-/S_8). However, the difference was in the final discharge product as well as the speed of disproportionation reactions. It was found that in a high dielectric solvent Li_2S was detected in the presence of polysulfide intermediates. On the other hand, in a low dielectric solvent Li_2S and hypothetically Li_2S_2/S_2^{2-} species were detected. These data highlight that the disproportionation steps are faster in poorly ion-solvating environments. Theoretical studies (DFT calculations) done by Wang et al. [61] show that selection of low dielectric constant solvents may bring up the potential of the lower discharge plateau.

On a similar topic, when studying the effect of solvents, Cuisinier et al. [62] reported that stabilization of sulfur radical $S_3^{\cdot-}$ in an electron pair donor (EPD) solvent may help Li_2S to be chemically reoxidized to the soluble species S_n^{2-}. This could potentially prevent the passivation of the positive electrode until the end of electrochemical reduction and full discharge capacity could be obtained.

6.2.4 Mechanistic Insight into Li_2S/Li_2S_2 Nucleation and Growth

In the literature, much attention has been paid to the mechanisms of the formation of Li_2S/Li_2S_2 species and the pathways of the redox reactions associated with it. However, only recently the insight into the kinetics of solid precipitates formation has started to gain more consideration. Lithium–oxygen batteries have been studied in the past in terms of nucleation and the growth mechanism of solid precipitates such as Li_2O_2 [63]. This gained knowledge can be adapted to the Li–S system. Pioneering work in understanding the nucleation and growth mechanism was presented by Fan, Chiang, and coworkers [64, 65]. The authors were studying the kinetics and morphology of Li_2S being electrodeposited on the surface of the MWCNT in potentiostatic as well as galvanostatic conditions, so that the effect of the overpotential and current density, respectively, were investigated [65]. In the first place, the authors proposed the mechanism of Li_2S nucleation. It was found that a critical value of overpotential is necessary for the nucleation of Li_2S to begin. This value is distinct for each solvent studied; however, for glyme molecules explored in this work, the overpotential value was very close, i.e. 100 mV for tetra and triglyme and 90 mV for diglyme. In other words, the absolute cell voltage needs to be below 2.06 V (assuming 2.15 V as an equilibrium potential) to initiate the nucleation. This observation is in a good agreement with the presence of a little voltage dip at the beginning of the lower discharge plateau, as indication of the nucleation of a solid phase. Nucleation requires higher driving force than growth to overcome the surface energy barriers. Growth, on the other hand, occurs at lower overpotentials by reduction of soluble polysulfides at the three-phase boundary between the existing nucleation site (i.e. already precipitated Li_2S), conductive surface of the carbon, and the solution phase. Growth in the thickness direction is likely to occur via surface diffusion at the Li_2S–electrolyte interface, rather than via bulk chemical diffusion through the Li_2S layer, because of Li_2S conductivity limitations. In a Li–S cell during galvanostatic operation different morphologies of precipitates can be obtained depending on the current rate, which in turn determines the value of the overpotential. High C-rates are accompanied by a larger overpotenial during the discharge. As a result, the nucleation rate is high, which produces high nuclei density and therefore a continuous morphology of many small crystallites on the carbon surface. On the contrary, at slower C-rates, lower nucleation rate leads to the formation of lower number of nucleation sites; however, much larger particles of deposit were found on the carbon surface, as large 2D growth at the three-phase boundary took place. An illustrative schematic of how the C-rate affects the morphology of the precipitates formed is shown in Figure 6.6.

2D growth of an insulating Li_2S precipitate causes blocking of the active surface, which leads to a rapid end of discharge once the surface is clogged. However, as reported by Gerber et al. [64], addition of benzo[*ghi*]peryleneimide (BPI) redox mediator (which is tuned to the potential of Li_2S electrodeposition) allows to electrodeposit Li_2S as a porous 3D structure on the carbon current collector during cell discharge.

A similar methodology as described previously was used, where potentiostatic and galvanostatic conditions were applied together with SEM analysis to study the mechanism of nucleation and growth of Li_2S electrodeposition [64]. Addition of BPI to the

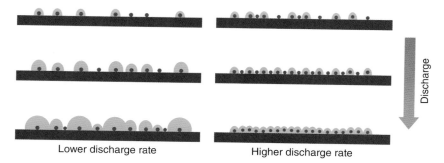

Figure 6.6 Schematic illustration of Li$_2$S precipitation depending on the discharge C-rate. Source: Ren et al. 2016 [66]. Reproduced with permission from Elsevier.

electrolyte led to an impressive sixfold increase in Li$_2$S formation capacity. Addition of a redox mediator did not affect the nucleation process. However, the growth mechanism was significantly modified. The growth trajectory of such 3D deposits involves reduction of BPI at the carbon surface to form a radical BPI·$^-$, which in turn diffuses and circulates in the catholyte solution. It then reduces the polysulfide species to Li$_2$S, which can deposit either on the carbon surface or on already existing Li$_2$S. This leads to the formation of a 3D morphology (see Figure 6.7a), successfully delaying the coverage of the electroactive surface of carbon with an insulating layer. Cobaltocene was also reported to be an effective redox mediator [67], which allowed not only for more Li$_2$S deposition but also a thicker Li$_2$S layer. Schematic illustration of the functioning of a redox mediator is shown in Figure 6.7b. Manthiram's group [68] reported that when utilizing N,O-co-doped carbon hollow fiber (NCHF) sheets as a current collector impregnated with Li$_2$S$_6$ solution, more uniform deposition of Li$_2$S was achieved with no obvious big particles or agglomerations. According to the authors, one of the reasons for improved performance and more uniform morphology of the deposit was attributed to the functional groups on NCHF, providing additional nucleation sites for precipitation.

Noh et al. [17] came to similar conclusions as Fan et al. [65] and reported that by increasing the discharge rate more amorphous Li$_2$S phase was created, while lower C-rate led to more crystalline Li$_2$S deposits. The authors also studied the effect of Li$_2$S morphology (resulting from different discharge C-rates) on prolonged cycling and capacity retention. It was found that higher discharge C-rates resulted in better stability and lower capacity fade due to higher reversibility of amorphous phase. In contrast, SEM images of the electrodes charged up to 2.3 V after various cycles at slow C-rate discharge exhibited large amounts of undecomposed Li$_2$S, explaining the reason for faster capacity fade due to the loss of active material.

In deciphering the complexity of the working mechanism of Li–S cells, theoretical modeling is a very useful tool being used more often recently. Ren et al. [66] developed a one-dimensional model for the discharge behavior of the Li–S cell, which incorporated, for the first time ever, surface nucleation and growth dynamics. Thanks to that the authors depicted the rate-dependent precipitation phenomenon and proposed that the growth of precipitated Li$_2$S during the discharge is suppressed at a higher discharge rate, which causes limitation to the active material utilization. A good agreement was achieved between their theoretical work and the experimental SEM studies they performed on discharge cathodes at various C-rates.

Figure 6.7 (a) Schematic representation of Li$_2$S deposition on a carbon cloth fiber with or without BPI additive used in the electrolyte, together with SEM photos. Source: Gerber et al. 2016 [64]. Reproduced with permission from ACS. (b) Schematic of redox reactions and Li$_2$S deposition in the presence of cobaltocene redox mediator. Source: From Ref [67].

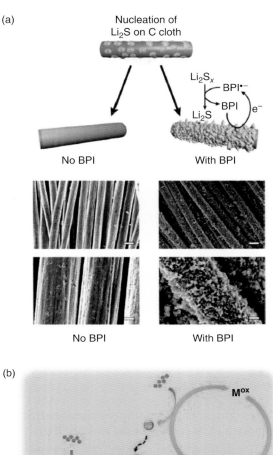

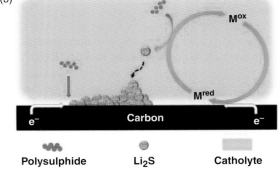

Previous works made an assumption on the formation of Li$_2$S solely, without taking into consideration the formation of Li$_2$S$_2$. The work of Lang et al. [46], where *in situ* AFM supported by *ex situ* XPS and Raman spectroscopy were used, gave a new insight into nucleation, growth, and the resulting morphology of both solid precipitates, Li$_2$S and Li$_2$S$_2$ (see Figure 6.8). The authors were studying the growth of insoluble products at the highly oriented pyrolytic graphite (HOPG)/electrolyte interface. It was found that at approximately 2 V the nanoparticles of insoluble Li$_2$S$_2$ nucleate. When the discharge proceeds, the nanoparticles grow and accumulate on the HOPG substrate. Li$_2$S starts to appear when the discharge potential decreases and at 1.83 V lamellar Li$_2$S deposits were observed, accompanied by rapid growth and diffusion, which caused the formation of Li$_2$S in the form of micron-sized islands. During the subsequent charge, Li$_2$S

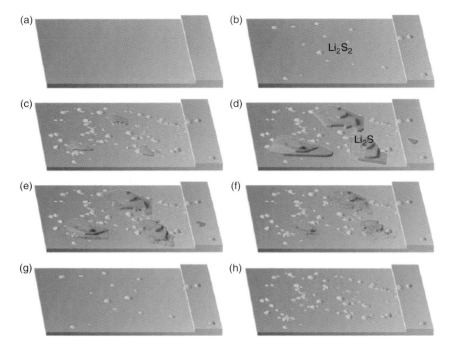

Figure 6.8 Schematic illustration of the nucleation and growth of Li_2S_2 (yellow) and Li_2S (blue) during discharge (a–d) and charge (e–h). Source: Lang et al. 2016 [46]. Reproduced with permission from John Wiley & Sons.

fully reoxidized whereas Li_2S_2 remained partially on the surface and accumulated over cycling, which, according to the authors, led to capacity fade.

6.2.5 Strategies to Limit Li_2S Precipitation and Enhance the Capacity

Formation of an insulating layer of Li_2S/Li_2S_2 that passivates the electrode is considered as one of the key reasons for rapid capacity fade. Because of that several attempts were undertaken to delay, minimize, or even hinder the solid products precipitation.

A simple approach adopted by several researchers to minimize the formation of an insoluble Li_2S/Li_2S_2 is to alter the cycling protocol [5, 69]. This could be done by limiting the discharge cutoff voltage to a level below which rapid voltage drop occurs, or to cycle the cell in a capacity control rather than voltage control mode. In general, by applying these protocols, much more stable capacity retention and significantly enhanced cycle life were obtained. As an indication of reduced Li_2S formation during prolonged cycling with different discharge cutoff voltages, resistance measured at the end of each discharge did not increase over >120 cycles [5]. While some cycling strategies are targeting to minimize the severe passivation of the cathode, the opposite approaches were also demonstrated, where the authors tried to confine the electrochemical reactions during cycling to the lower discharge plateau, i.e. to the reactions between soluble Li_2S_4 and insoluble Li_2S species [70]. This approach, however, aimed to tackle the problem of dissolution and diffusion of high-order polysulfides causing shuttle effect, rather than limiting the formation of solid precipitates. According to the authors, the presence of some amount of solid Li_2S on the cathode surface during all stages of cycling acts on the one hand as nuclei sites that facilitate further precipitation of Li_2S, but on the other hand also serves as an adsorbent and contributes to the adsorption of soluble polysulfides.

Another proposed strategy is the use of redox mediators [64, 67, 71]. Different redox mediators may play different roles in enhancing the cell performances, such as tailoring the growth of Li_2S into 3D structure [64, 67] or inhibiting active mass (polysulfide) dissolution into the electrolyte and controlling the deposition of Li_2S or Li_2S_2 [71].

Other additives to either the electrolyte or directly into the cathode were also studied. Zhang [72] implemented the concept of stabilizing polysulfide anions in the solution based on Pearson's hard and soft acids and bases (HSAB) theory. According to this sulfur intermediates can form complexes that may stabilize them from being completely reduced to the insoluble S_2^{2-} form. The author used quaternary ammonium compounds, either as a co-salt (tetrabutylammonium triflate) or as an ionic liquid ($PYR_{14}TFSI$) and reported on effectively suppressed disproportionation of polysulfide intermediates and enhanced capacity retention. Pan et al. [73] also used quaternary ammonium additives, this time however in DMSO-based electrolyte to promote dissolution of Li_2S. NMR studies revealed that strong interactions between NH_4^+ and S^{2-} through hydrogen bond coupled with solvation by DMSO solvent and NO_3^- is crucial for dissociating crystalline Li_2S. Wu et al. [53] reported on improved performance after employing CS_2 as an electrolyte additive which, according to the authors, stabilized the S_8^{2-} species and inhibited the appearance of further reduced species. *In situ* Raman and CV experiments confirmed that CS_2 addition effectively changed the sulfur reduction mechanism. Demir-Cakan et al. [5] used a V_2O_5 additive mechanically mixed with the mesoporous C/S composite into the cathode structure. Although the exact reason for the improved cyclability and the achieved capacity was not completely revealed by the authors, the experiments suggested that V_2O_5 acts like an internal redox mediator, capable of assisting the oxidation of Li_2S species to more soluble ones. *In situ* XRD data showed that no crystalline Li_2S was detected at any stage of discharge/charge whereas the impedance measured at the end of each discharge stayed fairly stable during prolonged cycling.

As already mentioned, Li_2S is considered as an insoluble species in most of the aprotic solvents. Nevertheless, in the XAS studies done by Cuisinier et al. [62] where DMA (*N,N*-dimethylacetamide), an EPD solvent, was used, very high sulfur utilization was achieved with 98% of S_8 being transformed into a stable intermediate solution. The reason behind this improvement is the capability of DMA molecules to stabilize the sulfur radical $S_3^{\cdot-}$. In addition, Li_2S can be partially solvated by DMA molecules thanks to the highly dissociative properties of the solvent, which may promote utilization of Li_2S, thus enhancing the delivered capacity.

6.2.6 Charge Mechanism and its Difficulties

The charge process, in a simplified description, involves oxidation of insoluble Li_2S or Li_2S/Li_2S_2 products into soluble polysulfide species, which are further oxidized to elemental sulfur. Contradictive opinions exist where it is believed that soluble high-order polysulfide species (mainly Li_2S_8) are the end of charge products and no traces of solid S_8 is detected. Oxidation of solid precipitates clears out the electroactive surface of the

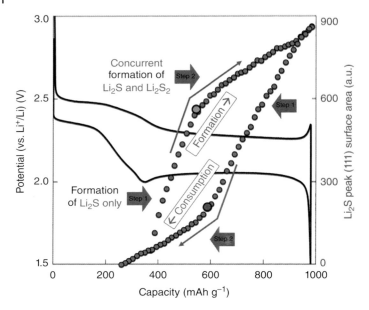

Figure 6.9 Evolution of a (111) XRD peak of a Li_2S compound during its formation (discharge; blue arrow) and disappearance (charge; red arrow). Source: Waluś et al. 2015 [36]. Reproduced with permission from John Wiley & Sons.

cathode. The mechanism of charge process is equally complex as the discharge one; how-ever, significantly much less attention in the literature has been paid to reveal the exact reaction pathways. It is also generally accepted that the charge is not an inverse process of the discharge, and a significant hysteresis between discharge/charge pathways is seen (Figure 6.9) when applying different *in situ* characterization techniques [36, 42].

The oxidation of Li_2S is known to be a difficult process, mainly due to its non-soluble character, while its insulating nature is obviously another obstacle. On the contrary, reduction of elemental sulfur, which is an even more insulating species (10^{-30} S cm^{-1}) but highly soluble in aprotic solvents, is a very easy process. Characteristic overpoten-tial (also known as an activation kick) at the beginning of charge is usually seen. The higher the kick, the more difficult it is to initiate the oxidation process. Different factors can modify the magnitude of this overpotential, with the most straightforward being the amount of Li_2S deposited, reflected by the capacity of the lower plateau of the preceding discharge. Some studies have presented the effect of the morphology of deposited Li_2S on the activation kick. It is believed that the passivating layer composed of amorphous or small crystallites of Li_2S is much easier to oxidize than scattered and bigger crystals. As previously mentioned in Section 6.2.4, the difference in Li_2S morphologies can be obtained when discharging the cell at different currents (C-rates). An example is shown in Figure 6.10, where different C-rate discharges were applied, while the charge pro-cess was maintained at the same constant rate of C/10. The lower the discharge current, significantly higher the activation kick.

The strategic approaches described in Section 6.2.5 to minimize the formation of Li_2S (such as cycling with limited cutoff voltage or capacity, redox mediators, additives, etc.),

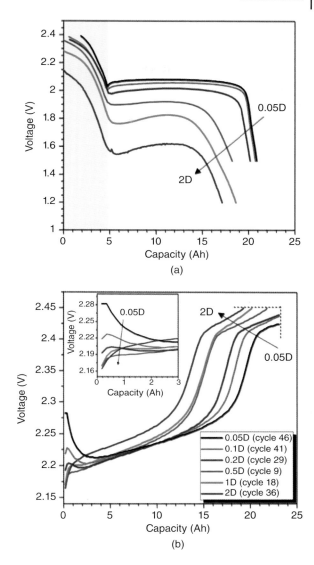

Figure 6.10 Discharge voltage profiles (a) of a 20 Ah Li/S cell (OXIS Energy) discharged at different C-rates and following charge profiles at a constant rate of C/10 (b). Zoomed-in image of the beginning of charge, i.e. activation of Li_2S.

when successfully applied to Li–S cell, resulted in decreased oxidation kick at the beginning of charge.

Studies on charge mechanisms are mainly conducted from two perspectives. One focus is on solid Li_2S/Li_2S_2 species and the following points are being debated: (i) whether a complete reoxidation of these solids takes place and what is the onset time for their disappearance, (ii) if solid Li_2S/Li_2S_2 accumulates with progressive cycling and cause the increase of the cell resistance, loss of active material, and rapid capacity fade. Another focus is to precisely detect the soluble polysulfide species present during the charge so that the reaction pathway could be proposed. Several different *in situ* characterization techniques (XRD, XAS, UV–Vis, EPR) were applied to decipher a charge working mechanism. Moreover, much work on revealing the mechanism of Li_2S

oxidation was conducted in a system where Li_2S-based cathodes were used in Li_2S/Li cells, and Section 6.3 provides more details about the fundamental understanding on the electrochemical processes.

In situ XRD studies in the majority of the cases [25, 37, 38] reported on complete reoxidation and disappearance of crystalline Li_2S species, meaning that no accumulation of the insulating layer on the cathode surface at the end of charge or during further cycling [36] took place. In contrast, Demir-Cakan et al. [5, 51] claimed that incomplete conversion of Li_2S at the end of charge was achieved, suggesting that Li_2S irreversible accumulation was one of the main causes of capacity fade.

Cuisinier et al. [45] in their *operando* XANES work observed monotonic consumption of Li_2S species during charge, and at the same time formation of shorter chain polysulfides. S_6^{2-} species were detected directly from the beginning of charge and their concentration increased progressively. S_4^{2-} species appeared as transient polysulfides, allowing oxidation of Li_2S to more stable S_6^{2-}. Toward the end of charge Li_2S disappeared completely together with the S_4^{2-} species, while the maximum of S_6^{2-} concentration coincided with the voltage rise, signaling the final oxidation of S_6^{2-} to elemental sulfur.

The latest work has revealed the presence of radical anion $S_3^{.-}$ and its role in the charge mechanism, which gave more insight into the pathways of the redox reactions [38, 56]. As explained by Wang et al. [56], the chemical equilibrium that establishes at the end of discharge between Li_2S/Li_2S_2 and $S_3^{.-}$ leads to the formation of S_4^{2-} species (see Figure 6.5). During charge, once the overpotential becomes sufficiently high, S_4^{2-} species will be directly oxidized to long chain polysulfides, i.e. S_6^{2-}/S_8^{2-}, due to the fast kinetics of semisolid-to-liquid conversion. As new S_6^{2-} species are produced, more $S_3^{.-}$ radicals are being formed and their presence drives the chemical reactions to produce more of S_4^{2-} to facilitate the conversion of Li_2S/Li_2S_2 precipitates. Consequently, more Li_2S_2/Li_2S products are consumed chemically, in addition to its electrochemical consumption.

It was also found that in EPD solvents such as DMA, oxidation of Li_2S during charge occurs with practically no overpotential kick at the beginning, as compared with the standard glyme solvents [62]. The reason for this is assigned to the stabilization of the $S_3^{.-}$ radical by the EPD solvent, which then facilitates the phase transfer from solid Li_2S to soluble polysulfides.

6.3 Li_2S-Based Cathodes: Toward a Li Ion System

6.3.1 General

Recently, lithium sulfide (Li_2S) has received much attention as an active material for being used in the cathode. With a very high theoretical specific capacity of 1166 mAh $g_{Li_2S}^{-1}$, far exceeding the values of the conventional positive electrodes used in Li ion cells, it offers a possibility for being used with non-lithium high capacity anode materials, such as silicon or tin-based compounds. Therefore, the problems encountered when using metallic Li as an anode in the cell, i.e. dendritic or mossy growth of Li affecting safety and life cycle, high reactivity requiring dry conditions for cell building, and low coulombic efficiency, could be eliminated or at least minimized. Moreover, in such a configuration, the battery would be built in a discharged state (as all the Li$^+$ ions would initially be in the cathode), thus eliminating the risk of self-discharge after manufacturing and during initial storage, which is known to be an issue in conventional Li–S cells [74].

The main hindrance from using Li$_2$S as an active material for cathode is its insulating nature, both electronic and ionic. To resolve this problem, active materials require an intimate contact with electronically conductive additives, mainly carbon-based ones. Addition of transition metals, such as Cu [75], Fe [76], and Co [77], was also reported in the early studies, however, with no major breakthrough. Li$_2$S is also considered as non-soluble in most of the aprotic solvents, which hinders its oxidation, as compared to the easiness of the reduction of sulfur. Finally, Li$_2$S is very sensitive to moisture, contact with which leads to the formation of LiOH and poisoning H$_2$S gases with a rotten egg odor. It could also be easily oxidized when exposed to oxygen. Because of this, cathode fabrication requires a protective atmosphere, such as a glove box, which is not a practical solution for large-scale application. In the vision of preparing confined/encapsulated carbon-active material structures by infiltrating the active material into the pores of carbon (like it was widely applied in the case of sulfur cathodes), Li$_2$S is not the ideal candidate owing to its high melting point of 950 °C [78]. In general, the loadings (mg cm^{-2}) of active material usually reported in the literature are much smaller for Li$_2$S-based cathodes (1–3 mg cm^{-2}) [79] than for S$_8$-based ones (even >6 mg cm^{-2}). In order to obtain relatively high capacity from a Li$_2$S-based cathode, often the cathodes had to have the content of the active material usually lower than 50 wt%, while higher ratios of Li$_2$S led to poorer utilization.

Despite the abovementioned disadvantages, research and development is ongoing and interesting findings have been obtained. To date, the research on Li$_2$S cathodes is still conducted in a configuration with Li metal, although successful, complete Li$_2$S-based Li ion systems have been developed and will be briefly described in Section 6.3.3.

6.3.2 Initial Activation of Li$_2$S – Mechanism of First Charge

The cell containing Li$_2$S-based cathode starts its operation from de-lithiation, i.e. charge process, during which the oxidation of Li$_2$S to soluble species occurs. It is well accepted that the initial charge of a Li$_2$S/Li cell is very unique with the activation barrier at the beginning (a circle in Figure 6.11a). Once the initial charge is accomplished, the Li$_2$S/Li system during the subsequent cycles performs similarly to the standard Li–S cells. Apart from the activation barrier, often a continuously higher overpotential during the whole initial charge is observed, with the voltage curve having a weakly reproducible shape (Figure 6.11c). The interesting thing is that regardless of the first charge profile, the proceeding discharge curve displays standard features of a Li–S cell with two well-known plateaus (Figure 6.11b,d).

This activation barrier originates from the phase nucleation of Li$_2$S, where an extra driving force is required to nucleate the new phase, i.e. soluble polysulfides [80]. The authors examined the contribution of thermodynamic and kinetics factors to the origin of this activation barrier and revealed the mechanism of the initial charge. A schematic illustration is presented in Figure 6.12.

The height of the activation barrier strongly depends on the current rate, and was reported to be even as high as 1.15 V for C/8 rate, whereas at C/2000 (almost zero current conditions) it was only 25 mV [80]. It is therefore believed that the contribution of

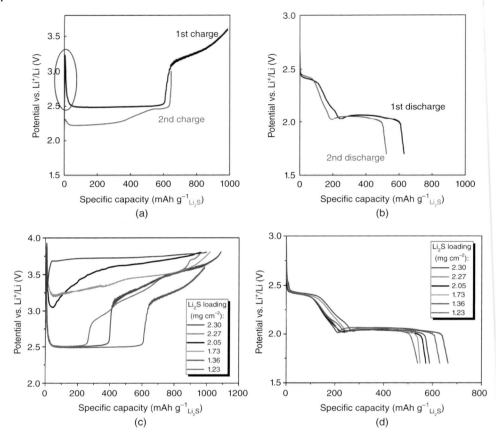

Figure 6.11 First and second charge (a) and discharge (b) profiles of a Li$_2$S/Li cell. Source: From Ref. [16]. Variety of different initial charge curves with different polarization (c) and subsequent discharge profiles (d). Source: From Ref [79].

all kinetic factors to the height of the activation barrier is concomitant with the phase nucleation process. At very low currents where kinetics factors have negligible effect on the electrochemical reaction, the origin of the activation barrier is more attributed to the thermodynamic effects and phase nucleation. On the other hand, with the increase in the current, the overpotential associated with kinetic factors increases exponentially and is the dominant factor in the magnitude of the barrier height. After thorough analysis, it was found that among three main kinetics factors, such as electronic conductivity of Li$_2$S, diffusivity of Li$^+$ ions in Li$_2$S, and charge transfer process at the surface of Li$_2$S particles, it is the charge transfer process that controls the overpotential of the activation barrier and is the main reason for its appearance. As shown in Figure 6.12, during Steps 1 and 2, Li$_2$S and a single phase of Li$_{2-x}$S (with high lithium deficiency on the surface due to the low ionic diffusivity of Li$^+$ in Li$_2$S) undergo a slow charge transfer process due to the difficulty in charge transfer between Li$_2$S and the electrolyte. Once polysulfides are created, it is assumed that the charge transfer between Li$_2$S and polysulfides is easier in the polysulfides environment than in pure electrolytes (Step 3). At the end of

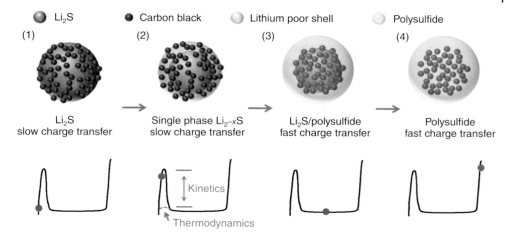

Figure 6.12 Schematic illustration of the proposed mechanism of the initial charge process of a Li$_2$S/Li cell. Source: Yang et al. 2012 [80]. Reproduced with permission from ACS.

charge, redox reactions are quite fast as the species involved in the reactions are in the liquid phase. The fact that there is no activation barrier at the beginning of charge in the subsequent cycles or when polysulfides are added into the electrolyte strongly supports the hypothesis that Li$_2$S activation process is the phase nucleation of Li$_2$S into polysulfides. Additional factors resulting in an overpotential during the activation process were presented by Son et al. [81], and it was attributed to the presence of a native layer on the surface of the Li$_2$S particle. This is related with high reactivity of Li$_2$S to moisture and oxygen, which leads to the formation of a stable native layer of LiOH in combination with an unstable S—H native layer. Jung and Kang [82] supported the argument about the presence of a thin insulating layer on the surface of Li$_2$S particles. This time, however, the authors reported on the presence of Li$_2$SO$_3$- or Li$_2$SO$_4$-like structures, obtained through *ex situ* XPS. Moreover, according to the authors, these insulating layers are responsible not only for the initial activation barrier but also for the continuous overpotential present along the charge curve. Another reason ascribed to enhancing the overpotential in the first charge, during both activation and along the whole charge process, is the micrometric size of the Li$_2$S particles as well as the morphology and structure of the cathode.

The mechanisms hidden behind the initial charge process of Li$_2$S-based cathode were far less studied than the charge mechanism of a standard sulfur cathode. Both processes are significantly different, as reflected by their voltage profiles (Figure 6.11a). Understanding what causes such significant differences may bring better understanding of the working mechanisms ruling Li–S cells in general. Different experimental techniques (including characterization as well as electrochemical evaluation) were applied to study the behavior of Li$_2$S/Li cells, particularly with the focus on the initial charge and the cause of the large overpotential. XRD technique applied via *ex situ*, *in situ*, and *operando* modes was used to study the structural transformation of a crystalline Li$_2$S active material upon charge and possible formation of sulfur at the end of the process [79, 80, 83]. A relatively good agreement between different reports was obtained, showing that peaks of crystalline Li$_2$S were progressively diminishing during the charge suggesting the formation of soluble species. Complete disappearance of Li_2S at the end of charge was confirmed after having reached 85–90% of the theoretical capacity [83]. The opinion about sulfur formation at the end of charge is more divided. Yang et al. [80] did not detect any traces of solid sulfur, while Jha et al. [83] did see the appearance of sulfur peaks when the cell was charged to 4.0 V. Waluś et al. [79] detected peaks of elemental sulfur formed on the electrodes being charged above 3.6 V, and the presence of β-sulfur was confirmed. Formation of another allotropic form of sulfur during charge has been previously seen in standard sulfur cathodes in Li–S configuration [25, 36, 37, 84]. It is interesting to observe that the potential of solid sulfur formation during initial charge of Li_2S cathode is significantly shifted toward higher voltage values as compared with standard sulfur cathode, i.e. >3.6 V vs. generally 2.4 V. *Operando* XAS by Gorlin et al. [85] confirmed the formation of solid sulfur at the end of charge as well. *Ex situ* XPS studies of Jung and Kang [82] performed on Li_2S electrodes charged to a different cutoff voltage showed that no traces of sulfur formation could be found when the electrode was charged to 3.5 V and some residuals of Li_2S were still detected. On the contrary, when charging the cathodes to 4.0 V, a complete disappearance of Li_2S-related peaks was observed and the appearance of sulfur was confirmed.

Galvanostatic interruption titration technique (GITT) was applied to Li_2S electrodes to follow the equilibrium, and under load potentials to find an explanation for the origin of the overpotential. The results showed [79] that the equilibrium potential during initial charge is fairly stable at about 2.4–2.5 V, and that it is much higher than the equilibrium potential during the second and further charges, as shown in Figure 6.13a–c. It was concluded that during the initial activation barrier, and in agreement with previous discussions, immediate formation of medium- and high-order polysulfides occurred, more likely through an electrochemical process. The amount of capacity related with the activation dip is very small, approximately $50\,mAh\,g^{-1}$, which stands for only 5–6% of the theoretical value (i.e. $874\,mAh\,g^{-1}$ if assuming the conversion reaction $4Li_2S \rightarrow Li_2S_4 + 6Li^+ + 6e^-$). During "Step 2," where the overpotential is very low (100–50 mV only), several electrochemical processes can occur: oxidation of Li_2S, which is of slow kinetics, together with the oxidation of medium-order polysulfides S_4^{2-}, S_6^{2-} to high-order polysulfides S_8^{2-}. Chemical disproportionation of Li_2S with high-order polysulfides cannot be excluded in this step, although it has not been confirmed experimentally in this work. At the end of charge very high polarization is observed, most likely to be caused by the very sluggish kinetics of Li_2S oxidation. Nevertheless, the equilibrium potential stayed almost unchanged, suggesting the presence of high-order polysulfides in the electrolyte. Finally in "Step 4," where the underload potential was kept at 3.6 V, equilibrium potential increased to 2.8 V, suggesting oxidation of long chain polysulfides to elemental sulfur. Jung and Kang [82] also applied GITT and observed very high polarization all along the charge profile, while the equilibrium potential stayed constant at ~2.4 V during most of the charge curve (Figure 6.13d). The authors assigned the presence of large polarization to the generally difficult conversion of Li_2S due to the presence of oxidized products (such as S—O, C—O, related bonds) on the surface of Li_2S particles.

The work of Gorlin et al. [85], where *operando* XAS was applied to identify the intermediate species during the charge process of a Li_2S-based cathode, gave a more complete picture of the evolution of polysulfide species of both nature, i.e. soluble and solid precipitates. Moreover, spatially resolved data were obtained, where the electrode

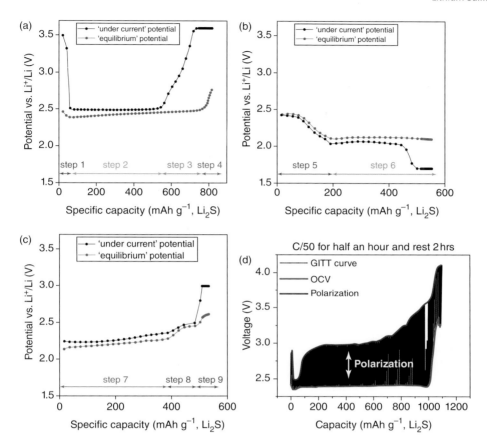

Figure 6.13 (a–c) GITT results applied to study overpotential phenomenon in Li$_2$S/Li cell during first charge–first discharge—second charge, at a constant rate of C/20 applied during 20 minutes, followed by 30 minutes of relaxation. Source: Waluś et al. 2015 [79]. Reproduced with permission from Elsevier. (d) GITT results of highly polarized initial charge of a Li$_2$S/Li cell. Source: From Ref. [82].

and separator/electrolyte were monitored. The results showed that formation of S$_8$ (detected in the separator, thus in the electrolyte) started very early, at approximately 10% SoC, and its concentration progressively increased throughout the entire charge process. At the same time, no soluble intermediate species were detected at any particular point of charge, or as suggested by the authors, they were potentially present in the electrolyte but in non-detectable concentrations. These results demonstrated that for charge processes occurring at a significant overpotential (>1.2 V above the equilibrium potential), Li$_2$S can be directly converted to S$_8$, which allows attainment of the full charge capacity, without the formation of intermediate species. The possible explanation for this observation is as follows: in the initial step of the charge, solid sulfur (S$_{8, \text{solid}}$) is electrochemically generated through direct oxidation of Li$_2$S, and then dissolves in the electrolyte (S$_{8, \text{solution}}$) and diffuses toward Li$_2$S particles and reacts chemically to produce S$_n^{2-}$ in non-detectable concentrations. Later on, S$_n^{2-}$ converts through a series of chain-growth/disproportionation reactions to polysulfide species that can be electrochemically oxidized to solid sulfur S$_{8, \text{solid}}$. Then again, S$_{8, \text{solid}}$ can dissolve in the electrolyte and react chemically with Li_2S in the cathode. This proposed mechanism involves both types of Li_2S activation. Part of the Li_2S particles that have good interfacial contact with the conductive substrate are electrochemically activated to form sulfur ($S_{8, solid}$), which is reflected by the appearance of an activation barrier during extraction of the initial $10\,mAh\,g^{-1}$ of the capacity. In parallel, other parts of Li_2S particles can be oxidized through a series of chemical reactions, leading to the formation of polysulfide species, which can be further electrochemically oxidized to $S_{8, solid}$. The second charge overpotential was significantly lower, in agreement with what is usually observed in the literature [79, 85, 86]. This time, intermediate polysulfides were detected at a relatively constant concentration and composition through the entire charge process. Their presence facilitates the chemical step of Li_2S oxidation. Since the second charge process of Li_2S cathode looks very similar to any charge cycle of a standard sulfur cathode, this result may be translated to understanding the charge mechanism of a standard Li–S configuration, i.e. that oxidation of Li_2S being formed at the end of discharge involves more of a chemical activation rather than the electrochemical one. Recently Koh et al. [87] studied the oxidation route of electrochemically insulated Li_2S in a specially designed cell. This work supported the hypothesis that Li_2S oxidation is a mix of chemical and electrochemical interactions. The results suggest, in the first place, that commercial Li_2S powder contains <2 wt% of polysulfide impurities, such as S^{2-}, S_2^{2-}, S_3^{2-}, S_4^{2-}, and traces of S_6^{2-}, all identified by UV spectroscopy. The CV scans of elucidated solutions proved that among them at least two are electrochemically active. These electroactive species can be oxidized to form long chain polysulfides, which can react with insoluble Li_2S to generate soluble medium chain length polysulfides, which practically participate in the charge transfer process. Therefore, the authors propose that the electrochemical oxidation of Li_2S occurs through chemical reactions coupled with a charge transfer process rather than through a direct charge (electron) transfer between solid Li_2S and conducting materials (such as carbon).

In the recent work of Vizintin et al. [86] the authors coupled two *in situ* techniques, UV–Vis and XAS, to study the mechanisms of the Li_2S oxidation depending on the overpotential of the initial voltage curve. Their work confirmed the hypothesis of direct transformation of Li_2S to sulfur without the formation of the intermediate soluble polysulfides, when charge occurs with a high overpotential along the entire process. On the other hand, in a charge process when a 2.5-V plateau is present, formation of polysulfides was detected. In this case, the authors suggested that partially dissolved sulfur that is formed at the beginning of charge reacts chemically with Li_2S and forms long chain polysulfides. The latter can partially react with Li_2S to form additional polysulfides, which lead to Li_2S activation through its dissolution, until it becomes electrochemically accessible. The authors also explained that the reason for such different voltage profiles could be attributed to the choice of the binder and its dispersion in the cathode, which in turn affects the ionic and electronic wiring of Li_2S particles. On top of this, the electrolyte amount used in the cell can significantly affect the mechanism of the initial oxidation. The cells with the lowest electrolyte amount displayed the profile with large overpotential, most likely due to wettability issues.

To minimize the initial activation barrier (as well as overpotential along the charge curve) several approaches have been proposed in the literature. The simplest one and easily applicable is altering/increasing charge cutoff voltage, so that during the initial

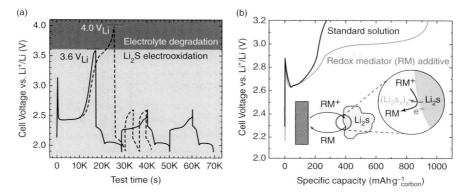

Figure 6.14 (a) Voltage profiles comparison showing the consequence of electrolyte degradation depending on the charge cutoff voltage. (b) Principles of redox mediators mechanism and differences in the voltage shape when mediators were/were not used. Source: Meini et al. 2014 [89]. Reproduced with permission from ACS.

step of charging the voltage limit will not be prematurely triggered by rapidly increasing potential due to the activation barrier [80]. As previously mentioned, reducing the charge rate also leads to a decrease in the initial activation barrier [80]. It was reported that by setting the charge cutoff voltage at a higher value (even 4.0 V), more Li$_2$S can be activated/oxidized and transformed to elemental sulfur, leading to higher discharge capacity in the subsequent discharge [79, 82]. Nevertheless, such high potentials during the initial charge (even 4.4 V [88]) may lead to unwanted electrolyte degradation. Therefore, redox mediators were proposed as a solution to facilitate activation of Li$_2$S-based cathodes by reducing the initial activation barrier and improve utilization of Li$_2$S active material without the necessity of charging to very high voltages [89]. The concept of using redox mediators is shown in Figure 6.14. Other redox mediator additives, such as LiI [89] and P$_2$S$_5$ [90], were also used.

As already mentioned, addition of polysulfides to the electrolyte is another easily scalable solution for reducing/eliminating the activation barrier, often reported in the literature [80, 91]. With the presence of polysulfides, the activation barrier is minimized most likely through chemical activation, due to the synergetic presence of the reduced and oxidized forms of Li$_2$S. In such configuration Li$_2$S can spontaneously oxidize to low-order polysulfides when in direct contact with high-order species, as explained by Xu et al. [91]. Other solutions applied to reduce the overpotential and improve utilization of Li$_2$S were undertaken at the cathode side, at both levels – active material by itself as well as cathode architecture. Section 6.3.3 provides more details about the development of Li$_2$S-based cathodes leading to improved performance.

6.3.3 Recent Developments in Li$_2$S Cathodes for Improved Performances

Several attempts have been made to improve the performance of Li$_2$S/Li cells by modifying the structure of Li$_2$S active material or the design of the complete cathode and cell. The literature presents examples where annealing treatment (between 500 and 850 °C) [81] or etching with acetonitrile [82] of commercially available Li$_2$S powder helped remove the insulating native layer present on the particles, leading to decreased overpotential during initial charge, which consequently led to higher capacity.

Another proposed approach is to decrease the particle size of Li_2S, which also led to a reduced activation barrier, better utilization of Li_2S, and in general improved performance. Some of the strategies can allow for great control of the particle size and shape distribution.

One route proposed by Cai et al. [92] was a simple and cost-effective high-energy dry ball milling of commercially available micrometer-sized Li_2S powders with carbon additives. Further addition of MWCNT to the nanocomposite allowed to improve cycling stability. The effect of ball milling was studied by other groups as well [14, 80, 93]. Another way to form nano-sized Li_2S–C composites in an easily scalable, solution-based method was proposed by Wu et al. [94] (Figure 6.15B), where commercially available Li_2S was dissolved in ethanol with polyvinylpyrrolidone (PVP), followed by pyrolization of PVP to form a carbon coating on the particles. This structure resulted in a good rate capability as well as cycling performance, maintaining a discharge capacity of > 835 mAh $g_{Li_2S}^{-1}$ after 100 cycles at 0.2 C. *In situ* formation of Li_2S–carbon composites was also proposed by several researchers [96–98] as an alternative way to form nanoparticles of active material well dispersed in the conductive agent matrix. Different strategies were demonstrated, such as lithiation of prior prepared S-composite with the use of stabilized lithium metal powder (SLMP) [97], chemical reduction of pre-sublimed sulfur by lithium triethylborohydride ($LiEt_3BH$) [99], and electrochemical reduction of soluble polysulfides [98]. Nevertheless, it is important to keep in mind that scalability of some of the composites may be difficult to achieve due to the complicated and expensive synthesis route, while the results obtained are not superior to the ones when using simpler methods of fabrication.

In order to improve the electronic conductivity of Li_2S particles, several strategies to form conductive coating directly onto Li_2S particles were developed. A dry coating process through mechanical grinding of sucrose (as a source of carbon) with Li_2S was presented; however, wet coating with polyacrylonitrile (PAN) dissolved in *n*-methyl-2-pyrrolidone (NMP) resulted in Li_2S particles homogenously coated with carbon flakes, which led to improved cycling performance [100]. Heat treatment of adsorbed PVP and acetylene black particles created an amorphous coating on Li_2S, which according to the authors [101] resulted in reduced shuttle phenomenon. Lithium sulfate (Li_2SO_4) has also been applied as a precursor for Li_2S formation through carbothermal reduction in the presence of a carbon source, such as graphene [88] or PVP, which led to the formation of nano-sized Li_2S coated with nitrogen-doped carbon [86]. Particle size and shape of Li_2SO_4 can be easily controlled by selection of the cosolvent used for Li_2SO_4 dissolution. More complicated methods for coating were also applied, including the work of Nan et al. [95], where Li_2S spheres with controlled nano-size were prepared, followed by chemical vapor deposition (CVD) method to convert them into Li_2S/C core–shell particles, as illustrated in Figure 6.15A.

An approach to improve ionic conductivity of Li_2S particle was described by Lin et al. [3], where nanoparticles of Li_2S were first produced by simple reaction of sulfur with lithium triethylborohydride in tetrahydrofuran (THF), and then exposed to P_2S_5, which resulted in the formation of a thin layer of a superionic conductor (Li_2S–P_2S_5 or Li_3PS_4) on the surface of Li_2S particles.

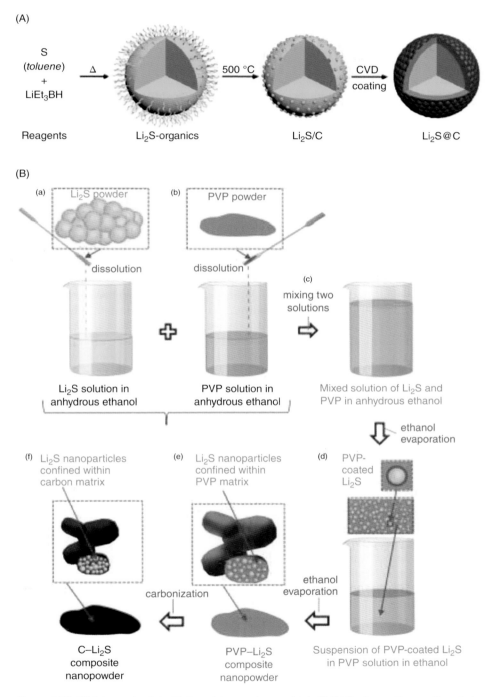

Figure 6.15 (A) Schematic of the Li$_2$S particles coating to obtain Li$_2$S@C spheres. Source: Nan et al. 2014 [95]. Reproduced with permission from ACS. (B) Preparation of Li$_2$S nanoparticles involving steric interactions and self-assembly phenomena. Source: From Ref. [94].

Table 6.1 Comparison of electrochemical performance of the recently developed Li$_2$S-based electrodes.

Cathode electrode	Preparation T (°C)	Loading (mg cm^{-2})	Rate	Initial discharge capacity (mAh g^{-1})	Stable discharge capacity (mAh g^{-1}) and cycles	Electrolyte additive	Cutoff voltage (V vs. Li$^+$/Li) of 1st charge	References
Li$_2$S/MWCNT paper	100	0.9	C/10	879	776 and 50 cycles	No	2.8 V at C/20	[104]
			C/5	843	705 and 100 cycles			
			C/2	794	676 and 100 cycles			
			1 C	729	634 and 100 cycles		3 V at C/20	
		1.8	C/10	764	705 and 50 cycles		3 V at C/20	
		3.6	C/10	590	576 and 50 cycles		3 V at C/40	
Li$_2$S/CNF paper		0.9	C/5	827	716 and 80 cycles		2.8 V at C/20	
Li$_2$S/carbon	300	1	C/5	400	370 and 50 cycles	No	4.1 V at C/20	[106]
Li$_2$S/carbon	900	0.54	C/2	340	280 and 50 cycles	Polysulfides	3 V at C/10	[96]
Carbon coated Li$_2$S	550	1.7	C/50	556	270 and 50 cycles	No	3.5 V at C/50	[100]
Li$_2$S/TiS$_2$	400	1	C/2	666	512 and 400 cycles	No	3.8 V at C/20	[107]
Li$_2$S/carbon black	110	1	C/10	600	480 and 50 cycles	Polysulfides	3.8 V at C/25	[80]
Li$_2$S/Sandwich carbon	50	1	1 C	620	520 and 100 cycles	No	4 V at C/20	[105]
Li$_2$S/carbon black	50	3	C/10	460	190 and 150 cycles	Redox mediators	3.6 V at C/20	[89]
Li$_2$S/CNF	50	1	C/10	800	620 and 80 cycles	P$_2$S$_5$	4 V at C/20	[90]
Li$_2$S/carbon	600	0.63	C/6	410	490 and 20 cycles	No	3.5 V at C/20	[108]
Carbon coated Li$_2$S	600	1.5	C/10	810	680 and 50 cycles	No	3.8 V at C/50	[109]
Li$_2$S/MWCNT	600	1	C/5	420	130 and 50 cycles	No	4 V at C/20	[110]
Li$_2$S/graphene	400	1	C/10	550	400 and 100 cycles	No	4 V at C/20	[111]
Li$_2$S/acetylene black	60	1.5	C/10	520	450 and 50 cycles	No	4.2 V at C/20	[112]
Li$_2$S/CNF	600	0.9	C/2	650	500 and 100 cycles	No	3.2 V at C/40	[113]
Li$_2$S/N-doped carbon	600	1	C/5	860	600 and 100 cycles	No	4 V at C/20	[114]
Li$_2$S/rGo	700	0.96	C/10	860	300 and 50 cycles	Polysulfides	3.5 V at C/20	[102]

Source: From Ref. [104].

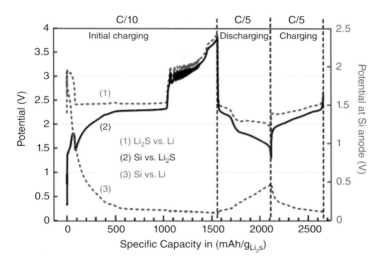

Figure 6.16 Voltage behavior measured in a three-electrode cell with Li reference electrode. Potentials of a full Li$_2$S/Si cell and each individual electrode are shown. Source: Jha et al. 2015 [83]. https://creativecommons.org/licenses/by/4.0/.

Reduced graphene oxide (rGO) has been widely applied as a conductive additive in creating the composites with Li$_2$S [99]. A simple solution-based method and thermal reduction have been used to form rGO–Li$_2$S nanocomposites, where nanoparticles of Li$_2$S were uniformly distributed on the rGO sheets [102]. A concept of freestanding and binder-free Li$_2$S-based electrodes was presented, where different current collector substrates such as rGO sheets [103], CNT, or carbon fiber paper [104] were used. In a similar manner, sandwiched cathode configuration composed of pristine Li$_2$S powder squeezed in between two layers of self-weaving, binder-free CNT electrodes was presented by Fu et al. [105].

A summary of cycling performances of recently developed Li$_2$S-composite cathodes can be found in Table 6.1.

The main advantage of developing Li$_2$S cathode is its potential for being used with metal anodes other than Li and surpassing the specific energy of commercial Li ion cells. To date, several examples of full Li$_2$S-based Li ion systems were presented, where a Li$_2$S composite cathode was paired with Si nanowires [12], Si/C composite [83, 93], Si thin film anode [99], and nanocomposite of Sn—C [115]. Utilization of a graphite anode in combination with standard carbonate electrolyte was also demonstrated and it was achievable thanks to nano-Li$_2$S composite cathode, allowing for non-ether electrolytes to be used [97]. When replacing Li metal by other metal Li-free anode, the operating voltage of the full cell is lower as compared with a Li$_2$S/Li equivalent system, and it depends on the voltage characteristics of the particular anode material, as demonstrated in Figure 6.16. Zheng et al. [97] demonstrated a full Li$_2$S/mesoporous carbon graphite cell delivering a reversible capacity of 600 mAh g$^{-1}_{Li_2S}$ during 150 cycles, at a current density of 168 mA g^{-1} and with 1.6 V average discharge voltage. Jha et al. [83] presented a Li$_2$S/Si full cell, which after 70 cycles at C/5 rate was still delivering a capacity of 280 mAh g$^{-1}_{Li_2S}$, while the Li$_2$S/Li counterpart cell remained at 400 mAh g$^{-1}_{Li_2S}$.

6.4 Summary

Lithium sulfide (Li_2S) is undoubtedly one of the main components of a Li–S system, either as an end of discharge product or a starting active material for cathode fabrication. It is considered as one of the main reasons for cell performance deterioration, since it forms an insulating passivation layer on the active surface of the cathode and blocks further reactions from occurring, causing premature end of the discharge. However, its formation, if assuming reduction of S_4^{2-} soluble species, is accompanied by exchange of $12e^-$, resulting in almost 75% of the theoretical capacity (1256 over 1675 mAh g_S^{-1}). Therefore, its formation is still a desired process. Recently, much effort has been made to better understand the kinetics of Li_2S nucleation and growth, and its dependence on different factors, which is the essential knowledge for further development and improvement. On the other hand, when Li_2S is used as an active material during the cathode fabrication, it is considered as an interesting alternative to move toward a Li-metal-free system. Moreover, initial oxidation of Li_2S is a very particular and still not completely understood process. This pushes the research toward applying more sophisticated characterization techniques (*in situ* and *operando*) to fully decipher the way Li_2S acts in Li–S cells and its role.

References

1 Bruce, P.G., Freunberger, S.A., Hardwick, L.J., and Tarascon, J.M. (2011). Li–O$_2$ and Li–S batteries with high energy storage. *Nature Materials* 11 (1): 19–29.

2 Song, M.K., Cairns, E.J., and Zhang, Y. (2013). Lithium/sulfur batteries with high specific energy: old challenges and new opportunities. *Nanoscale* 5 (6): 2186–2204.

3 Lin, Z., Liu, Z., Dudney, N.J., and Liang, C. (2013). Lithium superionic sulfide cathode for all-solid lithium–sulfur batteries. *ACS Nano* 7 (3): 2829–2833.

4 Xu, R., Lu, J., and Amine, K. (2015). Progress in mechanistic understanding and characterization techniques of Li–S batteries. *Advanced Energy Materials* 5 (16): 1500408.

5 Demir-Cakan, R. (2015). Targeting the role of lithium sulphide formation for the rapid capacity fading in lithium–sulphur batteries. *Journal of Power Sources* 282: 437–443.

6 Yan, J., Liu, X., and Li, B. (2016). Capacity fade analysis of sulfur cathodes in lithium–sulfur batteries. *Advanced Science* 3 (12): 1600101.

7 Cheon, S.-E., Ko, K.-S., Cho, J.-H. et al. (2003). Rechargeable lithium sulfur battery. *Journal of the Electrochemical Society* 150 (6): A796.

8 Yuan, L., Qiu, X., Chen, L., and Zhu, W. (2009). New insight into the discharge process of sulfur cathode by electrochemical impedance spectroscopy. *Journal of Power Sources* 189 (1): 127–132.

9 Barchasz, C., Leprêtre, J.-C., Alloin, F., and Patoux, S. (2012). New insights into the limiting parameters of the Li/S rechargeable cell. *Journal of Power Sources* 199: 322–330.

10 Li, Z., Huang, J., Yann Liaw, B. et al. (2014). A review of lithium deposition in lithium-ion and lithium metal secondary batteries. *Journal of Power Sources* 254: 168–182.

11 Wandt, J., Marino, C., Gasteiger, H.A. et al. (2015). *Operando* electron param-agnetic resonance spectroscopy – formation of mossy lithium on lithium anodes during charge–discharge cycling. *Energy & Environmental Science* 8 (4): 1358–1367.

12 Yang, Y., McDowell, M.T., Jackson, A. et al. (2010). New nanostructured Li_2S/silicon rechargeable battery with high specific energy. *Nano Letters* 10 (4): 1486–1491.

13 Yamada, T., Ito, S., Omoda, R. et al. (2015). All solid-state lithium–sulfur battery using a glass-type P_2S_5–Li_2S electrolyte: benefits on anode kinetics. *Journal of the Electrochemical Society* 162 (4): A646–A651.

14 Nagao, M., Hayashi, A., and Tatsumisago, M. (2012). High-capacity Li_2S–nanocarbon composite electrode for all-solid-state rechargeable lithium batteries. *Journal of Materials Chemistry* 22 (19): 10015.

15 Barchasz, C., Molton, F., Duboc, C. et al. (2012). Lithium/sulfur cell discharge mechanism: an original approach for intermediate species identification. *Analytical Chemistry* 84 (9): 3973–3980.

16 Walus, S. (2015). Lithium/sulfur batteries – development and understanding of working mechanisms. PhD thesis. Universite de Grenoble.

17 Noh, H., Song, J., Park, J.-K., and Kim, H.-T. (2015). A new insight on capacity fad-ing of lithium–sulfur batteries: the effect of Li_2S phase structure. *Journal of Power Sources* 293: 329–335.

18 Cañas, N.A., Fronczek, D.N., Wagner, N. et al. (2014). Experimental and theoretical analysis of products and reaction intermediates of lithium–sulfur batteries. *The Journal of Physical Chemistry C* 118 (23): 12106–12114.

19 Zhang, S.S. (2013). Liquid electrolyte lithium/sulfur battery: fundamental chemistry, problems, and solutions. *Journal of Power Sources* 231: 153–162.

20 Cañas, N.A., Hirose, K., Pascucci, B. et al. (2013). Investigations of lithium–sulfur batteries using electrochemical impedance spectroscopy. *Electrochimica Acta* 97: 42–51.

21 He, X., Ren, J., Wang, L. et al. (2009). Expansion and shrinkage of the sulfur com-posite electrode in rechargeable lithium batteries. *Journal of Power Sources* 190 (1): 154–156.

22 Elazari, R., Salitra, G., Talyosef, Y. et al. (2010). Morphological and structural stud-ies of composite sulfur electrodes upon cycling by HRTEM, AFM and Raman spectroscopy. *Journal of the Electrochemical Society* 157 (10): A1131.

23 Hagen, M., Dörfler, S., Fanz, P. et al. (2013). Development and costs calculation of lithium–sulfur cells with high sulfur load and binder free electrodes. *Journal of Power Sources* 224: 260–268.

24 Choi, Y.-J., Chung, Y.-D., Baek, C.-Y. et al. (2008). Effects of carbon coating on the electrochemical properties of sulfur cathode for lithium/sulfur cell. *Journal of Power Sources* 184 (2): 548–552.

25 Walus, S., Barchasz, C., Colin, J.F. et al. (2013). New insight into the working mech-anism of lithium–sulfur batteries: *in situ* and *operando* X-ray diffraction characteri-zation. *Chemical Communications* 49 (72): 7899–7901.

26 Diao, Y., Xie, K., Xiong, S., and Hong, X. (2013). Shuttle phenomenon – the irre-versible oxidation mechanism of sulfur active material in Li–S battery. *Journal of Power Sources* 235: 181–186.

27 Busche, M.R., Adelhelm, P., Sommer, H. et al. (2014). Systematical electrochemical study on the parasitic shuttle-effect in lithium–sulfur-cells at different temperatures and different rates. *Journal of Power Sources* 259: 289–299.

28 Mikhaylik, Y.V. and Akridge, J.R. (2004). Polysulfide shuttle study in the Li/S battery system. *Journal of the Electrochemical Society* 151 (11): A1969.

29 Yamin AG, H., Penciner, J., Sternberg, Y., and Peled, E. (1988). Lithium sulfur battery. *Journal of the Electrochemical Society* 135 (5): 1045.

30 Peled, E., Sternberg, Y., Gorenshtein, A., and Lavi, Y. (1989). Lithium–sulfur battery: evaluation of dioxolane-based electrolytes. *Journal of the Electrochemical Society* 136 (6): 1621–1625.

31 Sharma, R.A. (1972). Equilibrium phases in the lithium–sulfur system. *Journal of the Electrochemical Society* 119 (11): 1439–1443.

32 Wang, Y., Huang, Y., Wang, W. et al. (2009). Structural change of the porous sulfur cathode using gelatin as a binder during discharge and charge. *Electrochimica Acta* 54 (16): 4062–4066.

33 Ryu, H.S., Guo, Z., Ahn, H.J. et al. (2009). Investigation of discharge reaction mechanism of lithium|liquid electrolyte|sulfur battery. *Journal of Power Sources* 189 (2): 1179–1183.

34 Cañas, N.A., Wolf, S., Wagner, N., and Friedrich, K.A. (2013). In-situ X-ray diffraction studies of lithium–sulfur batteries. *Journal of Power Sources* 226: 313–319.

35 Nelson, J., Misra, S., Yang, Y. et al. (2012). In *operando* X-ray diffraction and transmission X-ray microscopy of lithium sulfur batteries. *Journal of the American Chemical Society* 134 (14): 6337–6343.

36 Waluś, S., Barchasz, C., Bouchet, R. et al. (2015). Lithium/sulfur batteries upon cycling: structural modifications and species quantification by *in situ* and *operando* X-ray diffraction spectroscopy. *Advanced Energy Materials* 5 (16): 1500165.

37 Kulisch, J., Sommer, H., Brezesinski, T., and Janek, J. (2014). Simple cathode design for Li–S batteries: cell performance and mechanistic insights by in operando X-ray diffraction. *Physical Chemistry Chemical Physics: PCCP* 16 (35): 18765–18771.

38 Lowe, M.A., Gao, J., and Abruña, H.D. (2014). Mechanistic insights into operational lithium–sulfur batteries by *in situ* X-ray diffraction and absorption spectroscopy. *RSC Advances* 4 (35): 18347.

39 Yang, G., Shi, S., Yang, J., and Ma, Y. (2015). Insight into the role of Li_2S_2 in Li–S batteries: a first-principles study. *Journal of Materials Chemistry A* 3 (16): 8865–8869.

40 Assary, R.S., Curtiss, L.A., and Moore, J.S. (2014). Toward a molecular understanding of energetics in Li–S batteries using nonaqueous electrolytes: a high-level quantum chemical study. *The Journal of Physical Chemistry C* 118 (22): 11545–11558.

41 Feng, Z., Kim, C., Vijh, A. et al. (2014). Unravelling the role of Li_2S_2 in lithium–sulfur batteries: a first principles study of its energetic and electronic properties. *Journal of Power Sources* 272: 518–521.

42 Patel, M.U., Demir-Cakan, R., Morcrette, M. et al. (2013). Li–S battery analyzed by UV/Vis in *operando* mode. *ChemSusChem* 6 (7): 1177–1181.

43 Kawase, A., Shirai, S., Yamoto, Y. et al. (2014). Electrochemical reactions of lithium–sulfur batteries: an analytical study using the organic conversion technique. *Physical Chemistry Chemical Physics* 16 (20): 9344–9350.

44 Hagen, M., Schiffels, P., Hammer, M. et al. (2013). *In-situ* Raman investigation of polysulfide formation in Li–S cells. *Journal of the Electrochemical Society* 160 (8): A1205–A1214.

45 Cuisinier, M., Cabelguen, P.-E., Evers, S. et al. (2013). Sulfur speciation in Li–S batteries determined by *operando* X-ray absorption spectroscopy. *The Journal of Physical Chemistry Letters* 4 (19): 3227–3232.

46 Lang, S.Y., Shi, Y., Guo, Y.G. et al. (2016). Insight into the interfacial process and mechanism in lithium–sulfur batteries: an *in situ* AFM study. *Angewandte Chemie* 55 (51): 15835–15839.

47 Paolella, A., Zhu, W., Marceau, H. et al. (2016). Transient existence of crystalline lithium disulfide Li_2S_2 in a lithium–sulfur battery. *Journal of Power Sources* 325: 641–645.

48 Suo, L., Hu, Y.S., Li, H. et al. (2013). A new class of solvent-in-salt electrolyte for high-energy rechargeable metallic lithium batteries. *Nature Communications* 4: 1481.

49 See, K.A., Leskes, M., Griffin, J.M. et al. (2014). *Ab initio* structure search and *in situ* ^{7}Li NMR studies of discharge products in the Li–S battery system. *Journal of the American Chemical Society* 136 (46): 16368–16377.

50 Huff, L.A., Rapp, J.L., Baughman, J.A. et al. (2015). Identification of lithium–sulfur battery discharge products through ^{6}Li and ^{33}S solid-state MAS and ^{7}Li solution NMR spectroscopy. *Surface Science* 631: 295–300.

51 Demir-Cakan, R., Morcrette, M., Gangulibabu et al. (2013). Li–S batteries: simple approaches for superior performance. *Energy & Environmental Science* 6 (1): 176.

52 Patel, M.U., Arcon, I., Aquilanti, G. et al. (2014). X-ray absorption near-edge structure and nuclear magnetic resonance study of the lithium–sulfur battery and its components. *ChemPhysChem: A European Journal of Chemical Physics and Physical Chemistry* 15 (5): 894–904.

53 Wu, H.L., Huff, L.A., and Gewirth, A.A. (2015). *In situ* Raman spectroscopy of sulfur speciation in lithium–sulfur batteries. *ACS Applied Materials & Interfaces* 7 (3): 1709–1719.

54 Yeon, J.T., Jang, J.Y., Han, J.G. et al. (2012). Raman spectroscopic and X-ray diffraction studies of sulfur composite electrodes during discharge and charge. *Journal of the Electrochemical Society* 159 (8): A1308–A1314.

55 Xu, R., Belharouak, I., Zhang, X. et al. (2014). Insight into sulfur reactions in Li–S batteries. *ACS Applied Materials & Interfaces* 6 (24): 2.

56 Wang, Q., Zheng, J., Walter, E. et al. (2015). Direct observation of sulfur radicals as reaction media in lithium sulfur batteries. *Journal of the Electrochemical Society* 162 (3): A474–A478.

57 Helen, M., Reddy, M.A., Diemant, T. et al. (2015). Single step transformation of sulphur to Li_2S_2/Li_2S in Li–S batteries. *Scientific Reports* 5: 12146.

58 Gao, J., Lowe, M.A., Kiya, Y., and Abruña, H.D. (2011). Effects of liquid electrolytes on the charge–discharge performance of rechargeable lithium/sulfur batteries: electrochemical and *in-situ* X-ray absorption spectroscopic studies. *The Journal of Physical Chemistry C* 115 (50): 25132–25137.

59 Zhang, B., Lai, C., Zhou, Z., and Gao, X.P. (2009). Preparation and electrochemical properties of sulfur–acetylene black composites as cathode materials. *Electrochimica Acta* 54 (14): 3708–3713.

60 Gorlin, Y., Siebel, A., Piana, M. et al. (2015). *Operando* characterization of intermediates produced in a lithium–sulfur battery. *Journal of the Electrochemical Society*
162 (7): A1146–A1155.

61 Wang, L., Zhang, T., Yang, S. et al. (2013). A quantum-chemical study on the discharge reaction mechanism of lithium–sulfur batteries. *Journal of Energy Chemistry*
22 (1): 72–77.

62 Cuisinier, M., Hart, C., Balasubramanian, M. et al. (2015). Radical or not radical:
revisiting lithium–sulfur electrochemistry in nonaqueous electrolytes. *Advanced
Energy Materials* 5 (16): 1401801.

63 Gallant, B.M., Kwabi, D.G., Mitchell, R.R. et al. (2013). Influence of Li_2O_2 morphology on oxygen reduction and evolution kinetics in $Li–O_2$ batteries. *Energy &
Environmental Science* 6 (8): 2518.

64 Gerber, L.C., Frischmann, P.D., Fan, F.Y. et al. (2016). Three-dimensional growth of
Li_2S in lithium–sulfur batteries promoted by a redox mediator. *Nano Letters* 16 (1):
549–554.

65 Fan, F.Y., Carter, W.C., and Chiang, Y.M. (2015). Mechanism and kinetics of Li_2S
precipitation in lithium–sulfur batteries. *Advanced Materials* 27 (35): 5203–5209.

66 Ren, Y.X., Zhao, T.S., Liu, M. et al. (2016). Modeling of lithium–sulfur batteries incorporating the effect of Li_2S precipitation. *Journal of Power Sources* 336:
115–125.

67 Kim, K.R., Lee, K.S., Ahn, C.Y. et al. (2016). Discharging a Li–S battery with
ultra-high sulphur content cathode using a redox mediator. *Scientific Reports* 6:
32433.

68 Qie, L. and Manthiram, A. (2016). Uniform Li_2S precipitation on N,O-codoped
porous hollow carbon fibers for high-energy-density lithium–sulfur batteries with
superior stability. *Chemical Communications* 52 (73): 10964–10967.

69 Zheng, J., Gu, M., Wang, C. et al. (2013). Controlled nucleation and growth process
of Li_2S_2/Li_2S in lithium–sulfur batteries. *Journal of the Electrochemical Society* 160
(11): A1992–A1996.

70 Su, Y.S., Fu, Y., Cochell, T., and Manthiram, A. (2013). A strategic approach to
recharging lithium–sulphur batteries for long cycle life. *Nature Communications* 4:
2985.

71 Liang, X., Hart, C., Pang, Q. et al. (2015). A highly efficient polysulfide mediator for
lithium–sulfur batteries. *Nature Communications* 6: 5682.

72 Zhang, S.S. (2013). New insight into liquid electrolyte of rechargeable lithium/sulfur
battery. *Electrochimica Acta* 97: 226–230.

73 Pan, H., Han, K.S., Vijayakumar, M. et al. (2017). Ammonium additives to dissolve
lithium sulfide through hydrogen binding for high-energy lithium–sulfur batteries.
ACS Applied Materials & Interfaces 9 (5): 4290–4295.

74 Ryu, H.S., Ahn, H.J., Kim, K.W. et al. (2006). Self-discharge characteristics of
lithium/sulfur batteries using TEGDME liquid electrolyte. *Electrochimica Acta*
52 (4): 1563–1566.

75 Hayashi, A., Ohtsubo, R., and Tatsumisago, M. (2008). Electrochemical performance of all-solid-state lithium batteries with mechanochemically activated
$Li_2S–Cu$ composite electrodes. *Solid State Ionics* 179 (27–32): 1702–1705.

76 Obrovac, M.N. and Dahn, J.R. (2002). Electrochemically active lithia/metal and
lithium sulfide/metal composites. *Electrochemical and Solid-State Letters* 5 (4): A70.

77 Zhou, Y., Wu, C., Zhang, H. et al. (2007). Electrochemical reactivity of Co–Li_2S nanocomposite for lithium-ion batteries. *Electrochimica Acta* 52 (9): 3130–3136.

78 Wu, F., Lee, J.T., Fan, F. et al. (2015). A hierarchical particle-shell architecture for long-term cycle stability of Li_2S cathodes. *Advanced Materials* 27 (37): 5579–5586.

79 Waluś, S., Barchasz, C., Bouchet, R. et al. (2015). Non-woven carbon paper as current collector for Li-ion/Li_2S system: understanding of the first charge mechanism. *Electrochimica Acta* 180: 178–186.

80 Yang, Y., Zheng, G., Misra, S. et al. (2012). High-capacity micrometer-sized Li_2S particles as cathode materials for advanced rechargeable lithium-ion batteries. *Journal of the American Chemical Society* 134 (37): 15387–15394.

81 Son, Y., Lee, J.-S., Son, Y. et al. (2015). Recent advances in lithium sulfide cathode materials and their use in lithium sulfur batteries. *Advanced Energy Materials* 5 (16): 1500110.

82 Jung, Y. and Kang, B. (2016). Understanding abnormal potential behaviors at the 1st charge in Li_2S cathode material for rechargeable Li–S batteries. *Physical Chemistry Chemical Physics* 18 (31): 21500–21507.

83 Jha, H., Buchberger, I., Cui, X. et al. (2015). Li–S batteries with Li_2S cathodes and Si/C anodes. *Journal of the Electrochemical Society* 162 (9): A1829–A1835.

84 Villevieille, C. and Novák, P. (2013). A metastable β-sulfur phase stabilized at room temperature during cycling of high efficiency carbon fibre–sulfur composites for Li–S batteries. *Journal of Materials Chemistry A* 1 (42): 13089.

85 Gorlin, Y., Patel, M.U.M., Freiberg, A. et al. (2016). Understanding the charging mechanism of lithium–sulfur batteries using spatially resolved *operando* X-ray absorption spectroscopy. *Journal of the Electrochemical Society* 163 (6): A930–A939.

86 Vizintin, A., Chabanne, L., Tchernychova, E. et al. (2017). The mechanism of Li_2S activation in lithium–sulfur batteries: can we avoid the polysulfide formation? *Journal of Power Sources* 344: 208–217.

87 Koh, J.Y., Park, M.S., Kim, E.H. et al. (2014). Understanding of electrochemical oxidation route of electrically isolated Li_2S particles. *Journal of the Electrochemical Society* 161 (14): A2133–A2137.

88 Li, Z., Zhang, S., Zhang, C. et al. (2015). One-pot pyrolysis of lithium sulfate and graphene nanoplatelet aggregates: *in situ* formed Li_2S/graphene composite for lithium–sulfur batteries. *Nanoscale* 7 (34): 14385–14392.

89 Meini, S., Elazari, R., Rosenman, A. et al. (2014). The use of redox mediators for enhancing utilization of Li_2S cathodes for advanced Li–S battery systems. *Journal of Physical Chemistry Letters* 5 (5): 915–918.

90 Zu, C., Klein, M., and Manthiram, A. (2014). Activated Li_2S as a high-performance cathode for rechargeable lithium–sulfur batteries. *Journal of Physical Chemistry Letters* 5 (22): 3986–3991.

91 Xu, R., Zhang, X., Yu, C. et al. (2014). Paving the way for using Li_2S batteries. *ChemSusChem* 7 (9): 2457–2460.

92 Cai, K., Song, M.K., Cairns, E.J., and Zhang, Y. (2012). Nanostructured Li_2S–C composites as cathode material for high-energy lithium/sulfur batteries. *Nano Letters* 12 (12): 6474–6479.

93 Agostini, M., Hassoun, J., Liu, J. et al. (2014). A lithium-ion sulfur battery based on a carbon-coated lithium–sulfide cathode and an electrodeposited silicon-based anode. *ACS Applied Materials & Interfaces* 6 (14): 10924–10928.

94 Wu, F., Kim, H., Magasinski, A. et al. (2014). Harnessing steric separation of freshly nucleated Li_2S nanoparticles for bottom-up assembly of high-performance cathodes for lithium–sulfur and lithium-ion batteries. *Advanced Energy Materials* 4 (11): 1400196.

95 Nan, C., Lin, Z., Liao, H. et al. (2014). Durable carbon-coated Li_2S core–shell spheres for high performance lithium/sulfur cells. *Journal of the American Chemical Society* 136 (12): 4659–4663.

96 Yang, Z., Guo, J., Das, S.K. et al. (2013). *In situ* synthesis of lithium sulfide–carbon composites as cathode materials for rechargeable lithium batteries. *Journal of Materials Chemistry A* 1 (4): 1433–1440.

97 Zheng, S., Chen, Y., Xu, Y. et al. (2013). *In situ* formed lithium sulfide-microporous carbon cathodes for lithium-ion batteries. *ACS Nano* 7 (12): 10995–11003.

98 Fu, Y., Zu, C., and Manthiram, A. (2013). *In situ*-formed Li_2S in lithiated graphite electrodes for lithium–sulfur batteries. *Journal of the American Chemical Society* 135 (48): 18044–18047.

99 Zhang, K., Wang, L., Hu, Z. et al. (2014). Ultrasmall Li_2S nanoparticles anchored in graphene nanosheets for high-energy lithium-ion batteries. *Scientific Reports* 4: 6467.

100 Jeong, S., Bresser, D., Buchholz, D. et al. (2013). Carbon coated lithium sulfide particles for lithium battery cathodes. *Journal of Power Sources* 235: 220–225.

101 Liu, J., Nara, H., Yokoshima, T. et al. (2014). Carbon-coated Li_2S synthesized by poly(vinylpyrrolidone) and acetylene black for lithium ion battery cathodes. *Chemistry Letters* 43 (6): 901–903.

102 Han, K., Shen, J., Hayner, C.M. et al. (2014). Li_2S-reduced graphene oxide nanocomposites as cathode material for lithium sulfur batteries. *Journal of Power Sources* 251: 331–337.

103 Wang, C., Wang, X., Yang, Y. et al. (2015). Slurryless Li_2S/reduced graphene oxide cathode paper for high-performance lithium sulfur battery. *Nano Letters* 15 (3): 1796–1802.

104 Wu, M., Cui, Y., and Fu, Y. (2015). Li_2S nanocrystals confined in free-standing carbon paper for high performance lithium–sulfur batteries. *ACS Applied Materials & Interfaces* 7 (38): 21479–21486.

105 Fu, Y., Su, Y.-S., and Manthiram, A. (2014). Li_2S-carbon sandwiched electrodes with superior performance for lithium–sulfur batteries. *Advanced Energy Materials* 4 (1): 1300655.

106 Hassoun, J., Sun, Y.K., and Scrosati, B. (2011). Rechargeable lithium sulfide electrode for a polymer tim/sulfur lithium-ion battery. *Journal of Power Sources* 196 (13): 343–348.

107 Seh, Z.W., Yu, J.H., Li, W. et al. (2014). Two-dimensional layered transition metal disulphides for effective encapsulation of high-capacity lithium sulphide cathodes. *Nature Communications* 5: 5017–5024.

108 Guo, J., Yang, Z., Yu, Y. et al. (2013). Lithium–sulfur battery cathode enabled by lithium–nitrile interaction. *Journal of the American Chemical Society* 135 (2): 763–767.

109 Suo, L., Zhu, Y., Han, F. et al. (2015). Carbon cage encapsulating nano cluster Li_2S by ionic liquid polymerization and pyrolysis for high performance Li–S batteries. *Nano Energy* 13: 467–473.

110 Wu, F., Magasinski, A., and Yushin, G. (2014). Nanoporous Li$_2$S and MWCNT-linked Li$_2$S powder cathodes for lithium–sulfur and lithium-ion battery chemistries. *Journal of Materials Chemistry A* 2 (17): 6064–6070.

111 Wu, F., Lee, J.T., Magasinski, A. et al. (2014). Solution-based processing of graphene–Li$_2$S composite cathodes for lithium-ion and lithium–sulfur batteries. *Particle and Particle Systems Characterization* 31 (6): 639–644.

112 Liu, J., Nara, H., Yokoshima, T. et al. (2015). Li$_2$S cathode modified with polyvinylpyrrolidone and mechanical milling with carbon. *Journal of Power Sources* 273: 1136–1141.

113 Ye, F., Hou, Y., Liu, M. et al. (2015). Fabrication of mesoporous Li$_2$S–C nanofibers for high performance Li–Li$_2$S cells cathode. *Nanoscale* 7 (21): 9472–9476.

114 Chen, L., Liu, Y., Ashuri, M. et al. (2014). Li$_2$S encapsulated by nitrogen-doped carbon for lithium sulfur batteries. *Journal of Materials Chemistry A* 2 (42): 18026–18032.

115 Hassoun, J. and Scrosati, B. (2010). A high-performance polymer tin sulfur lithium ion battery. *Angewandte Chemie* 49 (13): 2371–2374.

7

Degradation in Lithium–Sulfur Batteries

Rajlakshmi Purkayastha

OXIS Energy, E1 Culham Science Centre, Abingdon, Oxfordshire OX14 3DB, UK

7.1 Introduction

Degradation refers to the process of breakdown or deterioration that occurs when an object is put to use. For a lithium–sulfur (Li–S) battery, this breakdown of ability is measured in the deterioration of the usable capacity of the cell. For batteries in particular this refers to the breaking down/chemical transformation/deterioration of internal components, which leads to changes in the measured output characteristics, i.e. voltage and current over a period of time. Since fluctuations of voltage and current are possible due to changes in operating conditions as well, a better quantity to measure deterioration over time is the capacity of the battery. The capacity can be defined as the total available charge within the battery, and is expressed in units of ampere-hour. The capacity is directly correlated to the active mass present within the battery, but is dependent on all components operating efficiently. Thus the degradation of the internal components will lead to the charge either becoming inaccessible (e.g. loss of connection of active material from the circuit) or being consumed in parasitic side reactions (e.g. solid–electrolyte interphase (SEI) formation, polysulfide shuttling). Thus the loss of capacity or capacity fade is a direct measure of the degradation processes within the battery. When the cell is used, depending on the use case of operation including ambient environment and rates of power, the capacity one can draw from the cell can change from cycle to cycle. As the cell degrades, the capacity obtained deviates from the expected value. When this process manifests over a time of operation we refer to it as aging of the cell.

Why does a lithium–sulfur battery degrade? The reasons are manifold and are not fully understood. In order to understand how degradation manifests, we examine the lifetime characteristics of a lithium–sulfur cell cycled under constant current conditions. In Figure 7.1, we present the plot for the lifetime capacity of a 3.4 Ah lithium–sulfur cell.

The cell is cycled at 30 °C in the voltage limits of 1.5–2.45 V. The cell is cycled at constant current, with a discharge rate of 0.2 C and a charge rate of 0.1 C. Initially, the capacity seems lower than the rated capacity and rises slowly to the peak capacity. This is because initially, wetting of the separator and other components is not complete, and takes place gradually over a number of cycles [1]. This behavior is dependent on the viscosity of the electrolyte and other components, including separator properties,

Lithium–Sulfur Batteries, First Edition. Edited by Mark Wild and Gregory J. Offer.

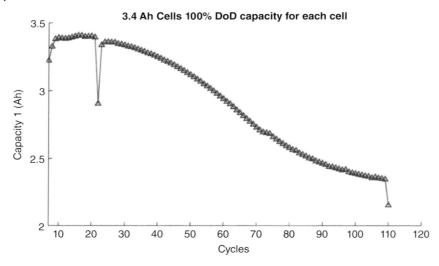

Figure 7.1 Discharge capacity of a cell cycled at a constant C-rate of 0.2 C during discharge and 0.1 C during charge at 30 °C.

and as such varies from cell to cell. Some work, for example, suggests that activation is dependent on the morphology and electrode structure [1]. Around cycle 20, a gradual decline in the capacity of the cell begins. The decline is almost linear, and the cell reaches the value of 80% of beginning of life (BoL) capacity around 70 cycles. This limit, 80% BoL, is used by most lithium ion battery manufacturers [2] and has been adopted by most researchers and developers of alternative battery technologies, such as lithium–sulfur. For continued use in homes etc. this limit may be dropped to as low as 60%.

In Figure 7.2, we examine the voltage profiles of certain cycles in greater detail. Figure 7.2a contains plots of voltage vs. capacity across different discharge cycles. The Li–S discharge voltage profile is a two-step profile. The initial part of the profile, which we refer to as the "upper plateau," starts around 2.45 V for this particular cell, due to voltage limits used, but can also start from voltages as high as 2.8 V. The changes in profiles for cells cycled in voltage windows can be found elsewhere [3] but the effects are largely the same. The voltage decreases in a monotonic manner, until it reaches a transition point. The voltage then becomes flat, and the second half of the curve is referred to as the lower plateau. The phenomena behind the shape of this curve has been described in detail in many other references [4], including earlier in this book. We shall describe it in brief.

The Li–S system works by a series of electrochemical reactions between lithium and sulfur. At the start of discharge, sulfur is present in a "higher order" form S_8, either as a solid or dissolved as a liquid. A series of electrochemical reactions convert this form into middle-order (S_4^{2-}, S_6^{2-}) polysulfides and lower order (S_2^{2-}, S^{2-}) polysulfides. The overarching reaction is

$$16Li^+ + 2S_8 \rightarrow 8Li_2S \text{ (solid)} \tag{7.1}$$

Li_2S or lithium sulfide is very sparingly soluble and precipitates out almost immediately when formed. The transition point between the upper and lower plateau is also

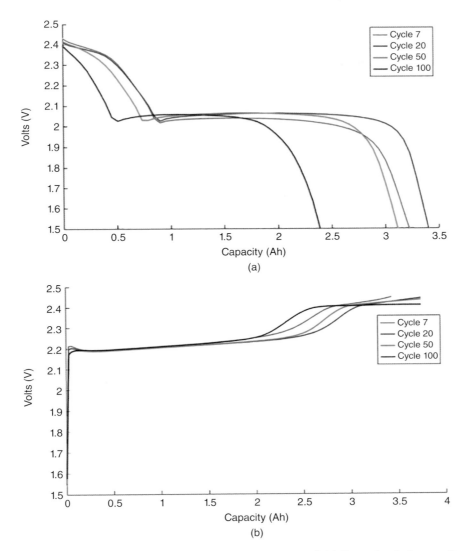

Figure 7.2 Voltage profiles of certain cycles of the previous cell. (a) Shows the discharge voltage profiles at cycles 7, 20, 50, and 100. (b) Shows the voltage profiles at cycles 7, 20, 50, and 100.

known as the "supersaturation point," [4] as lithium sulfide starts precipitating out of solution at that point. In fact, it is the presence of Li_2S which is thought to cause this behavior. This is better described in Chapter 6, and more details can also be found here [4–6]. The entire lower plateau is an equilibrium of sorts between the middle and lower order species [7], and toward the end of discharge, as the lithium sulfide precipitation reaction begins to dominate, the voltage drops with no gain in capacity. For reasons later described in this chapter discharge is stopped at 1.5 V.

The charging process, while analogous to the discharge, is not completely reversible in profile, as can be seen in Figure 7.2b. This is because the initial portion of the charging curve is dominated by the rate of dissolution of the lithium sulfide precipitates. The nature of the previous discharge also has an impact, as the type of precipitate formed might change [8, 9] although new work suggests that this might not be the case [10]. The charge curve then has a "lower plateau" region followed by a rise into the higher plateau region.

The changes in the features of the voltage profile give us some hints as to the processes underlying the loss of capacity seen over lifetime cycling. For discharge there is an initial increase in capacity followed by a steady decrease. In cycle 7, the cell has not reached its full capacity yet, but all of the characteristic features can be seen. In cycle 20, we can see that the upper plateau voltage lies almost on top of the voltage curve of cycle 7, but the lower plateau of cycle 20 seems longer, i.e. the extra capacity gain comes from the lower plateau. The voltage of the cell in cycle 7 has also not reached its full potential as compared to cycle 20. By cycle 50 the capacity has started to decrease, with the losses coming from both the upper and lower plateaus. By cycle 100, the cell has lost more than 20% of its capacity. The upper plateau capacity has decreased sharply, with the voltage almost becoming linearly decreasing. The lower plateau, while maintaining the same voltage, shows a marked reduction in capacity. In terms of a resistance increase over cycles, it therefore would seem that the polarization is not a linear or fixed quantity. This indicates that several different effects occur simultaneously.

In Figure 7.3 we present a close-up of the initial and final stages of the voltage profile during charge. The overpotential at the start of charge initially, as seen in Figure 7.3a, has a higher voltage but over time this voltage decreases. The voltage profile at the end of charge shows a slightly more complex behavior. As the cell ages, if there is an increase in the internal resistance of the cell, the value of the voltage will shift higher. The lithium–sulfur charge profile does not follow this behavior. Initially for cycle 7, the voltage proceeds to rise upward until it reaches the voltage cutoff of 2.45 V. By cycle 20, the profile while similar, now takes longer to reach the upper cutoff. In fact, it does not reach the cutoff voltage of 2.45 V at all, but is stopped because of having reached a time limit. By cycle 50, the profile at the top of charge has become flatter and by cycle 100 the profile flattens at 2.4 V almost continuously at the top of charge. This process of the voltage flattening at the top of charge is known as "shuttle." As mentioned earlier, during discharge, the sulfur species is reduced from higher order polysulfides to lower order polysulfides, all the way down to lithium sulfide. During charge, the reverse process occurs, with the lithium sulfide oxidizing to middle- and higher order polysulfides. Very quickly, the electrolyte solution contains a mixture of different sulfide species, including S_4^{2-}, S_6^{2-}, and S_8^{2-}. As the concentration of middle-order polysulfides increases at the cathode, due to the combination of a favorable concentration gradient and attracting charged lithium metal anode surface, the middle- and higher order polysulfides start drifting toward the anode. At the anode surface, they are reduced back to lower order polysulfides. Again, due to favorable concentration gradients, these diffuse back to the cathode where they are oxidized again. This starts a looping parasitic reaction, wherein species are endlessly oxidized at the cathode and reduced at the anode. This is why the voltage profile flattens out after a number of cycles as complete oxidation of all polysulfide species to S_8^{2-} is not allowed to take place. We shall explore the effects of shuttle in greater detail in Section 7.2.2, which describes the degradation processes at the anode.

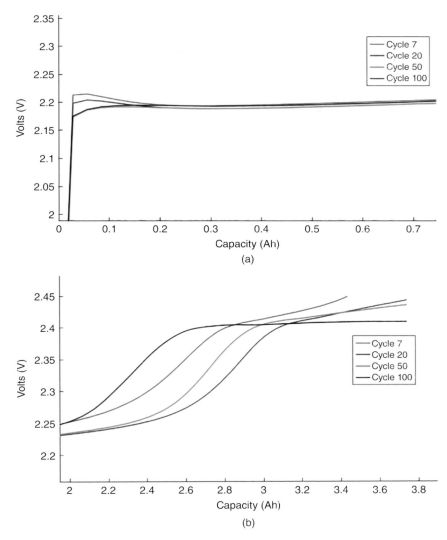

Figure 7.3 Close-up of features observed in a charge curve. (a) Shows the voltage profile during the initial part of the charge curve while (b) shows the voltage profile toward the end of charge for a cycled 3.4 Ah OXIS Long life cell.

The rest of this chapter is divided as follows. We first proceed to highlight all known degradation processes taking place in the cell components, due to ambient conditions and due to cell build and geometry. We then describe analytical and computational models designed to study capacity fade. We then list a number of methods by which degradation can be measured, e.g. resistance curves. We finally list a number of methods used to curtail degradation within the lithium–sulfur cell in order to achieve better capacity and cycle life. This becomes useful in defining limits of operation for the user as well.

7.2 Degradation Processes Within a Lithium–Sulfur Cell

In Figure 7.4 we see a schematic of different degradation processes within a lithium–sulfur cell. Before we proceed to describe all the known processes of degradation, a caveat must be issued. Lithium–sulfur is a technology in development, which means there is no fixed method of cell build or particular materials only used to make these cells. Even the lithium and the sulfur used to make these cells can be made in different ways, e.g. melted sulfur or patterned lithium. While some degradation processes, such as shuttle, are mechanistic phenomena common to all lithium–sulfur cells, a lot of degradation phenomena are linked to the types of components used and their interactions, e.g. electrolyte sulfur ratio. Therefore, some degradation phenomena will be more active in particular cells than others. This is important to keep in mind while examining the results of any life cycle analysis as the components and ambient conditions might limit the active degradation phenomena.

7.2.1 Degradation at Cathode

The lithium–sulfur cathode comprises of sulfur, the active material, mixed with a conducting carbon in order to form an electronic pathway, and a binder, to give it mechanical stability. The cathode must be sufficiently porous to allow for the infiltration of electrolyte, which allows the ionic transport of ions.

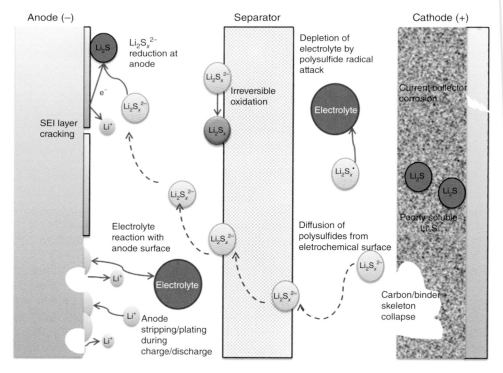

Figure 7.4 Schematic of degradation processes within a lithium–sulfur cell. Source: Wild et al. 2015 [4]. Reproduced with permission from RSC.

In order to maximize energy density, the amount of sulfur must be high compared to the binder and the carbon. At the same time, the entire cathode itself must be quite thick, in order to allow for complete utilization of the lithium as well. In order to be economically viable a deliverable capacity of 8 mAh cm^{-2} is recommended [11], which is equivalent to a sulfur loading of 6 mg cm^{-2}. At these loadings though, the structure of the cathode is unstable. The issue with a sulfur cathode is that unlike a standard lithium metal oxide electrode found in a lithium ion cell, solid sulfur dissolves into liquid polysulfides at the start of discharge. Toward the end of discharge solid lithium sulfide precipitates out, and not necessarily at the same locations where the sulfur was originally present [12]. A binder with large elasticity is therefore required in order to accommodate all these volume changes and still maintain the integrity of the cathode structure [3, 12, 13]. The densities change from 1.03 to 1.67 g cm^{-3}, as the sulfur converts to lithium sulfide, leading to a volume expansion of upwards of 75% in total [3]. While the expansion occurs through polysulfide formation, it is still a substantial change over the life of the cell.

The binder binds the electrode to the current collector, and it directly affects the power capability of the sulfur cathode [14]. Without robust adhesion a large polarization can develop at the interface of the electrode and current collector, and if the binder is unable to maintain contact with the carbon network, the electronic pathway is compromised. The binder therefore is also important for the cohesion of the cathode [14]. As it swells and shrinks with age, and reacts over time with the electrolyte, it loses its properties, rendering the cathode unable to deliver the same reversible capacity. While nano-carbon is good for developing highly conductive pathways, it hinders the mechanical integrity of the cathode [14].

Finally, higher sulfur loading has a deleterious effect on the cycle life simply because the greater depth to which de-plating/plating of lithium occurs, the more likely it is for dendrites to grow through the separator [15].

The other issue is the poor conductivity of both sulfur and lithium sulfide [12, 14, 16]. In order to allow for the sulfur to dissolve a large amount of carbon must be introduced into the system. As a result of this, a large amount of electrolyte is required to wet the porous carbon, reducing the energy density of the cell [16]. Reduction of sulfur particle size does not seem to have a long-term benefit on cycling [16]. Methods such as infiltration of the carbon material with molten sulfur, while showing some benefit, is insufficient to prevent long-term capacity fade [16]. Lithium sulfide provides a bigger challenge, as it deposits across both the cathode and the anode interface. By rendering the surface inert it effectively stops discharge from proceeding, causing large polarization to develop.

This combination of poor conductivity and mechanical problems causes higher loading cathodes to be unable to deliver higher power, with the structure collapsing and degrading when cycled at high C-rates. The sulfur suffers from mass transport limitations and sulfur utilization plummets as well at all C-rates, as can be seen in Figure 7.5 [1, 14, 16]. In fact, it has been found that areal specific capacity only increases with thickness for some time before falling again. The spatial distribution of the polysulfides leads to omitted pathways for diffusion, and the electrolyte uptake as well is not able to support very high current densities [16]. Lithium ion transport is therefore not seen as the limiting factor [14]. In order to maximize the energy density, along with increasing the cathode thickness the electrolyte is either kept the same or

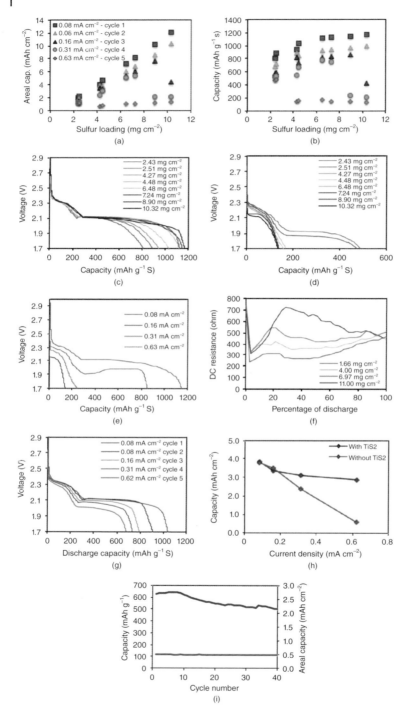

Figure 7.5 The effect of sulfur loading on discharge capacity cycle life. Source: Hagen et al. 2014 [15]. Reproduced with permission from Elsevier.

decreased, resulting in an effective decrease in the electrolyte sulfur ratio. As a result viscosities in the electrolyte increase considerably, especially around the saturation point [14]. Not only does polarization increase but so does lithium sulfide precipitation [17]. Conversely though, it has been shown that higher electrolyte loadings can lead to more rapid capacity fade, which is ascribed mainly to lithium sulfide growth [18]. Since local surface kinetics can be a limiting factor, more electrolyte promotes better transport of PS and Li ions, leading to greater lithium sulfide formation at the cathode.

During the initial part of discharge, sulfur rapidly forms polysulfides (PS). The solubility of polysulfides has a strong impact on both capacity and capacity fade. If the solubility is high, the discharge can take place more rapidly and higher capacity is achieved. However, they also then diffuse to the anode surface and start the shuttling reaction. A method to stop these is to create a physical barrier to prevent them from leaving the cathode. However, these PS species are an intrinsically polar species. They have a terminal sulfur, which bears the negative charge [16]. As a result, physical confinement does not seem to work as well as chemical interactions. As the PS species diffuse out of the cathode they become inaccessible for further reactions and therefore begin capacity loss [14, 19]. It is the PS dissolution that leads to shuttling, which ultimately leads to low coulombic efficiencies and self-discharge [14]. However, without the dissolution of PS, the reduction of sulfur can only occur on the carbon–sulfur interface, leaving the bulk sulfur unutilized [13].

Irreversible degradation processes occur due to the formation of anionic polysulfide radicals that result from sulfur reduction [13]. These can react with many organic polymers and therefore suitable tests need to be done. Finally, at high temperatures aluminum and sulfur can react violently with each other, and precautions must be taken, e.g. coating the current collector with carbon, in order to keep these two components separate.

Lithium sulfide is a bottleneck in the use of lithium–sulfur cells. While its formation and types are described in greater detail elsewhere in this book, we briefly look at its impact on cell cycling.

Lithium sulfide has a very low solubility in most electrolytes and as a result precipitates out of the electrolyte. Because it is a poor conductor, it blocks areas of conducting carbon, essentially shutting off the electronic conducting pathway. Initially, it can help create a protective SEI on the anode surface and protect it from pitting. But its insulating properties create a large polarization on the anode surface as well, reducing usable capacity as a whole.

There have been a series of studies on the nucleation and growth of Li_2S [17, 20]. They find that the initial dip at the saturation point deepens at lower electrolyte/sulfur ratios showing that the amount and type of Li_2S might be different. If the density of Li_2S nuclei has been reduced, the passivation effects are delayed, with fewer larger Li_2S particles. The deposition of Li_2S appears to be rate dependent. At lower discharge rates, the transport of ions to the crystal surface is able to keep up with diffusion through the crystal, and larger crystals are preferred. Diffusion through the crystalline phase is always slower as there is a shortage of defects compared to the amorphous phase. At high rates, nucleation is preferred, leading to the formation of amorphous Li_2S. Crystalline Li_2S, because of this relatively insulating nature, reduces the area through which current can flow. Thus, later on as the relative current densities through the remainder of the cathode increase, amorphous Li_2S is formed. Modeling studies [21] confirm

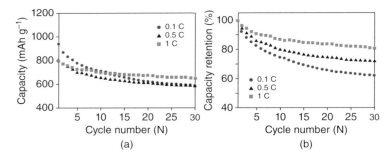

Figure 7.6 The effect of constant current cycling at different rates on cycle life of a lithium–sulfur cell. Source: Orsini et al. 1998 [22]. Reproduced with permission from Elsevier.

this fact, showing that initially particle growth is favored followed by nucleation as the surface area increases. Thus, a bimodal distribution of particles is found at the end of discharge. Therefore, it is suggested that the nucleation rate for Li_2S should be promoted to stop capacity fade, as at a higher initial nucleation rate the discharge capacity increases.

Amorphous Li_2S appears to be more reversible than crystalline Li_2S, showing lower capacity fade. This is in line with other studies [20] showing that while capacities obtained at lower rates are smaller, the overall fade rate is lower than for slow rates (Figure 7.6). Thus, severe capacity fade is explained by the accumulation of irreversible Li_2S in the cathode with cycling [9].

The formation of Li_2S is better understood by examining the initial part of charge, as the charge profile can reveal the type of Li_2S formed. In [9, 20] it was found that the charge following a slow discharge had a higher overpotential "hump" than after a fast C-rate. Scanning electron microscopy (SEM) images show that the large Li_2S particles formed during slow discharge do not disappear on charge. This has also been found in experimental work on the thickness of the cell [9]. While most of the thickness growth is ascribed to mossy lithium, there is some part that is governed by the formations of Li_2S deposits.

Thus, looking at irreversible loss of capacity due to loss of active sulfur, it is possible to separate this into three different components [18]. The first is due to loss into the electrolyte, the second is due to precipitation of Li_2S onto both electrodes, and the third is due to incomplete conversions between the PS.

7.2.2 Degradation at Anode

One of the key features of lithium–sulfur battery chemistry is the use of the lithium metal anode. This almost immediately gives the cell a boost in capacity as compared to the traditional graphite intercalation anodes used in lithium ion batteries because of lithium's relatively larger theoretical charge density (3861 vs. 372 mAh g^{-1} in graphite and 744 mAh g^{-1} in graphene). However, the use of lithium comes with caveats. Lithium has a much lower melting point, and its use makes the battery prone to issues such as thermal runaway. In brief, the main issues are those of dendrite formation, cycling efficiency, safety, and SEI formation, which is true for all lithium metal systems [19]. For lithium–sulfur in particular, the presence of shuttle and the formation of insulating Li_2S

on the surface cause additional problems. Gaseous and solid products can be formed on the surface, which use up active material and cause capacity fade [13].

The main issue with lithium metal anodes is the formation of dendrites: needle-like protrusions, which if allowed to extend through the separator will cause the battery to short circuit [23]. In the case of lithium–sulfur, this problem is in part mitigated by the presence of polysulfides, but there is a continuous stripping and deposition of lithium, and the deposits are mossy or dendrite-like in manner [19]. Dendrite formation occurs due to the lithium/electrolyte interface having local surface nonuniformities or imperfections [24, 25]. This causes a concentration variance across the surface of lithium, which causes lithium to plate preferentially and grow as dendrites. This is part of a larger inefficient process, where lithium plates and de-plates preferentially across certain sites on the anode rather than uniformly. As the dendrites grow, on de-plating it is possible for lithium to become isolated to form "dead" lithium, which contributes to low coulombic efficiency [24].

Chazalviel [26] used a simple electrochemical system consisting of two metallic electrodes to study the initiation conditions for dendrite growth. At high current density large-scale polarization causes an imbalance in the distribution of charge, inducing a large local electric field that causes dendritic growth to initiate. However, this would preclude the formation of dendrites at lower current densities, which is not what is seen experimentally. Other workers have therefore argued that the lithium surface state should also be considered in the formation of dendrites [27, 28]. Gireaud et al. [29] conducted a series of studies to analyze the lithium stripping and plating behavior to understand the effect of the lithium surface defects on dendritic growth. Even lithium with a smooth surface will show the presence of grain boundaries. The chemical reactivity of the grain boundary is higher than the grain itself, and therefore will be the preferred sites for SEI formation, on reaction with the electrolyte. Thus, when high current densities are passed through lithium these sites of "high interfacial energy" or slip lines will be the preferred sites for lithium stripping. At low current densities pits form on the surface of the lithium, with the viscosity of the electrolyte influencing the shape of the pits. The plating experiments indicated that at both high and low current densities, plating of lithium onto a pitted surface produced a mossy structure that was localized in the pits. No deposits were seen on smooth surfaces. In essence, conditions leading to a more homogenous current density lead to mossy deposits while nonhomogeneous currents due to surface imperfections lead to dendritic growth. Finally, the authors conducted cycling studies in order to see the variation in morphology over time. There are two distinct regions seen when cycling a lithium foil. There are regions of dendritic growth. But there is also the appearance of "bulk" lithium behavior, with grain boundaries. Thus, over many cycles these grain boundaries will grow deeper and form sites for preferential growth of dendrites. Therefore, they believe there is no benefit of surface treatment vis-à-vis dendrite growth. Finally, they conduct an experiment wherein the applied pressure on the electrode is varied, and they find that the efficiency values increase up to 90% with an increase in applied pressure. The increasing efficiency values can be directly correlated with the compactness of the lithium deposits.

Orsini et al. [22, 30] did experiments to observe the morphology of the lithium deposited on different substrates. Cycling of a lithium battery at low currents (C/5) saw the formation of mossy lithium. After initially appearing porous and crystallized, after a number of cycles the texture of the moss becomes relatively denser. On increasing the current density, dendrites were formed; however, they also observed the formations of "3D" or tangled dendrites – a system of deposits that grow in an intertwined manner. Wood et al. [24] observed using visualization techniques that most of the electrodeposition occurs on the dendrite surface rather than the surrounding bulk, which indicates that the kinetics of nucleation is sluggish compared to that of dendrite growth. During stripping of lithium the processes in order of importance were found to be electro-dissolution of dendrites, then planar bulk lithium, and then pitted surfaces. In the reverse process, electrodeposition occurs onto dendrites occurred first, followed by nucleation of new dendrites.

The SEI becomes critical in the suppression of dendrite growth as it can form a mechanical barrier to the growth of dendrites. Desirable properties for the SEI therefore include high ionic conductivity to allow lithium to plate, and high mechanical and elastic strength so that it can stretch to accommodate large strains without cracking [23]. Since the SEI itself adds resistance to charge transfer across the anode, it should be as thin as possible. If the SEI cracks or spalls off, then more electrolyte is consumed in the formation of new SEI, which leads to decrease in transport properties of the ions (due to increased viscosity of the electrolyte). The thickness of the SEI varies with the system but it is on the order of several tens of angstroms. The thickness itself can be nonuniform due to surface variations, and this might cause the formation of dendrites. Li et al. [23] summarize the effects of the SEI in the following manner. The uneven surface of the electrode will cause nonuniform deposition of lithium under the surface of the SEI and therefore exert stress on the SEI. If the SEI restrains the growth underneath (good elastic properties) or if the growth is not continuous, the morphology of the deposited lithium is mossy or particulate type. There might be instances where the lithium grows out of the surface due to cracking in the SEI and the electrolyte reacts to form new layers.

Han et al. [31] performed a study to look at the effect of sulfur loading on the corrosion behavior and SEI formation on lithium metal anodes. In this work, the effect of polysulfides on corrosion of the anode was studied by changing the amount of sulfur used to make the cathode. While the initial cycles do not show much difference in terms of thickness of layers of SEI formed on the anode, by 50 cycles, the thickness of the film is seen to be directly proportional to the sulfur loading (increasing from 98 μm with a 5 mg sulfur loading to 235 μm with a 10 mg sulfur loading). The SEI nominally consists of two sections: a bottom layer mainly comprised of lithium sulfide and lithium disulfide and a top layer that contains reaction products of the electrolyte with the lithium anode. The earlier hypothesis was that the topmost layer prevents the formation of additional sulfides, but that does not seem to be the case. Researchers believe this is due to the formation of lithium dendrites, due to uneven current densities across the anode. These dendrites change in volume during formation and dissolution, which destroys the structure of the film and changes the morphology of the anode. The incipient porosity makes it easy for the polysulfides to react with the lithium to form more precipitates on the anode surface. Cycling tests confirmed the detrimental effect of the sulfur loading, with the largest sulfur loading showing the fastest capacity fade.

Shuttle has been described in detail earlier in this chapter as well as in other sections of this textbook. For completion we present a "diagnostic" view of shuttle in Figure 7.7.

Figure 7.7c shows the charge capacity increasing with time: this indicates the presence of the shuttle reaction. Conversely, it directly correlates with a drop in cycling efficiency

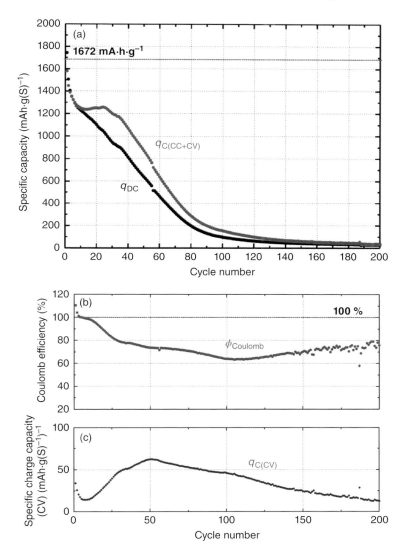

Figure 7.7 Evaluation of the discharge capacity (black), charge capacity (red), potentiostatic charge capacity (violet), and the coulombic efficiency (green) during 200 cycles at a rate of 0.1 C. Source: Busche et al. 2014 [3]. Reproduced with permission from Elsevier.

as seen in the image above it. Shuttle interferes with the deposition of lithium during the charge cycle. However, it reaches a maximum due to the irreversible consumption and decomposition of polysulfides, which is why the coulombic efficiency improves, but the overall capacity of the cell fades quite rapidly [3].

7.2.3 Degradation in Electrolyte

The electrolyte is a vital component of the lithium–sulfur battery. Unlike the lithium ion battery where transport occurs mainly via the lithium ion, in a lithium–sulfur battery the presence of polysulfide (PS) species necessitates a different approach. Degradation in electrolytes occurs due to stability issues, low conductivity, and high solubility of polysulfides. Most electrolytes are attacked from the "S" center of PS, but not ethers. PS anions and radicals react with common electrolyte systems such as esters, carbonates, and phosphates [13]. Ethers have low viscosity, leading to better wettability [32]; the ethers being used are polyethylene glycol dimethyl ether (PEGDME) and tetraethylene glycol dimethyl ether (TEGDME) [32]. Polysulfides show a high solubility in polar aprotic media, which explains the popularity of these solvents [3]. Among ether molecules there is most interest in glyme molecules as they have low volatility, no flammability, and are environmentally compatible [1]. Linear and cyclic ethers such as dimethyl ether (DME) and 1,3-dioxolane (DOL) are most commonly used. They have synergetic properties. DME has better PS solubility and reaction kinetics and is more stable to Li metal. Cyclic DOL forms a more stable SEI with lower solubility and kinetics. However, the solvents are still gradually depleted due to reaction with PS and Li metal [13]. The equations are

$$R\!-\!O\!-\!R + Li_2S_n \rightarrow R\!-\!OLi + R\!-\!S_nLi$$

$$R\!-\!O\!-\!R + 2Li \rightarrow 2R\!-\!OLi + R\!-\!R + \cdots$$

This problem is made worse with high sulfur loads [15]. More fresh surface is created, especially when the fraction of DME is low. For high sulfur loadings a ratio of almost $60\,\mu l\,mg^{-1}$ is recommended [15].

The choice of solvents is also dictated by permittivity, with high permittivity and donor numbers required to solvate PS. While high viscosity can cause sluggish transport it can also stop the disproportionation of PS [19]. Given the many properties required from an electrolyte for the Li–S system, a single solvent will most probably not work.

Carbonate electrolytes that are widely used in lithium ion batteries and are not ideal solvents for lithium–sulfur batteries as they allow dendritic deposition of lithium metal, which leads to low coulombic efficiencies (CE) [33]. On the other hand, ether-based electrolytes are used commonly with LiTFS, LiFSI, and LiTFSI as salts, as they can improve dendrite formation and reduce separator breach. As lithium–sulfur operates in a low voltage range, the anti-oxidation capacity of the electrolyte system is not an issue. Conventional salts such as $LiPF_6$, $LiBF_4$, and LiBOB are not suitable as they form hydrofuran (HF) [13].

An important role of the electrolyte is the formation of the SEI. The electrolyte reacts with the lithium metal surface to form a protective layer that allows for transport of lithium ions without further parasitic reactions. However, because lithium plating and de-plating is not uniform, cracks can develop in the SEI, leading to new areas being exposed to the electrolyte. This finally leads to "drying out" of the cell due to depletion of electrolyte. In a study [34] it was found that electrolyte decreased for a DOl:DME system after 35 cycles, and the loss in electrolyte directly correlates to capacity fade. This is another reason why electrolyte loadings are kept so high, as electrolyte-heavy cells always have longer cycle life than "electrolyte-lean" cells [34, 35].

Ideally, solid-state electrolytes would be used in lithium–sulfur batteries, because as mentioned in the previous section, their high shear modulus can prevent dendrite growth and allow for more uniform lithium plating. However, they suffer from poor lithium ion conductivity and over time contact with electrodes is poor, leading to a large polarization [19, 33].

Additives are commonly used to improve the performance of lithium–sulfur batteries and the most common one is lithium nitrate ($LiNO_3$), which acts as a shuttle inhibitor. It is believed that this additive reacts with metallic lithium in order to form insoluble Li_xNO_y products. This forms a "robust" SEI on the lithium anode, which prevents it from reacting further with the electrolyte as well as PS [3, 36]. However, $LiNO_3$ is also believed to act as a catalyst and converts highly soluble PS to slightly soluble elemental sulfur near the end of the charging process (around 2.5 V) [36]. By disassembling a cycled cell and combining the used electrodes with pristine ones, it was shown that while pre-cycled Li still showed shuttle, the pre-cycled sulfur cathode still showed shuttle suppression (This can be seen in Figure 7.8). The mechanism is believed to be

$$2NO_3 - 2e^- \rightarrow 2NO_3$$
$$2NO_3 + S_n^{2-} \rightarrow 2NO_3^- + n/8\, S_8$$

Lithium nitrate is also thought to have an effect on dendrite shape, with the formation of sharper, thinner dendrites favored when present. This is not good as these are more likely to pierce the separator. However, when both lithium nitrate and PS are present, the morphology is altered to a more favorable one [37]. It is thought that PS aids in the formation of a layer of Li_2S. This acts as a protective layer and increases CE by preventing further reaction. The competition between lithium nitrate and Li_2S ultimately leads to a beneficial SEI. The authors claim that the interfacial resistance is negligible at low C-rates.

However, lithium nitrate is thought to take part in several corrosive processes as well. Below 1.8 V the additive undergoes a reduction reaction. It is thought to occur on the sulfur cathode, leading to the formation of Li_2O [36]. Capacity fading therefore accelerates quite dramatically if the cell is consistently taken below 1.7 V when lithium nitrate is added [13, 36]. Lithium nitrate also causes gas formation in the cell [9, 38]. Experiments with gas chromatography and mass spectroscopy show that N_2 is formed in high concentrations and that lithium nitrate is responsible for its production. This severely hampers its use in large voltage and temperature ranges. Lithium nitrate is

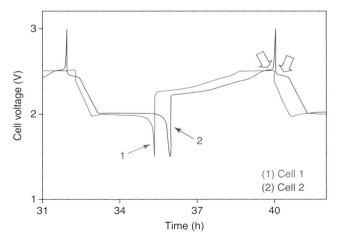

Figure 7.8 Voltage profile of cell with and without shuttle. Source: Zhang 2013 [13]. Reproduced with permission from Elsevier.

gradually consumed on the lithium anode surface [13], which leads to a decrease in its protective abilities. Finally, it is a strong oxidative agent that can lead to potential safety issues when a high concentration is used at high temperatures [13].

Gas products can be detected in lithium–sulfur cells throughout cycling, and can lead to swelling if not kept in check [9, 34]. At the BoL mostly hydrogen and nitrogen are detected, with hydrogen thought to arise from decomposition of the solvent. After extended cycling, methane and ethane and at very high cycle numbers, CO_2 and ethene can be detected [34]. The authors note that a cell that developed short circuits seems to develop more ethene. Hydrogen reduces over time while nitrogen increases, with a linear relationship between the lithium nitrate concentration and nitrogen gas.

One of the big issues with lithium–sulfur battery design is the ratio of the amount of electrolyte to the amount of active sulfur in the system or the E:S ratio. This ratio sets the upper bound on PS dissolution [27]. There is a decrease in conductivity with increase in PS concentration, which occurs with low E:S ratios. The choice of solvents affects the rate capabilities of the LiS cells due to effects on the kinetics and transport of the ions. For example, the viscosity of the electrolyte increases with increasing ether chain length, which gives rise to a lower ionic conductivity with longer solvent molecules [27]. There is significantly higher polarization and lower capacity and rate capability in LiS cells with reduced electrolyte [17] as electrodeposition of Li_2S is also adversely affected.

As the cell cycles, there is progressive wetting of the different components. If the E:S ratio is low an "activation" cycle is seen with a gradual rise in capacity as the electrodes flood completely. A low E:S ratio can therefore help by limiting the production of PS [1]. The viscosity of the electrolyte changes as well when the cell cycles due to the dissolution and precipitation of PS. The resistance of the solution is proportional to its viscosity. At lower E:S ratios, the increase of viscosity can cause a significant rise in internal resistance to reduce the discharge capacity quite significantly [15], with work showing that only ratios at $7 : 1\ \mu l\ mg^{-1}$ and above give reasonable results.

In Figure 7.9 a plot of different rates of capacity fade at different electrolyte loadings are shown [18]. Smaller loadings have low but stable capacity while large loadings show high capacity but more rapid fade. The authors believe that for the case of large electrolyte loading, more sulfur is being used up as Li_2S that is not being oxidized back to PS, which is confirmed by SEM images. The thicker nonconductive layers increase resistance of mass transport, which causes capacity fade.

7.2.4 Degradation Due to Operating Conditions: Temperature, C-Rates, and Pressure

The capacity of a lithium–sulfur cell is strongly influenced by both temperature and C-rate [39]. Lithium–sulfur is known to have a limited rate capability for a multitude of reasons. The main reasons are the insulating nature of the end products [19] and the sluggish transport of the various polysulfide species in solution (Various results at different rates can be seen in Figure 7.10 and 7.11).

In Figure 7.12 the effect of electrolyte volume by discharging at different C-rates can be seen. At high electrolyte volumes the viscosity of the solution is reduced, leading to better mass transport and rate capabilities. As the C-rate is increased, the high plateau remains more or less the same. However, the lower plateau shows a decrease in capacity

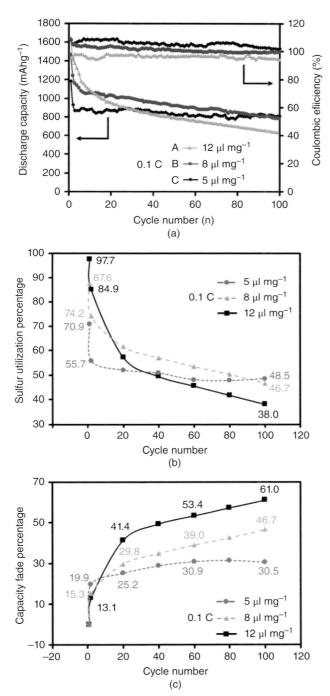

Figure 7.9 Sulfur utilization as a function of electrolyte loading [18].

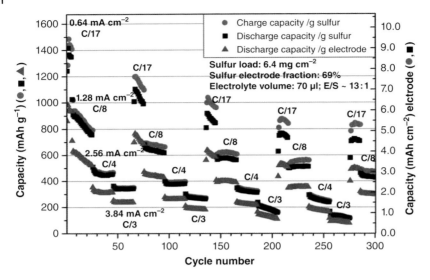

Figure 7.10 Effect of C-rate on capacity of lithium–sulfur cells over multiple cycles. Source: Hagen et al. 2015 [38]. Reproduced with permission from John Wiley & Sons.

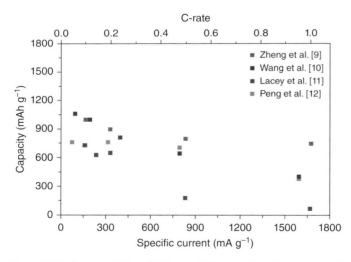

Figure 7.11 Rate capability of different lithium–sulfur cells as reported in literature. Source: Poux et al. 2016 [35]. Reproduced with permission from the Journal of the Electrochemical Society.

and a lower overall voltage [3, 35]. This indicates that utilization of sulfur becomes increasingly incomplete at higher rates [21].

Results from modeling studies show that more long chain PS reduces on the anode at low C-rates. There will be a reduction of side reactions on the anode as a passivating layer of Li_2S is formed [21]. Similar to results for electrolyte loading, batteries cycled at higher C-rates undergo a steep drop before they show stable cycling [21] while cells cycled at slower C-rates show a more gradual decrease in capacity. In [3] the authors' studies confirmed this fact with the relative capacity fade being 1.15% at 0.05 C and 0.88% for 0.5 C. There is an increase to 1.8% at 1 C. This is mainly due to the strong impact of shuttle. At lower rates, there is more time for the concentration gradients to build up

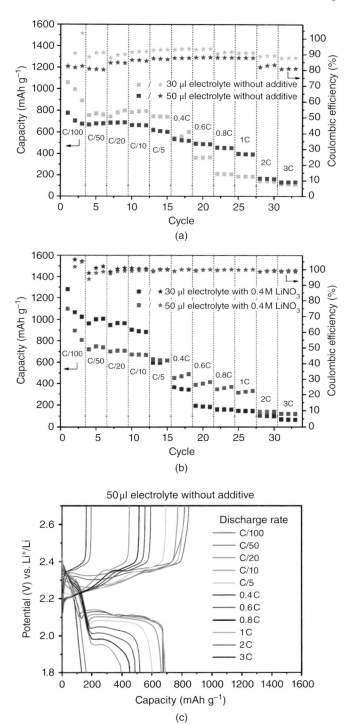

Figure 7.12 The rate capability of lithium–sulfur cells with and without the lithium nitrate additive with different volumes of electrolyte. Source: Rosso et al. 2002 [28]. Reproduced with permission from Elsevier.

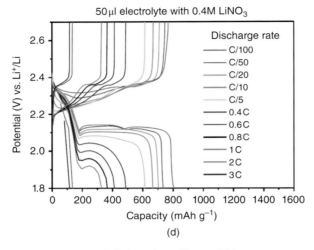

(d)

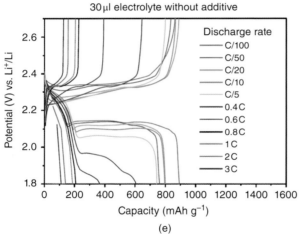

(e)

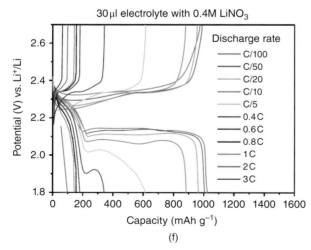

(f)

Figure 7.12 (*Continued*)

across the cell, which encourages shuttle, with lower charge rates proving detrimental for the cell. At high C-rates the capacity loss is mainly due to uneven plating of the lithium anode [9]. Slow C-rates aggravate material loss while high C-rates affect stability of the cell [35].

Finally, a routine method of testing the rate capability of batteries is to perform a "C-rate screen." In this the cells are cycled at gradually increasing rates. This might be alright for lithium ion systems, but in lithium–sulfur, the impact of previous cycles means that this method of testing needs to be modified. [35] shows that the order of the C-rate screen has an impact, with gradually decreasing C-rates showing better performance than gradually increasing C-rates. Hence, a lot of data in the literature might be overstating the case of the poor performance of lithium–sulfur at high C-rates. In order to mitigate this, it would be recommended to insert a few cycles at a "standard" C-rate. This makes sure that all species are brought back to the same starting configuration before a new C-rate is applied.

There is not a lot of work done on temperature effects in a lithium–sulfur system. A few studies [3] show that an increase of temperature strongly deteriorates the cell capacity. An increase in temperature reduces the viscosity of the electrolyte and accelerates all reactions in the system as well. Therefore, for a few initial cycles, it is possible to obtain higher capacities than at lower temperatures. Over time, however, the shuttle reaction is accelerated at high temperatures. In fact, Mikhaylik and Akridge [40] show that this can lead to a self-sustaining cycle, wherein shuttle causes a temperature rise, which further accelerates the shuttle reaction. This leads to the lithium–sulfur equivalent of thermal runaway. This model is described in more detail later.

After a few cycles, at all cycling rates the discharge capacities are significantly lower than those measured at room temperature [3]. It is interesting to note the deviations that occur as the cells cycle. This may indicate that because of so many competing processes, cells that start off essentially similar can have a very wide variation of outcomes. This can also highlight one of the outcomes of using a solution-based chemistry with so many different species; it can lead to a "butterfly" effect of sorts with small initial changes leading to large variations (This is very clearly elucidated in Figure 7.13).

7.2.5 Degradation Due to Geometry: Scale-Up and Topology

An important issue to raise is that most of the data available in the literature is for coin cells. This is not useful because different degradation mechanisms may become important as cells are scaled up. For example, in [41] it was found that the areal current density on the electrodes is higher in pouch cells compared to coin cells at the same C-rate. This leads to more severe shuttle as well as more rapid dendrite growth. In a coin cell, the presence of a relatively thick lithium metal anode makes shuttle the major method of degradation. However, mossy growth and "powdering" of the anodes becomes more critical as we increase currents when we scale up. This powdering can increase lithium ion diffusion, consume electrolyte, and reduce coulombic efficiency. The study in [41] showed that the initial capacity loss in the first five cycles was due to precipitation of nonconducting Li_2S. After 70 cycles, there is an increase in polarization, which is measured by the voltage of the lower plateau. SEM images confirm that at this point of high polarization, there is a change in anode surface morphology, with mossy growth occurring with lots of dead lithium on the surface.

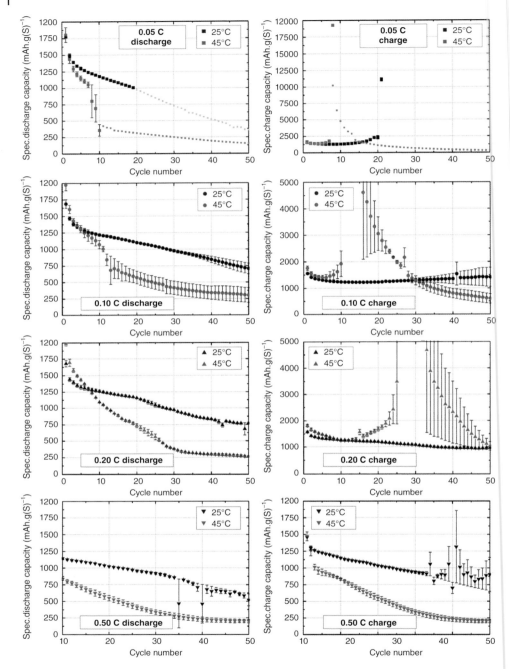

Figure 7.13 Charge and discharge capacities for lithium–sulfur cells cycled at different C-rates at two different temperatures. The error bars represent variations between the cells. Source: Busche et al. 2014 [3].

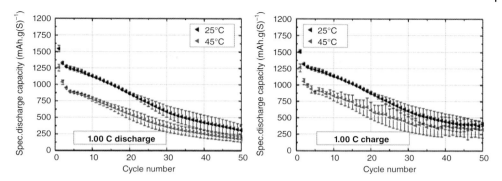

Figure 7.13 (*Continued*)

Another more insidious issue is the relative electrolyte-to-sulfur ratio. As we have mentioned earlier, this ratio is critical to the development of commercially viable cells. However, coin cells have a large reservoir of electrolyte, which can lead to unusually high E:S ratios. This means that even increasing the size of pouch cells themselves can lead to inversion of results (This phenomenon can be seen in Figure 7.14).

In the figure above, we show an example of such behavior. Results for a larger format cell are shown in Figure 7.12a and those for the small format with the same chemistry are shown in Figure 7.12b. The trend with temperature was found to have been inverted, for reasons not quite clear at present. Therefore, while it is likely that trends continue to hold, topological effects might be strong enough to reverse this. Hence, cycle life testing at all scales is always recommended.

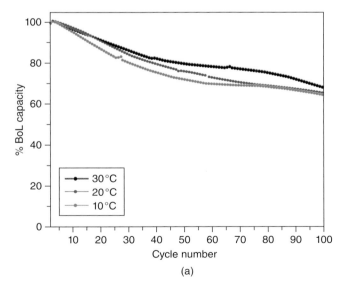

(a)

Figure 7.14 (a) OXIS prototype cell with greater than 30 layers (b) OXIS prototype cell with less than 10 layers. Source: Image courtesy Laura O'Neill, OXIS Energy.

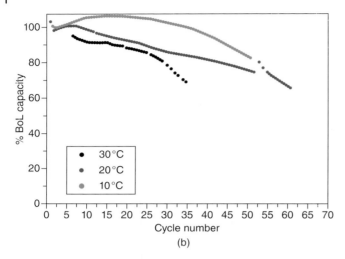

(b)

Figure 7.14 (*Continued*)

It therefore becomes instructive to normalize measured values by a common factor such as weight, in order to tease out the effect that scaling up might have. In Figure 7.15, we see the peak power performance for two kinds of cells. While the total power will be higher for the larger cell, it is interesting to note that scaling up seems to improve the performance of the cell when specific power values are compared.

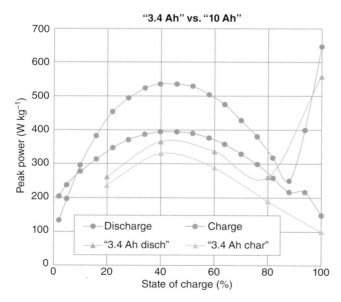

Figure 7.15 Comparison of specific peak power of two cells with similar chemistries but different formats. Source: Courtesy of Sylwia Waluś, OXIS Energy.

7.3 Capacity Fade Models

Experimental studies of capacity fade primarily concentrate on isolating individual degradation processes in order to describe mechanistic processes. Translating these studies into operational battery systems with several processes occurring in tandem can prove more laborious. This is where modeling of capacity fade proves more useful. Modeling of capacity fade can be done by either identifying the physical processes behind degradation and modeling them or by using statistics to analyze data and develop models or by a combination of the processes [42].

Most models of lithium–sulfur have been developed to describe the mechanism of lithium–sulfur in discharge, and usually contain no more than one discharge cycle. An example of this is the model by Kumaresan et al. [7] involving the step-wise reduction of S_8 to S^{2-}. They also include the dissolution of solid sulfur into the electrolyte solution as well as the precipitation of lithium sulfide. While their model is only for discharge and does not have any explicit capacity fading included into it, by changing the material parameters it is possible to get different capacity curves. For example, changing the dissolution rate of sulfur will reduce the sulfur utilization and therefore the capacity. Similarly, changing the precipitation rate of lithium sulfide also helps increase or decrease the capacity of the curve.

These ideas were carried on further by Ghaznavi and Chen [43–45] who performed a series of sensitivity studies using this model. The key highlights were that changing the material properties as well as charging characteristics leads to changes in the shape of the voltage plateau, and could be used as an indicator of which parameter changes over time will lead to capacity fade. However, the model in particular does not represent a battery losing capacity, and does not charge back properly, indicating the need for the development of a more comprehensive model.

Mikhaylik and Akridge [40] published a comprehensive, if slightly simplified, model, of the effect of the polysulfide shuttle on the capacity of the lithium–sulfur cell. They show that when the shuttle current is much higher than the charging current, the charge voltage never reaches its full value because of the dominating parasitic current within the battery. The residual high plateau polysulfides also affect the discharge capacity because of the alternative pathway of reduction present at the anode, and the results are calibrated with experimental results. A heat model is also developed, wherein the heat generated in the battery is accounted to be generated purely due to the polysulfide shuttle, and the impact on the constant itself is modeled as an Arrhenius function. While this modeling is highly simplistic, it is instructive when looking at experimental results where an acceleration of the polysulfide shuttle is seen at elevated temperatures. Finally, self-discharge experiments were carried out in order to validate the values used for the shuttle constant.

Moy et al. [46] carried out experiments on modeling the shuttle current, which clearly show that the use of $LiNO_3$ as an additive to the electrolyte prevents shuttle from occurring, most likely due to the formation of an electronically insulating layer at the anode surface. They also develop a simple linear 1D model for the shuttle current, assuming that only S_8^{2-}, S_6^{2-}, and S_4^{2-} are the species present, with the products being reduced linearly. When a shuttle current is present, flow of ions occurs due to concentration gradients set up between the anode and the cathode. By equating currents and diffusive fluxes across the cell (essentially charge and mass balances over short periods of time) it is possible to come up with an equation for the shuttle current. The results seem to match the experimental results quite well. The authors note that in their experiments, the shuttle current values decrease over time and attribute it to the steady formation of insoluble products at the anode. Thus the shuttle current itself can be used as a measure of capacity fade. They assume that both S_2^{2-} and S^{2-} ions are formed. The irreversible loss of capacity can therefore be related to the average value of the rate of decay of the shuttle constant. Since lithium nitrate itself decays over time, this method of analysis can be used to predict capacity fade for systems using this additive.

Hofmann et al. [47] have developed a thermodynamically consistent model for the lithium–sulfur cell, which includes polysulfide shuttling. While polysulfide shuttling in itself can be classified as reversible degradation, it does result in precipitation of lithium sulfide at the anode, which causes irrecoverable capacity fade. Their model is simplistic, as it only incorporates only four reactions and four species, and uses kinetics rather than the Butler–Volmer equation to model the electron transfer. However, the model is able to cycle, and more importantly it is able to model capacity fade in a reasonable manner. Like the previous models, it also includes the effects of precipitation of lithium sulfide and therefore decrease in porosity as well as incomplete sulfur utilization due to limitations in sulfur dissolution. The cell degradation due to cycling is calibrated using a "heuristic equation": i.e. a parameter, the anode reactive surface, is modified in order to obtain the correct capacity. The model is unable to capture capacity fade due to clogging by pores and increased impedance; a 3D model will be required for that.

Risse et al. [48, 49] have developed a nice set of stochastic Markovian models in order to predict the capacity fade of a lithium–sulfur cell.

In [48] this work, the authors borrow a model from biological systems that works remarkably well for lithium–sulfur. Essentially, the sulfur is broken down into four potential states: "sleeping," living and stable, living and unstable, and dead. The sleeping sulfur represents a reservoir that can be drawn upon, and the dead phase represents the phase that is lost irreversibly. Both the stable and unstable processes essentially represent the sulfur that takes part in the reversible processes. The model was fitted against systems with different cathodes, and does capture the different cathode behaviors very well, with the effect of different C-rates and degradation phenomena represented by different fractions of each of these four states. The only issue with this model is that it depends only on the previous state, and there is no interconvertibility between the states, i.e. a living phase does not become sleeping again, i.e. there is no accounting for "reversible" capacity loss. So while the previous history is represented, there will be a greater effect of cathode degradation processes. This should still work (and it does) because the anode processes ultimately act as the sulfur is being consumed. It is a very instructive model because of the simplicity with which it does capture a lot of complex behavior [48]. Figure 7.16 is a nice illustration of the types of curves for capacity fade. A is an ordinary battery aging curve that determines long-term stability of the system. Curves B and D describe a film formation period at the electrodes and an additional activation of material, respectively. Curve C represents a superposition of all processes.

[49] is an extension of the previous work of Risse et al., where the stable and unstable living processes have been rechristened as fast and slow capacity fading processes. They show that the fast capacity fade processes occur mainly due to formation of SEI at the anode surface, and prove this by comparing electrochemical impedance spectroscopy (EIS) results. The fade rates are faster therefore in the first few cycles and slow down

Figure 7.16 A schematic of different types of capacity fade curves observed in lithium–sulfur systems [32].

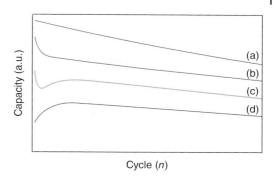

later, which is proved by validation. It is a good attempt to describe the two living stages in a more concrete manner than just stable and unstable.

In [50] Yoo et al. have done exactly as suggested earlier, modifying reaction parameters in order to determine the effect of PS solubility on capacity fade. They have analyzed shuttle with electrolytes of different PS solubility, with verification with experimental curves. They show that it is the combination of high PS solubility and a high reduction rate at the anode that leads to more rapid capacity fade, as seen in the case of Figure 7.17a. The other Figure 7.17b–d shows the impact of varying the rates. Figure 7.17b has a high polysulfide diffusivity and slow reaction rate. Figure 7.17c has low polysulfide diffusivity and fast reduction rate. Figure 7.17d has low polysulfide diffusivity and slow reaction rate which leads to the best cycling performance. The effect on the charge–discharge curves show that the high-voltage plateau is diminished after significant shuttle due to incomplete conversion to S_8 species (Figure 7.17).

7.3.1 Dendrite Models

Li et al. [23] have done an excellent analysis of dendrite growth in lithium systems. They divide the models into two different categories: physics-based models and Brownian/statistical models. In physical models, Barton and Bockris [51] in their model found that surface tension was one of the driving forces of dendrite growth, and was correlated to the overpotential through the pressure difference inside and outside the dendrite tip. The velocity at which the dendrite grows depends on the current density and the molar volume of the dendrite. The overpotential is a function of the surface tension and this relationship can be used to derive an optimal radius at which the growth rate is maximized. Brownian models essentially look at the morphology of the dendrite growth, with deposition probabilities based on current densities and overpotentials used to predict the growth of the dendrites in varying conditions. However, these are for smaller systems and would be computationally expensive to implement in an entire battery model.

The other important component in dendrite prevention is the separator. There are two aspects of the separator. The first is the mechanical strength of the separator. The second is the size of the pores. In a study by Jana et al. [52] a computational analysis was done on the effect of pore size and orientation through the separator on the growth of dendrites. The results show that the growth of dendrites through a separator depends not only on pore size but also on pore orientation. At large values of the pore radius even at low current densities the dendrites grow quite stably and are able to grow

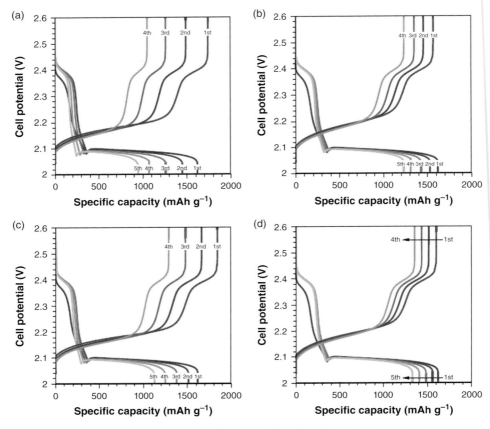

Figure 7.17 Charge and discharge curves of lithium–sulfur cells with varying relative degrees of polysulfide solubility and anode reaction (shuttle) rate. Source: Yoo et al. 2016 [50]. Reproduced with permission from Elsevier.

across the entire separator. Monroe and Newman [53] developed a model for growth of dendrites in the lithium–polymer system under galvanostatic charging. An extended model [54, 55] was applied to a polymer separator and for different shear moduli and Poisson's ratio, the displacements along different points of the separator can be predicted. The model indicates that surface tension has a much smaller contribution to the stability of the electrode as compared to the stresses due to deformation and compression. Deformation contributes to instability but increasing the shear modulus limits the impact of compressive forces, and the shear modulus should roughly be of the same order of magnitude as the elastic modulus in order to keep the electrode stable. Ferrese et al. [56, 57] found that the presence of a very stiff separator with a modulus of 16 GPa prevents excessive redistribution of lithium due to deformation. They use pressure-modified Butler–Volmer kinetics, but as lithium has a low yielding stress, the pressure developed is not very high. Hence the relatively straight profile is due to the stiffness of the separator itself rather than the modification of current densities.

A topic of interest is the morphology of plated lithium. While dendritic lithium has been studied extensively due to problems associated with short circuits and thermal runaway, lithium can also plate in a "mossy" or porous form. As noted in the experimental work, it is at lower current densities (and lower overpotentials) that mossy lithium is formed. Most models, however, tend to look at dendrite nucleation and growth in isolation. Ely et al. [58] developed a phase field model to look at lithium plating across a broad substrate. They find that current density and wettability of the substrate play an important role. At high wettability and low current, even deposits are formed. On increasing the current, while the initial deposition is uniform, protrusions start to develop into dendrites. In substrates with low wettability you have isolated islands depositing, with single dendrites showing a tendency to detach from the substrate. At low current densities, these islands tend to coalesce, and at higher densities they form long dendritic structures. Previous modeling of lithium deposition [37] showed that lithium deposition is biased toward the edges of a planer electrode due to greater presence of electrolyte in that region, which has also been confirmed experimentally.

7.3.2 Equivalent Circuit Network Models

Equivalent circuit network (ECN) models are an attempt to represent information obtained from mass spectroscopy as an electrical circuit. These are most useful for system engineers, who can design interactions between the cells and the battery management system (BMS) using these representations. They also help in reducing complex electrochemical processes into smaller, simpler units of interpretation. Most models have a bulk resistance term for the electrolyte, which is then coupled with multiple R–C elements in order to simulate the capacitive effects. Most of the proposed equivalent circuit models in the literature are always composed of a few (two/three/even four) Randles circuits (R/C) connected in series. However, the interpretation of the physical meaning of each R/C is more complicated, and several discrepancies between interpretations exist. Moreover, in the representation of these circuits for capacity fade, the values of these circuit elements will change with time, and the exact change ascribed to each element can be very subjective.

Deng et al. [59] carried out a series of EIS studies in order to study capacity fade within the battery. The first term, R_{el}, is a resistor in series, which represents the resistance due to diffusion through the bulk electrolyte. The second term, a parallel $R_l//CPE_l$ element, represents the effect of the lithium sulfide surface films formed on the anode. The second and final circuit $R_r//CPE_r$ represents the effect of the reactions at the sulfur pores. There is an additional composite polymer electrolyte (CPE) term after the resistance and two R//CPE circuit elements. The resistance in the high-frequency regime is the electrolyte resistance, and there are two additional semicircles in the high-frequency and middle-frequency regime. Instead of having a Warburg element represent diffusion limitations in the low frequency regime, it is replaced by a CPE element in order to deal with fitting discrepancies. The authors conducted EIS tests on a battery at various states of charge. The pattern remains the same, with two semicircles, and a long diffusion-related line. Similar to their previous assertions they conclude that the high-frequency semicircle cannot be associated with charge-transfer resistance because they believe that the insoluble products could not be formed on the higher plateau. The resistance is therefore ascribed to the resistance encountered as electrons conduct from the current collector toward the reaction sites. They neglect the contribution of the anode resistance due to the presence of polysulfides. Finally, impedance measurements were carried out after cycling in order to look at capacity fade over time. The results showed that the middle-frequency curve increases rapidly initially, followed by a slow increase, showing a change in the charge-transfer resistance. This is most likely due to the precipitation of insoluble lithium sulfides that accumulate over time, as their solubility is very low. It is the change of this value that primarily seems to cause capacity fade in the battery studied.

Canas et al. [60] performed a similar study but obtained a different circuit. The circuit consists of the single resistance and four R//CPE elements. They attribute the first R/CPE element with charge transfer occurring at the anode surface (small high-frequency semicircle). The second R/CPE element (medium high-frequency) is assigned to the charge transfer of sulfur intermediates and the third R//CPE (medium frequency) element is attributed to the formation and dissolution of sulfur and dilithium sulfide. The fourth and final R//CPE element is attributed to diffusion processes.

During discharge, the electrolyte resistance starts to increase as sulfur dissolves and polysulfides are formed, and after reduction to lithium disulfide the value of resistance reduces as well. The charge-transfer process in the anode (R1//CPE1) also follows a similar pattern, indicating the reaction of polysulfides with the anode surface to form SEI. The charge-transfer resistance at the cathode surface (R2//CPE2) decreases drastically at first, most likely due to the dissolution of sulfur, increasing the porosity and the subsequent reaction rates. R3 starts to appear toward the end of the high plateau indicating the effect of dissolution of the remaining sulfur in the cathode, and then toward the end of discharge when lithium disulfide starts to form. The diffusion process becomes hard to fit due to lack of data points, but is done through R4//CPE4.

During charging, there is a decrease in the absolute values of the resistances. The linking of the circuit elements with the processes changes slightly with the diffusion element R4 now linked to the dissolution of dilithium sulfide and formation of sulfur as well. Studies were done to look at the effect of cycling. The charge transfer of the cathode decreases considerably in the first few cycles, at a rate of loss that is similar to the capacity fade. The decrease in charge-transfer resistance is most likely due to the development of better electronic conduction. R3 gradually increases with cycling, most likely due to the formation of a film of reaction products at the interface of the cathode and the separator.

Kolosnitsyn et al. [61] proposed a new method of using a pulsed method of application of current, in order to get over difficulties of taking measurements at very low frequencies (which are the long-time requirements, which might violate the condition of stability over the course of the measurement). The internal resistance of the cell over different cycles was plotted, with a continual increase in impedance seen as the battery was cycled. The capacity is plotted as a function of the number of cycles. There is a sharp initial decrease, most likely due to the formation of SEI and insoluble sulfides. The rate of fade then stabilizes.

7.4 Methods of Detecting and Measuring Degradation

In lithium ion batteries it is common to use accelerated aging, wherein cells are cycled at relatively high rates in order to elucidate the degradation mechanisms. However, as described earlier, the mechanisms active in lithium–sulfur change with C-rates, and therefore this method is not directly applicable to lithium–sulfur. As a result of the difference in mechanisms between solution- and intercalation-based chemistry, newer methods have to be thought of in order to measure and track degradation.

7.4.1 Incremental Capacity Analysis

Incremental capacity analysis (ICA) refers to a method commonly used in lithium ion batteries wherein the evolution of the differential of the voltage vs. the capacity is analyzed as the cell cycles, in order to identify processes causing degradation [62]. The process cannot be directly transferred from lithium ion to lithium–sulfur as the degradation mechanisms are not as fully understood. Moreover, the non-monotonic nature of the lithium–sulfur curve provides its own trouble. The peaks in the IC curves represent electrochemical reaction equilibrium for the cathode. Previous work detected two peaks, approximately around 2.45 and 2.1 V, which are most likely related to the formation of higher and lower order PS. Degradation is detected by the movement of the peak in the upper plateau, which decreases and moves toward higher voltages. The charging curves provide better indicators for aging, which indicates that the cells concerned were aging primarily through the shuttle effect [62].

7.4.2 Differential Thermal Voltammetry

Differential thermal voltammetry uses the same principle as ICA, but instead of the capacity it considers the differential of temperature with voltage. As a result, it is able to detect temperature-related aging processes quite well [63]. When this method is applied to lithium–sulfur processes the appearance of shuttle is detected quite rapidly and at values where it has not fully manifested in the voltage profile [64]. The method shows promise for the detection of shuttle and possible mitigation at a reasonably early stage of cycling.

7.4.3 Electrochemical Impedance Spectroscopy

For a battery to function a lot of different physical phenomena occur such as electrochemical reactions, mass, and electron transfer, and many times they occur in the same location and at the same point in time. In order to understand the operation of the battery better there is a need for an analytical technique that will help separate these different processes. Impedance spectroscopy is a useful technique for analysis of these phenomena. Moreover, it is a nondestructive and easy-to-record technique. One of the principal assumptions behind impedance spectroscopy is that the properties of any electrode-material system are time invariant, and therefore one can determine these properties and how they change in differing conditions of temperature, voltage current, and so forth. The process involves applying a single-frequency voltage or current across the system and measuring the response. The response of the system is frequency dependent, and by studying the response of the system across a range of applied frequencies, it is possible to calculate the effective impedance across the battery system. From EIS results, ECN models can be developed, which have been described in Section 7.4.3. Li_2S deposits can possibly be detected by impedance spectroscopy, by looking at the impedance curves that develop along the discharge and charge curve in the mid-frequency domain [65]. The primary issue with EIS has always been interpretation. Because of the number of species and interfaces involved in a lithium–sulfur system it is difficult to separate processes occurring in a similar way. Details of the variability of interpretation have been described in Section 7.3.

7.4.4 Resistance Curves

One of the characteristic indicators of aging in the lithium–sulfur battery is the resistance curve.

In Figure 7.18 we see the evolution of the resistance curve for a 10 Ah OXIS cell. Figure 7.18a shows the evolution of the resistance for the charge curve, while Figure 7.18b shows the evolution during a discharge curve.

These resistances are measured by a galvanostatic interruption titration technique (GITT). In this technique, rest periods are inserted in the middle of constant current

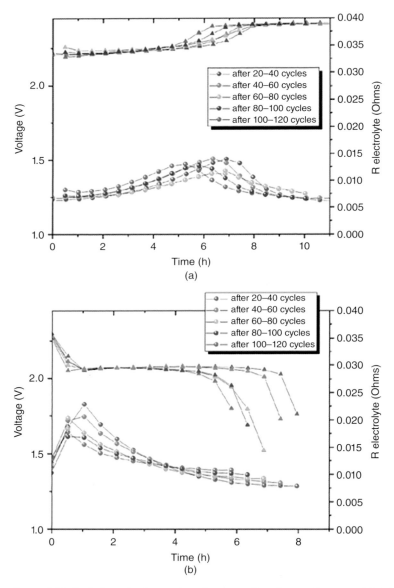

Figure 7.18 Evolution of resistance in cells over cycling. Source: Courtesy of Sylwia Waluś.

cycling. The instantaneous change in voltage at the start of rest is used to calculate the ohmic resistance of the system. Over time, as the cell ages, its capacity drops, which is indicated by the truncation of both the voltage and resistance curves with age. However, it is interesting to note that the resistance of the charge curve does not change much while in the discharge curve the values actually decrease. Hence, while this cell ages due to shuttle, the composition of species within the cell does not change much over time. It is likely that the measurement technique here has some overpotentials in it, as it is expected that the resistance should grow over time.

7.4.5 Macroscopic Indicators

Finally, as cell sizes grow bigger, it is possible to use macroscopic indicators, such as thickness or temperature of the cell to detect its state of health. Consider Figure 7.19.

The thickness of the cell was measured with a laser as described in [9]. The authors found that while the thickness of the cell has some degree of reversibility, due to processes such as stripping and plating of lithium, over time the overall thickness of the cell increases. The effect is found to be most apparent at lower C-rates, where the precipitation and dissolution of lithium sulfide are quite active. However, it is the plating of the lithium that seems to contribute most to the irreversible thickness rise, and it is likely that the uneven deposition of lithium leads to the formation of mossy and dead lithium, which contributes to an overall thickness increase.

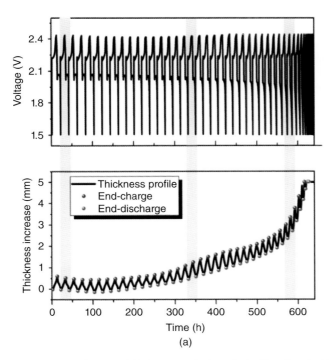

Figure 7.19 Voltage and thickness evolution of a 21 Ah cell cycled in a slow current regime of 0.106 C for both charge and discharge (a). End of charge and discharge thickness values, cycle to cycle thickness increase, and capacity retention as a function of cycle number (b). Walus et al. 2017 [9]. Reproduced with permission from Elsevier.

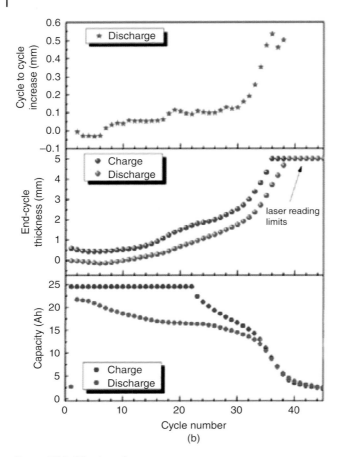

Figure 7.19 (*Continued*)

Similar to lithium ion cells, it is likely that the average temperature of the cell will rise over time. However, to measure this significant rise in temperature cycling of larger cells, which generate more heat, needs to be carried out.

7.5 Methods for Countering Degradation

Given the myriad ways in which a lithium–sulfur cell can degrade, the methods proposed to prevent these are also many. Figure 7.20 very succinctly highlights the different areas that need to be addressed in order to improve the cycle life and capacity of Li–S systems.

At the cathode, the restriction of PS diffusion is the main strategy to prevent capacity fade [66]. By containing the PS, they do not diffuse toward the anode, preventing shuttle and irrecoverable loss of sulfur. 3D conductive frameworks can help prevent the loss of conductivity that comes with the collapse of the cathode structure over cycling, especially for high-capacity cathodes [12]. Functional binders, especially cross-linked

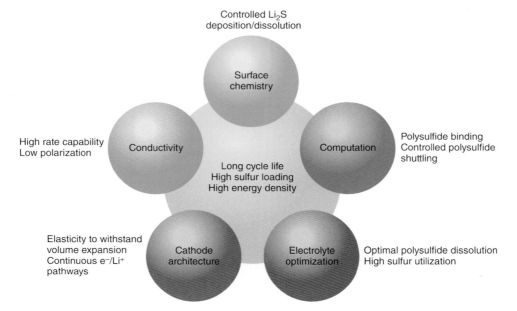

Figure 7.20 Summary of all the methods for improving capacity and performance of Li–S cells. Source: Xu et al. 2013 [25]. Reproduced with permission from RSC.

binders, can also help with protection of the cathode structure [12, 66]. "Engineered cathodes" [16] by Nazar and coworkers show that nanostructured mesoporous carbon–sulfur composites can provide better reversibility by delaying diffusion out of the cathodes. Porous carbon metal oxides and conducting polymers can be used for the capture of PS in the cathode [13]. Another suggestion is to replace a single thick cathode with a layered cathode [13]. There is a carbon cloth in the middle of the cathode that allows for better conductivity. Other ideas include freestanding cathode, layer-by-layer fabrication of 3D cathodes, and cross-linking cathodes [16]. All of these show promise but the techniques are involved and do not show a dramatic improvement in cycle life. Other methods of trapping polysulfides include adsorbing or encapsulating sulfur in various forms, metal complexes, functional binders, and physical barriers to suppress diffusion [67]. A possible technique is to bypass the sulfur cathode directly and use Li_2S cathodes, which can help reduce shuttling [68]. However, the initial discharge and charge has proved to be quite difficult, with further cycles showing similar issues to current Li–S cells. P_2S_5 has been used as a protective layer to protect the lithium anode and promote dissolution of PS [13]. Ionic liquids are seen as another alternative as they stabilize PS molecules against the Li ions and prevent shuttling [13]. However, their high density and cost make them commercially unviable.

A lot of efforts however are concentrated on protection of the lithium metal anode. Lee et al. [69] developed a system based on prior work with lithium air batteries, wherein a composite protective layer (CPL) was coated onto the lithium anode. The alumina particles provide mechanical strength and the copolymer allows fast lithium ion transport to occur through the layer. The effect of the CPL is twofold. The authors estimate that the shear modulus at an indentation of 25 nm is around twice that of lithium metal, which would indicate that dendrite formation is mechanically suppressed. Secondly, the CPL protects the formation of uneven layers of SEI across the anode surface. For an exposed anode, the unevenness in the SEI causes preferential plating of lithium to occur, which leads to the development of lithium dendrites. However, the CPL encourages uniform de-plating and plating across the anode, as can be seen from the results with the high C-rates (which would normally see some dendrite formation due to high polarization).

Another idea is to use polymer electrolytes with a shear modulus greater than that of lithium (109 MPa) [23, 33]. In [70] the electrolyte was composed of a block copolymer that blocked dendrite growth, but had highly conducting nano-channels present, which allowed the conduction of Li ions. However, in order to improve the overall conductivity of the electrolyte the operating temperatures will have to be raised to around 60 C, which might not be recommended for the lithium–sulfur system. Another way to increase the mechanical strength is to use inorganic fillers such as silica particles [71], which form a cross-linked structure as a result of aggregation of silica particles. This also helped increase the conductivity of the electrolyte. The authors of the study believed that the silica particles also helped remove impurities from the system, which inhibited dendrite formation. Another option is LiPON, which has a shear modulus of approximately 77 GPa, and can also suppress dendrite formation mechanically. In a similar vein, Zheng et al. [72] developed a layer of hollow carbon nanospheres, which can be deposited onto a copper current collector. A reduction of impedance is seen via EIS experiments as the carbon layer helps in the formation of a stable layer of SEI. Furthermore, if hollow nanotubes are used the layer of SEI formed is thin, thus reducing the charge-transfer impedance.

One of the methods of suppressing dendrite growth formation is the use of very concentrated electrolyte solution. In the work by Suo et al. [73] a ultrahigh salt concentration of up to 7 M used in Li–S batteries suppressed the formation of dendrites. It is likely that this very high concentration effectively nullifies variations along the surface of the lithium anode and thus promotes even deposition. Qian et al. [74] demonstrated that it was possible to achieve stable cycling of the anode at high rates by using a highly concentrated electrolyte, 4 M LiFSI-DME in particular. The SEI formed with this electrolyte was compact, and did not degrade after a large number of cycles. The impedance rise also indicates that this layer has a good ionic conductivity, which is maintained over a large number of cycles. In typical carbonate-based electrolytes, the lithium deposits in needle-like structures while in the LiFSI-DME electrolytes the lithium deposits in a nodule-like structure, with rounded instead of sharp edges, which prevents them from growing into the separator. Even though the viscosity of the electrolyte is significantly higher the batteries still performed better in terms of rate capability and stability. This was ascribed to the higher concentration of lithium ions, which allowed for high rate stripping and plating, and an improved electrolyte reductive stability due to reduced availability of the reactive solvent and the sacrificial reduction of anions. The only issue with this is the weight: the energy density of the lithium–sulfur cell will be significantly impacted with these concentrations. Park et al. [75] however show results that seem to indicate that increased lithium ion concentration in the electrolyte does not lead to retarded dendritic growth. Their results indicate that it is the viscosity of the solvent that plays a role in dendrite growth, with higher viscosities leading to more dendritic growth. They used a statistical diffusion-limited aggregate (DLA) model for validation. The more viscous the solvent the more collisions will occur in a span of time, and therefore the growth of dendrites becomes more probable. They found that in a more viscous solution longer dendrites were formed, while in a less viscous solvent thick, dense dendrites of shorter height were formed. This was validated by experiments.

Additives such as Ru^+ or Cs^+ ions can prevent the growth of dendrites by forming a "self-healing" electrostatic shield in the vicinity of the dendrite tips [23]. The beauty of this method is that unlike other additives, these are not consumed and therefore last for the duration of cycling. Other additives include halide salts as halide ions can reduce the activation energy barrier for lithium diffusion at the anode interface, and can suppress dendrite growth even at high current densities [23]. As lithium nitrate has been discussed earlier we abjure from any further discussion here.

Other methods for the anode include charging at low current densities to mitigate dendrite growth and pulse charging in order to alter surface morphology and allow for the scattering of polysulfides at the anode interface [23]. Micro-patterning of the anode can help in even deposition as lithium preferentially deposits and strips from pits rather than from dendrites [76].

Finally, we briefly mention the use of recovery cycles here, although it is described in detail elsewhere in this book. Briefly put, when a lithium–sulfur cell is not discharged down to its full capacity and then charged up again, there is an incomplete conversion of lithium sulfide species back up to sulfur. Thus, there is a gradual loss of capacity over time, which manifests as a drop in voltage. However, if a complete "recovery" cycle at full depth of discharge is performed, then the capacity recovers significantly, as all the sulfur trapped as lithium sulfide is released back into the system [77]. Conversely, in [35] they found that keeping a potentiostatic step at the end of discharge helped in full conversion to Li_2S, which causes an increase in life cycle. Hence the complete conversion of species to the end products is a common, and reversible, cause of degradation.

7.6 Future Direction

Lithium–sulfur is a challenging system. It is important to note that in order to make this system viable sulfur loads need to be at least $2\,mg\,cm^{-2}$ or higher [38]. In order to be truly groundbreaking this value has to shift to greater than $5\,mg\,cm^{-2}$. Because the scientific community is still working on the basic reaction mechanisms, the capacity fade of these cells is not being given as much attention. While maximizing the capacity does mean that fewer cycles will be required, ultimately, performance has to remain unchanging for at least 50–100 cycles before products using this technology can become viable. Therefore, it would be wise to keep cycle life in mind while designing a lithium–sulfur system.

In terms of the components that seem to cause the most trouble, it is a toss-up between the cathode and the anode. As of now, more attention has been paid to the anode, because cathode collapse always follows anode collapse. However, given the need for higher cathode loadings, we suspect that it will be the cathode structure that will be the dominant cause for capacity fade. Thus, it would be important to keep the cyclability of the cathode in mind while designing new structures and greater sulfur loadings.

References

1 Carbone, L., Peng, J., Agostini, M. et al. (2017). Carbon composites for a high energy lithium sulfur battery with a glyme based electrolyte. *ChemElectroChem* 4 (1): 209–215.

2 Warner, J. (2015). *The Handbook of Lithium-Ion Battery Pack Design: Chemistry, Components, Types and Terminology*. Elsevier.

3 Busche, M., Adelhelm, P., Sommer, H. et al. (2014). Systematical electrochemical study on the parasitic shuttle effect in lithium sulfur cells at different temperatures and different rates. *Journal of Power Sources* 259: 289–299.

4 Wild, M., O'Neill, L., Zhang, T. et al. (2015). Lithium sulfur batteries, a mechanistic review. *Energy and Environmental Science* 8 (12): 3477–3494.

5 Marinescu, M., Zhang, T., and Offer, G.J. (2016). A zero dimensional model of lithium–sulfur batteries during charge and discharge. *Physical Chemistry Chemical Physics* 18 (1): 584–593.

6 Walus, S., Barchasz, C., Colin, J.-F. et al. (2013). New insight into the working mechanism of lithium–sulfur batteries: in-situ and operando X-ray diffraction characterization. *Chemical Communications* 49: 7899–7901.

7 Kumaresan, K., Mikhaylik, Y., and White, R.E. (2008). A mathematical model for a lithium–sulfur cell. *Journal of the Electrochemical Society* 155 (8): A576–A582.

8 Fan, F.Y., Carter, W.C., and Chiang, Y.-M. (2015). Mechanism and kinetics of Li_2S precipitation in lithium–sulfur batteries. *Advanced Materials* 27 (35): 5203–5209.

9 Walus, S., Offer, G., Hunt, I. et al. (2017). Volumetric expansion of lithium–sulfur cell during operation – fundamental insight into applicable characteristics. *Energy Storage Materials* https://doi.org/10.1016/j.ensm.2017.05.017.

10 Zhang, T., Marinescu, M., Walus, S. et al. (2018). What limits the rate capability of Li–S batteries during discharge: charge transfer or mass transfer? *Journal of the Electrochemical Society* 165 (1): A6001–A6004.

11 Eroglu, D., Zavadil, K.R., and Gallagher, K.G. (2015). Critical link between materials chemistry and cell level design for high energy density and low cost lithium sulfur transportation battery. *Journal of the Electrochemical Society* 162 (6): A982–A990.

12 Pang, Q., Liang, X., Kwok, C.Y. et al. (2016). A comprehensive approach toward stable lithium–sulfur batteries with high volumetric energy density. *Advanced Energy Materials* 7 (6): 1601630.

13 Zhang, S.S. (2013). Liquid electrolyte lithium/sulfur battery: fundamental chemistry, problems, and solutions. *Journal of Power Sources* 231: 153–162.

14 Sun, K., Cama, C.A., Huang, J. et al. (2017). Effect of carbon and binder on high sulfur loading electrode for Li–S Battery Technology. *Electrochimica Acta* 235: 399–408.

15 Hagen, M., Fanz, P., and Tubke, J. (2014). Cell energy density and electrolyte/sulfur ratio in Li–S cells. *Journal of Power Sources* 264: 30–34.

16 Pang, Q., Liang, X., Kwok, C.Y., and Nazar, L.F. (2016). Advances in lithium–sulfur batteries based on multifunctional cathodes and electrolytes. *Nature Energy* 1: https://doi.org/10.1038/nenergy.2016.132.

17 Fan, F.Y. and Chiang, Y.-M. (2017). Electrodeposition kinetics in Li–S batteries: effects of low electrolyte/sulfur ratios and deposition surface composition. *Journal of the Electrochemical Society* 164 (4): A917–A922.

18 Yan, J., Liu, X., and Li, B. (2016). Capacity fade analysis of sulfur cathodes in lithium–sulfur batteries. *Advanced Science* 3 (12): 1600101.

19 Bruce, P.G., Freunberger, S.A., Hardwick, L.J., and Tarascon, J.-M. (2012). Li–O$_2$ and Li–S batteries with high energy storage. *Nature Materials* https://doi.org/10.1038/NMAT3191.

20 Noh, H., Song, J., Park, J.-K., and Kim, H.-T. (2015). A new insight on capacity fading of lithium–sulfur batteries: the effect of Li$_2$S phase structure. *Journal of Power Sources* 293: 329–335.

21 Ren, Y.X., Zhao, T.S., Liu, M. et al. (2016). Modeling of lithium–sulfur batteries incorporating the effect of Li$_2$S precipitation. *Journal of Power Sources* 336: 115–12522.

22 Orsini, F., Du Pasquier, A., Beaudoin, B. et al. (1998). In situ scanning electron microscopy (SEM) observation of interfaces within plastic lithium batteries. *Journal of Power Sources* 76 (1): 19–29.

23 Li, Z., Huang, J., Liaw, B.Y. et al. (2014). A review of lithium deposition in lithium-ion and lithium metal secondary batteries. *Journal of Power Sources* 254: 168–182.

24 Wood, K.N., Kazyak, E., Chadwick, A.F. et al. (2016). Dendrites and pits: untangling the complex behaviour of lithium metal anodes though operando video microscopy. *ACS Central Science* 2 (11): 790–801.

25 Xu, W., Wang, J., Ding, F. et al. (2013). Lithium metal anodes for rechargeable batteries. *Energy and Environmental Science* 7: 513–537.

26 Chazalviel, J.-N. (1990). Electrochemical aspects of the generation of ramified metallic electrodeposits. *Physical Review A* 42: 7355.

27 Rosso, M., Gobron, T., Brissot, C. et al. (2001). Onset of dendritic growth in lithium/polymer cells. *Journal of Power Sources* 97-98: 804–806.

28 Rosso, M., Chassaing, E., Chazalviel, J.-N., and Gobron, T. (2002). Onset of current-driven concentration instabilities in thin cell electrodeposition with small inter-electrode distance. *Electrochimica Acta* 47 (8): 1267–1273.

29 Gireaud, L., Grugeon, S., Laruelle, S. et al. (2006). Lithium metal stripping/plating mechanisms studies: a metallurgical approach. *Electrochemistry Communications* 8: 1639–1649.

30 Orsini, F., Du Pasquier, A., Beaudoin, B. et al. (1998). In situ SEM study of the interfaces in plastic lithium cells. *Journal of Power Sources* 81–82: 918–921.

31 Han, Y., Duan, X., Li, Y. et al. (2015). Effects of sulfur loading on the corrosion behaviors of metal lithium anode in lithium–sulfur batteries. *Materials Research Bulletin* 68: 160–165.

32 Li, G., Li, Z., Zhang, B., and Lin, Z. (2015). Developments of electrolyte systems for lithium–sulfur batteries: a review. *Frontiers in Energy Research* https://doi.org/10.3389/fenrg.2015.00005.

33 Lang, J., Qi, L., Luo, Y., and Wu, H. (2017). High performance lithium metal anode: progress and prospects. *Energy Storage Materials* 7: 115–129.

34 Schneider, H., Weiss, T., Scordilis-Kelley, C. et al. (2017). Electrolyte decomposition and gas evolution in a lithium–sulfur cell upon long-term cycling. *Electrochimica Acta* 243: 26–32.

35 Poux, T., Novak, P., and Trabesinger, S. (2016). Pitfalls in Li–S rate-capability evaluation. *Journal of The Electrochemistry Society* 163 (7): A1139–A1145.

36 Zhang, S.S. (2016). A new finding on the role of LiNO$_3$ in lithium–sulfur battery. *Journal of Power Sources* 322: 99–105.

37 Li, W., Yao, H., Yan, K. et al. (2015). The synergetic effect of lithium polysulfide and lithium nitrate to prevent lithium dendrite growth. *Nature Communications* https://doi.org/10.1038/ncomms8436.

38 Hagen, M., Hanselmann, D., Ahlbrecht, K. et al. (2015). Lithium–sulfur cells: the gap between the state-of-the-art and the requirements for high energy battery cells. *Advanced Energy Materials* https://doi.org/10.1002/aenm.201401986.

39 Stroe, D.I., Knap, V., Swierczynski, M., and Schaltz, E. (2017). Thermal behaviour and heat generation modeling of lithium–sulfur batteries. ECS Meeting, Abstract MA2017-01 529.

40 Mikhaylik, Y.V. and Akridge, J.R. (2004). Polysulfide shuttle study in the Li/S battery system. *Journal of the Electrochemical Society* 151 (11): A1969–A1976.

41 Cheng, X.-B., Yan, C., Huang, J.-Q. et al. (2017). The gap between long lifespan Li–S coin and pouch cells: the importance of lithium metal anode protection. *Energy Storage Materials* 6: 18–25.

42 Guo, J., Li, Z., and Pech, M. (2015). A Bayesian approach for Li-Ion battery capacity fade modelling and cycles to failure prognostics. *Journal of Power Sources* 281: 173–184.

43 Ghaznavi, M. and Chen, P. (2014). Sensitivity analysis of a mathematical model of lithium–sulfur cells. Part I: applied discharge current and cathode conductivity. *Journal of Power Sources* 257: 394–401.

44 Ghaznavi, M. and Chen, P. (2014). Sensitivity analysis of a mathematical model of lithium–sulfur cells. Part II: precipitation reaction kinetics and sulfur content. *Journal of Power Sources* 257: 402–411.

45 Ghaznavi, M. and Chen, P. (2014). Sensitivity analysis of a mathematical model of lithium–sulfur cells. Part III: electrochemical reaction kinetics, transport properties and charging. *Electrochimica Acta* 137: 575–585.

46 Moy, D., Manivannan, A., and Narayanan, S.R. (2015). Direct measurement of polysulfide shuttle current: a window into understanding the performance of lithium–sulfur cells. *Journal of the Electrochemical Society* 162 (1): A1–A7.

47 Hofmann, A.F., Fronczek, D.N., and Bessler, W.G. (2014). Mechanistic modelling of polysulfide shuttle and capacity loss in lithium–sulfur batteries. *Journal of Power Sources* 259: 300–310.

48 Risse, S., Angioletti-Uberti, S., Dzubiella, J., and Ballauff, M. (2014). Capacity fading in lithium/sulfur batteries: a linear four-state model. *Journal of Power Sources* 267: 648–654.

49 Risse, S., Canas, N.A., Wagner, N. et al. (2016). Correlation of capacity fading processes and electrochemical impedance spectra in lithium/sulfur cells. *Journal of Power Sources* 323: 107–114.

50 Yoo, K., Song, M.-K., Cairns, E.J., and Dutta, P. (2016). Numerical and experimental investigation of performance characteristics of lithium/sulfur cells. *Electrochimica Acta* 213: 174–185.

51 Barton, J.L. and Bockris, J.O.'.M. (1962). The electrolytic growth of dendrites for ionic solutions. *Proceedings of The Royal Society A* 268 (1335).

52 Jana, A., Ely, D.R., and Garcia, R.E. (2015). Dendrite-separator interactions in lithium-based batteries. *Journal of Power Sources* 275: 912–921.

53 Monroe, C. and Newman, J. (2003). Dendrite growth in lithium/polymer systems: a propagation model for liquid electrolytes under galvanostatic conditions. *Journal of The Electrochemical Society* 150 (10): A1377–A1384.

54 Monroe, C. and Newman, J. (2004). The effect of interfacial deformation on electrodeposition kinetics. *Journal of The Electrochemical Society* 151 (6): A880–A886.

55 Monroe, C. and Newman, J. (2005). The impact of elastic deformation on deposition kinetics at lithium/polymer interfaces. *Journal of The Electrochemical Society* 152 (2): A396–A404.

56 Ferrese, A., Albertus, P., Christensen, J., and Newman, J. (2012). Lithium redistribution in lithium-metal batteries. *Journal of The Electrochemical Society* 159 (10): A1615–A1623.

57 Ferrese, A. and Newman, J. (2014). Modeling lithium movement over multiple cycles in a lithium-metal battery. *Journal of The Electrochemistry Society* 161 (6): A948–A954.

58 Ely, D.R., Jana, A., and Garcia, R.E. (2014). Phase field kinetics of lithium electrodeposits. *Journal of Power Sources* 272: 581–594.

59 Deng, Z., Zhang, Z., Lai, Y. et al. (2013). Electrochemical impedance spectroscopy study of a lithium/sulfur battery: modelling and analysis of capacity fading. *Journal of the Electrochemical Society* 160 (4): A553–A558.

60 Canas, N.A., Hirose, K., Pascucci, B. et al. (2013). Investigations of lithium–sulfur batteries using electrochemical impedance spectroscopy. *Electrochimica Acta* 97: 42–51.

61 Kolosnitsyn, V.S., Kuzmina, E.V., and Mochalov, S.E. (2014). Determination of lithium sulfur batteries internal resistance by the pulsed method during galvanostatic cycling. *Journal of Power Sources* 252: 28–34.

62 Knap, V., Kalogiannis, T., Purkayastha, R. et al. (1980). Transferring the incremental capacity analysis to lithium–sulfur batteries. ECS Meeting Abstracts, Abstract MA2017-01.

63 Wu, B., Yufit, V., Merla, Y. et al. (2015). Differential thermal voltammetry for tracking of degradation in lithium-ion batteries. *Journal of Power Sources* 273: 495–501.

64 Xiao, X., Zhang, T., Marinescu, M., and Offer, G. Differential thermal voltammetry for tracking degradation in lithium sulfur batteries. (Submitted).

65 Varzi, A., Passerini, S., Ganesan, A., and Sahijumon, M. (2016). Graphene derived carbon confined sulfur cathodes for lithium–sulfur batteries: electrochemical impedance studies. *Electrochimica Acta* https://doi.org/10.1016/j.electacta.2016.08.030.

66 Fu, C. and Guo, J. (2016). Challenges and current development of sulfur cathode in lithium–sulfur battery. *Current Opinion in Chemical Engineering* 13: 53–62.

67 Li, Z., Deng, S., Ke, H. et al. (2017). Explore the influence of coverage percentage of sulfur electrode on the cycle performance of lithium–sulfur batteries. *Journal of Power Sources* 347: 238–246.

68 Vizintin, A., Chabanne, L., Tchernykova, E. et al. (2017). The mechanism of Li_2S activation in lithium–sulfur batteries: can we avoid the polysulfide formation? *Journal of Power Sources* 344: 208–217.

69 Lee, H., Lee, D.J., Kim, Y.-J. et al. (2015). A simple composite protective layer coating that enhances the cycling stability of lithium metal batteries. *Journal of Power Sources* 284: 103–108.

70 Balsara N. (2008). Polymer Electrolytes for High Energy Density Lithium Batteries. http://www1.eere.energy.gov/vehiclesandfuels/pdfs/merit_review_2008/exploratory_battery/merit08_balsara.pdf (accessed 07 August 2018).

71 Li, Y., Fedkiw, P.S., and Khan, S.A. (2002). Lithium/V6O13 cells using silica nanoparticle-based composite electrolyte. *Electrochimica Acta* 47: 3853–3861.

72 Zheng, G., Lee, S.w., Liang, Z. et al. (2014). Interconnected hollow carbon nanospheres for stable lithium metal anodes. *Nature Nanotechnology* 152: 1–6.

73 Suo, L., Hu, Y.S., Li, H. et al. (2013). A new class of solvent-in-salt electrolyte for high-energy rechargeable metallic lithium batteries. *Nature Communications* 4: 1481.

74 Qian, J., Henderson, W.A., Xu, W. et al. (2015). High rate and stable cycling of lithium metal anode. *Nature Communications* 6 (6362): 1–9.

75 Park, M.S., Ma, S.B., Lee, D.J. et al. (2014). A highly reversible lithium metal anode. *Nature* 4 (3815): 1–8.

76 Li, Y., Jiao, J., Bi, J. et al. (2017). Controlled deposition of Li metal. *Nano Energy* 32: 241–246.

77 Marinescu, M., O'Neill, L., Zhang, T. et al. (2018). Irreversible vs reversible capacity fade of lithium–sulfur batteries during cycling. The effects of precipitation and shuttle. *Journal of the Electrochemical Society* 165: A6107–A6118.

Part III

Modeling

For an introduction to the basic electrochemical theory, see Chapter 1, Electrochemical theory and physics. Part III begins with Chapter 8, a review of the current state of the art in physics-based modeling of Li–S systems. 0D models up to multi-scale models are discussed along with the limitations and knowledge that can be retrieved from simulating the underlying mechanisms of the lithium–sulfur cell. Chapter 7 includes a review of models that specifically focus on degradation mechanisms.

Chapter 9 looks at the development of state estimation models for application in battery management systems to determine real-time state of charge and state of health estimations, an important enabler for application engineers. Predictive modeling techniques are taken from computer science, and better informed by the development of physics-based models and knowledge of the underlying mechanisms (Part II).

Modeling of batteries stems from an understanding of the underlying mechanisms derived from analytical studies (Part II) and materials research (Part I) and can be practically applied in the control systems and battery management systems of applications (Part IV). Working predictive models of differing battery technologies are used by application engineers when developing products and choosing components to optimize battery performance at the systems level. Capturing the performance of a lithium–sulfur battery in an accurate predictive model will lead to better understanding and uptake of the technology by engineers and reduce the testing burden for cell manufacturers.

Lithium–Sulfur Batteries, First Edition. Edited by Mark Wild and Gregory J. Offer.
© 2019 John Wiley & Sons Ltd. Published 2019 by John Wiley & Sons Ltd.

8

Lithium–Sulfur Model Development

Teng Zhang[1], Monica Marinescu[2] and Gregory J. Offer[2]

[1] *University of Surrey, Department of Mechanical Engineering Sciences, 388 Stag Hill, Guildford GU2 7XH, UK*
[2] *Imperial College, Department of Mechanical Engineering, London, UK*

8.1   Introduction

Modeling Li–S cells presents many challenges due to their complex mechanisms including electrochemical reactions, ionic transport, and morphology change. So far, Li–S modeling research is still at the early stages and no single Li–S model has been able to capture all aspects of a Li–S cell's complex behaviors. A useful Li–S model should at least retrieve the key features during charge and discharge qualitatively. In terms of physical mechanisms, a basic Li–S model should consider the following:

Cathode electrochemistry: the reduction and oxidation of sulfur is a multistep process involving multiple intermediate polysulfide species. Chemical reactions also occur among different polysulfides in the form of disproportionation and association reactions [1, 2]. Owing to the complexity of the electrochemical and chemical reactions involved, Li–S models employ simplified, linear reaction mechanisms involving a few representative polysulfide species. The thermodynamics of the cathodic reactions are typically described with the Nernst equation, whereas the charge-transfer kinetics is modeled using the Butler–Volmer equation.

Precipitation and dissolution: the formation and dissolution of the final discharge product, Li_2S, is a key feature in Li–S cells that effectively controls the amount of dissolved ionic species in the electrolyte as well as the reaction potentials during charge/discharge [3]. The insulating nature of Li_2S also leads to electrode surface passivation and change in charge-transfer kinetics. Furthermore, the accumulation of Li_2S could cause pore-blocking in the cathode, which increases transport resistance. Li–S models need to capture the rates of Li_2S precipitation and dissolution in order to predict the basic behaviors of Li–S cells [4].

Shuttle: polysulfide shuttle significantly affects a Li–S cell's charge behavior and degradation rate at low currents. The shuttle process also manifests as self-discharge during battery storage and discharge [5, 6]. While shuttle is a complex process involving both transport of polysulfides between electrodes and parasitic reactions at the anode, this process can be modeled at the simplest level as a chemical reaction between high- and low-order polysulfide species.

As we shall see later in this chapter, a basic Li–S model capturing the above phenomena is able to reproduce the key features of a Li–S cell. More advanced Li–S models should further include the following:

Ionic transport: the electrolyte of a Li–S cell contains multiple dissolved polysulfides whose concentrations vary significantly during charge/discharge. Modeling ionic transport in Li–S cells – including the polysulfide transport during shuttle – is therefore difficult given no practical modeling framework for multicomponent transport in a concentrated electrolyte. Existing Li–S models are therefore based on simplified dilute-solution theory to describe ionic transport in Li–S cells. Modeling ionic transport is further complicated by the hierarchical nature of the cathode morphology: while Li$^+$ transport and shuttle occur primarily at the cell scale, most of the polysulfides diffuse and react inside the porous carbon/sulfur particles at the particle scale. Multi-scale modeling frameworks are recently being developed to describe ionic transport at both the cell and the particle scales [7, 8].

Anode electrochemistry: anode is where most of the Li–S degradation mechanisms take place. Repeated stripping and plating of lithium during battery cycling roughens the anode surface morphology and exposes fresh lithium to electrolyte. This subsequently accelerates parasitic reactions between lithium and the electrolyte, giving rise to continued solid–electrolyte interface (SEI) growth, formation of dead lithium, and anode thickness increase [9]. The parasitic shuttle reactions also occur at the anode between lithium and various polysulfides, leading to irreversible precipitation of Li$_2$S here. Capturing the anodic reactions is important for modeling Li–S cell degradation and volume changes.

Temperature dependence: like other battery chemistries, Li–S cell behaviors are sensitive to temperature and the operation of Li–S cells generates heat. For example, the shuttle rate is known to increase significantly with temperature, and the shuttle process itself is the most important heat generation mechanism during low current charge [5]. Most of the electrochemical and chemical processes in Li–S cells, such as charge transfer and Li$_2$S precipitation, are also temperature dependent. Thermally coupled Li–S models are therefore essential for predicting battery performance in real applications.

Modeling the range of complex reactions, transport, and the phase-change phenomena described above requires relevant physical parameters, most of which are difficult to obtain experimentally. For example, measuring the diffusion coefficient of a polysulfide species is challenging due to the difficulty in isolating individual polysulfide species. Likewise, it is difficult to determine the equilibrium potential and exchange current density of an individual polysulfide oxidation/reduction reaction because multiple electrochemical and chemical reactions occur simultaneously during cell charge/discharge. Consequently, most parameters in existing Li–S models are assumed quantities, and the models only capture Li–S behaviors qualitatively rather than quantitatively. Advanced Li–S models considering additional mechanisms require correspondingly more assumed parameters.

Despite the challenges in modeling and parameterizing complex Li–S mechanisms, existing Li–S models with various levels of simplifications have proved to be useful in understanding Li–S cells' cycling characteristics, performance bottlenecks, and degradation mechanisms [3, 4, 6–8, 10–17]. In this chapter, we introduce the basic elements of Li–S modeling with the use of a simple, zero-dimensional (0D) model. We first describe the mathematical formulation of the 0D model and its charge/discharge predictions in Section 8.2. In Section 8.3, we extend the 0D model to include voltage loss mechanisms arising from electrolyte resistance, surface passivation, and transport limitation. We also briefly discuss 1D and higher dimensional models in Section 8.4.

8.2 Zero-Dimensional Model

The large number of unknown physical properties and the unclear reaction mechanisms of Li–S cells obscure the interpretability and validity of Li–S models. It is therefore desirable to develop a simple Li–S model with a minimal number of parameters and simple reaction mechanisms to reproduce the key features of a Li–S cell charge/discharge. The 0D model represents one of the simplest Li–S models. It considers leading-order effects of electrochemical reactions, polysulfide shuttle, and precipitation/dissolution of Li_2S in the cathode while ignoring ionic transport. The 0D model qualitatively captures the charge/discharge characteristics of a Li–S cell at low current when ionic transport is not limiting. In the following, we introduce the mathematical formulation of the 0D model and demonstrate the model-predicted charge/discharge behaviors of a Li–S cell.

8.2.1 Model Formulation

8.2.1.1 Electrochemical Reactions

Li–S cells exhibit an upper and a lower voltage plateau during charge/discharge. To capture this feature with the simplest possible reaction mechanism, the model assumes a two-step reaction chain with a single electrochemical reaction dominating each voltage plateau:

$$S_8^0 + 4e^- \leftrightarrow 2S_4^{2-} \tag{8.1}$$

$$S_4^{2-} + 4e^- \leftrightarrow S_2^{2-} + 2S^{2-} \downarrow \tag{8.2}$$

This simple reaction mechanism was first introduced by Mikhaylik and Akridge [5] to model the equilibrium potential of a Li–S cathode. Additional reaction steps can be added into the model to represent more realistic reaction mechanisms, at the expense of more computation time and fitting parameters.

To describe the equilibrium potentials for the two reactions, we use the Nernst equations of the following form [6, 17]:

$$E_H = E_H^0 + \frac{RT}{4F} \ln \left(f_H \frac{S_8^0}{(S_4^{2-})^2} \right) \tag{8.3}$$

$$E_L = E_L^0 + \frac{RT}{4F} \ln \left(f_L \frac{(S_4^{2-})}{(S_2^{2-})(S^{2-})^2} \right) \tag{8.4}$$

where E_H^0 and E_L^0 are the standard potentials for Eqs. (8.1) and (8.2), respectively, R is the gas constant, F is the Faraday constant, T is the temperature, and S^{2-}, S_2^{2-}, S_4^{2-}, and S_8^0 are the masses of sulfur dissolved in the electrolyte in the respective forms. The constants f_H and f_L convert mass in grams into concentrations:

$$f_H = \frac{n_{S4}^2 M_{S8} v}{n_{S8}} \tag{8.5}$$

$$f_L = \frac{n_S^2 n_{S2} M_{S8}^2 v^2}{n_{S4}} \tag{8.6}$$

where v is the volume of electrolyte in the system, M_{S8} is the molar mass of sulfur, and the n represents the number of atoms per S_8^0 molecule and S^{2-}, S_2^{2-}, S_4^{2-} ions. The use of species concentrations rather than activities in the Nernst equations implies the assumption of dilute-solution theory.

The currents related to the two electrochemical reactions are described by the Butler–Volmer equation [6]:

$$i_H = -2i_{H,0} a_r \sinh\left(\frac{n_e F \eta_H}{2RT}\right) \tag{8.7}$$

$$i_L = -2i_{L,0} a_r \sinh\left(\frac{n_e F \eta_L}{2RT}\right) \tag{8.8}$$

Here, n_e is the number of electrons transferred in each reaction, which equals four for both reactions, $i_{H,0}$ and $i_{H,0}$ are the exchange current densities that are assumed to be constant, η_L and η_H are the activation overpotentials, and a_r is the active electrochemical surface area in the cathode. The activation overpotential is given by the difference between the terminal voltage of the cell and the Nernst potential:

$$\eta_H = V - E_H \tag{8.9}$$

$$\eta_L = V - E_L \tag{8.10}$$

In writing Eqs. (8.9) and (8.10), it is assumed that the anode potential – which is set to be the zero reference at equilibrium – does not vary significantly during charge/discharge such that the cell's equilibrium potential is dominated only by the cathodic reactions [10]. It is also assumed that the voltage drop across the electrolyte due to its Ohmic resistance is negligible. Charge conservation dictates that the measured cell current I is given by the combined contribution of the two reactions:

$$I = i_H + i_L \tag{8.11}$$

8.2.1.2 Shuttle and Precipitation

The 0D model treats both shuttle and precipitation as chemical reactions. Shuttle involves the diffusion of larger polysulfides from the cathode to the anode, where they react with metallic lithium and other polysulfide species to produce smaller polysulfides. As a result, polysulfide species are transformed without the production of electric current. The present model does not consider ionic transport and contains only one type of higher order polysulfide, S_8^0. The shuttle is therefore represented by a simple conversion between elemental sulfur and S_4^{2-} : $S_8^0 \rightarrow S_4^{2-}$.

The rate of shuttle is described with a first-order reaction rate:

$$\text{shuttle rate} = k_s S_8^0 \tag{8.12}$$

with k_s being the shuttle rate constant.

Following White and coworkers [10], the precipitation of Li$_2$S is described with a phenomenological nucleation and growth model:

$$\text{precipitation rate} = \frac{1}{v\rho_s}k_pS_p(S^{2-} - S_*^{2-})$$

(8.13)

Here, S_p is the mass of precipitated sulfur, ρ_s is its density, and S_*^{2-} is the saturation mass of S^{2-} in solution. Precipitation occurs when the concentration of S^{2-} exceeds the saturation concentration, while dissolution occurs when the S^{2-} concentration drops below the saturation point. The proportionality of the precipitation rate to the mass of precipitated sulfur imitates the initial nucleation process where the nucleation rate depends on the amount of precipitates already formed.

8.2.1.3 Time Evolution of Species

The electrochemical reactions during charge/discharge, together with the chemical reactions that take account of shuttle and precipitation, can be described with the following time-dependent governing equations for the five sulfur species in the system:

$$\frac{dS_8^0}{dt} = -\frac{n_{S8}M_{S8}}{n_eF}i_H - k_sS_8^0$$

(8.14)

$$\frac{dS_4^{2-}}{dt} = \frac{n_{S8}M_{S8}}{n_eF}i_H + k_sS_8^0 - \frac{n_{S4}M_{S8}}{n_eF}i_L$$

(8.15)

$$\frac{dS_2^{2-}}{dt} = \frac{n_{S2}M_{S8}}{n_eF}i_L$$

(8.16)

$$\frac{dS^{2-}}{dt} = 2\frac{n_{S2}M_{S8}}{n_eF}i_L - \frac{1}{v\rho_s}k_pS_p(S^{2-} - S_*^{2-})$$

(8.17)

$$\frac{dS_p}{dt} = \frac{1}{v\rho_s}k_pS_p(S^{2-} - S_*^{2-})$$

(8.18)

8.2.1.4 Model Implementation

The differential algebraic equations introduced above can be solved for the 12 unknowns: the Nernst potentials E_H, E_L, the two reaction currents i_H, i_L, the cell voltage V, the overpotentials η_H, η_L, and the mass of the five species S^{2-}, S_2^{2-}, S_4^{2-}, and S_8^0 and S_p. The Jacobian for the system is calculated analytically and the system is solved in MATLAB using a second-order solver. The model parameters are given in Table 8.1. The parameters associated with the reaction and precipitation kinetics serve as fitting parameters whose values are dependent on the measured cell. The initial conditions for all variables are calculated self-consistently from chosen values of V, S_8^0, and S_p for discharge, and V, S_p, and S^{2-} for charge.

8.2.2 Basic Charge/Discharge Behaviors

The 0D model calculates the evolution of sulfur species during charge/discharge and the corresponding cell potential. The simulated cell voltage during a 0.5 C discharge is shown in Figure 8.1a for different precipitation parameters (k_p and S_*^{2-}), along with the corresponding amounts of precipitated sulfur in Figure 8.1b. The model retrieves the

Table 8.1 Parameters for the 0D model.

Notation	Name	Units	Value
Cell properties			
a_r	Active reaction area per cell	m^2	0.96
f_H	Dimensionality factor L	g L mol^{-1}	0.73
f_L	Dimensionality factor H	$\text{g}^2 \text{L}^2 \text{mol}^{-1}$	0.067
v	Electrolyte volume per cell	L	0.0114
m_s	Mass of active sulfur per cell	g	2.7
ρ_s	Density of precipitated sulfur	g L^{-1}	2×10^{-3}
Kinetic properties			
E_H^0	Standard potential H	V	2.35
E_L^0	Standard potential L	V	2.195
$i_{H,0}$	Exchange current density H	A m^{-2}	10
$i_{l,0}$	Exchange current density L	A m^{-2}	5
Shuttle and precipitation parameters			
S_*^{2-}	S^{2-} saturation mass	g	1×10^{-4}
k_p	Precipitation rate	s^{-1}	100
k_s	Shuttle rate	s^{-1}	2×10^{-4}
Operational parameters			
I	External current	A	Variable
T	Temperature	K	298
Variables			
E_H, E_L	Nernst potentials	V	
i_H, i_L	Current contributions	A	
η_H, η_L	Overpotentials	V	
V	Cell voltage	V	
$S_8^0, S_4^{2-}, S_2^{2-}, S^{2-}$	Mass of dissolved sulfur	g	
S_p	Mass of precipitated sulfur	g	

typical upper and lower plateaus of a Li–S cell's discharge profile. From the figure, it is clear that the shape of the lower voltage plateau is sensitive to the thermodynamics and kinetics of precipitation. In the absence of precipitation, the cell voltage continues to decrease throughout the lower plateau as a result of increasing amount of smaller polysulfides (S_2^{2-}, S^{2-}) dissolving into the electrolyte. Precipitation removes these smaller polysulfides from the electrolyte, thereby raising the cell potential based on the Nernst equations (Eqs. (8.3) and (8.4)). Based on Figure 8.1a, a larger precipitation

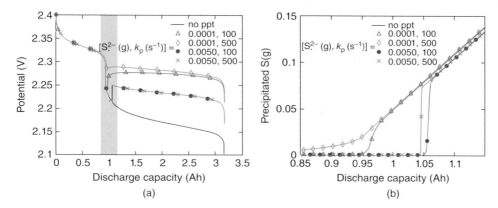

Figure 8.1 (a) Simulated constant-current discharge curves at 1.7 A (0.5 C) for various precipitation parameter values (no ppt – no precipitation), (b) and the corresponding evolution of the precipitated species. The *x*-axis in (b) corresponds to the shaded interval in (a).

rate corresponds to a higher cell voltage in the lower plateau. On the other hand, increasing the saturation mass not only leads to a drop in the lower plateau voltage but also a delay in the onset of the lower plateau. From Figure 8.1b, this delay is clearly related to the delayed onset of precipitation: larger saturation mass allows more S^{2-} to form before they could precipitate, thereby delaying the effect of precipitation on the cell voltage. The increased amount of dissolved S^{2-} also lowers the cell potential during discharge, as described by the Nernst equations. These simulated discharge behaviors demonstrate the significant role of precipitation in determining the shape of a Li–S cell's discharge curve.

The 0D model also captures the charge profile of a Li–S cell at different currents. As shown in Figure 8.2a, the charge capacity of the cell is reduced at a higher charge rate of 1 C, but is seemingly infinite at the lower rate of 0.1 C. The mechanism behind the observed behaviors is understood by examining the mass profiles of the species shown in Figure 8.2b,c. During the fast charge, the dissolution of precipitated sulfur, S_p (in the form of Li_2S), is not fast enough compared to the rates of current-dependent electro-chemical reactions, Eqs. (8.1) and (8.2). By the time all S_4^{2-} is consumed by reaction Eq. (8.1), there is still unutilized S_p in the form of Li_2S and consequently the charge ends with a reduced capacity. Furthermore, as the dissolution bottleneck forces the early formation of the high plateau species S_8^0, the boundary between the high and low plateaus becomes obscured during fast charge. During slow charge, on the other hand, all S_p can be converted into S_4^{2-} due to the slower electrochemical reactions; yet, only a fraction of S_4^{2-} can be converted back into S_8^0. This is because at low charge current the rate of shuttle becomes comparable to the rate of S_4^{2-} oxidation (Eq. (8.1)), causing continuous consumption of S_8^0 and the production of S_4^{2-}. As a result, the amounts of S_8^0 and S_4^{2-} reach a dynamic equilibrium toward the end of charge that gives rise to a flat upper plateau. While these simulation results are only qualitative, they provide a mechanistic understanding of the commonly observed Li–S behaviors during fast and slow charging.

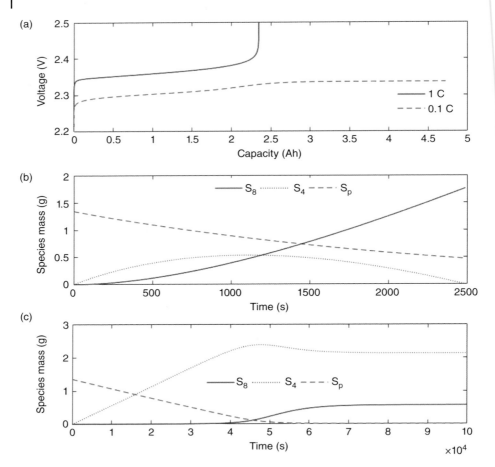

Figure 8.2 (a) Cell voltage during a 1 and a 0.1 C charge. (b, c) The corresponding sulfur species mass during the 1 and 0.1 C charge.

8.3 Modeling Voltage Loss in Li–S Cells

The 0D model reproduces some basic features of the Li–S charge/discharge curves, but it does not capture the charge/discharge voltage accurately. This limitation is mainly due to (i) the simple two-step reaction mechanism assumed in the model that has fewer reaction steps than the complex actual reaction mechanisms in Li–S cells, and (ii) additional voltage drop contributions from the electrolyte resistance, electrode surface passivation, and transport limitation not yet being included in the model. In this section, we demonstrate how the 0D model can be further extended to capture these major voltage loss mechanisms.

To better reproduce the shape of Li–S charge/discharge curves, in this section the 0D model is modified to include additional reaction steps proposed by White and coworkers [10]:

$$Li \leftrightarrow Li^+ + e^- \tag{8.19}$$

$$0.5S_8^0 + e^- \leftrightarrow 0.5S_8^{2-} \tag{8.20}$$

$$3/2S_8^{2-} + e^- \leftrightarrow 2S_6^{2-} \tag{8.21}$$

$$S_6^{2-} + e^- \leftrightarrow 3/2S_4^{2-} \tag{8.22}$$

$$0.5S_4^{2-} + e^- \leftrightarrow 2S_2^{2-} \tag{8.23}$$

$$0.5S_2^0 + e^- \leftrightarrow 2S^{2-} \tag{8.24}$$

$$Li^+ + S^{2-} \leftrightarrow Li_2S \downarrow \tag{8.25}$$

While still representing a simplified reaction mechanism, Eqs. (8.19)–(8.25) include three additional cathodic reactions with two additional sulfur species, S_8^{2-} and S_6^{2-}, as well as the anodic reaction of lithium dissolution and deposition. These reactions require the corresponding Nernst equations, Butler–Volmer equations, and mass conservation equations to include additional reaction potentials, reaction currents, and species quantities. The detailed mathematical formulation for the 0D model with the extended reaction mechanism can be found in [3]. We hereon focus on integrating various voltage loss mechanisms into the 0D model.

8.3.1 Electrolyte Resistance

Ohmic resistance contributes to a significant voltage loss mechanism in Li–S cells. The ohmic resistance of a Li–S cell is dominated by the resistance of its electrolyte, which can be measured by high-frequency electrochemical impedance spectroscopy. As shown in Figure 8.3, the electrolyte resistance first increases and then decreases during discharge, with a peak appearing at the transition between the voltage plateaus. This volcano-shaped resistance profile is a characteristic feature of Li–S cells, which stems from the variation of electrolyte concentration during discharge. In the higher plateau, lithium ions and polysulfides dissolve into the electrolyte, which increases the electrolyte viscosity and resistance. In the lower plateau, the onset of Li$_2$S precipitation reduces the ionic concentration, thereby lowering the electrolyte viscosity and resistance.

Figure 8.3 Voltage and Ohmic resistance during a 0.34 A discharge of a Li–S pouch cell provided by OXIS Energy Ltd.

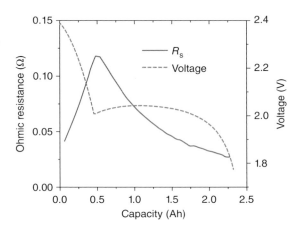

While a quantitative relation between polysulfide concentration and electrolyte conductivity has not been established, it was frequently observed that electrolyte conductivity decreases with increasing number of dissolved ions beyond an optimum ionic concentration [3]. With this observation, we employ a simple phenomenological expression to describe the concentration–conductivity relation of the Li–S electrolyte:

$$\sigma = \sigma_0 - b|c_{\text{Li}} - c_{\text{Li},0}| \tag{8.26}$$

where σ is the electrolyte conductivity, c_{Li} is the Li$^+$ concentration, σ_0 is the maximal conductivity at Li$^+$ concentration of $c_{\text{Li},0}$, and b is a fitting constant. For simplicity, we have assumed that the electrolyte conductivity varies with the total cation concentration rather than the concentration of each individual polysulfide species. Beyond the optimum concentration, $c_{\text{Li},0}$, which we assume is the initial Li$^+$ concentration in the cell, the electrolyte conductivity decreases linearly with slope b. The electrolyte resistance relates to its conductivity via

$$R_s = \frac{l}{\sigma A} \tag{8.27}$$

where l is the cell layer thickness and A is the electrode geometric area.

In Figure 8.4, the simulated electrolyte resistance based on Eq. (8.26) qualitatively agrees with the measured trend during discharge. The simulated discharge voltage incorporating the Ohmic IR_s drop also better resembles the typical discharge profile of a Li–S cell. The peak in electrolyte resistance contributes to the appearance of a dip between the higher and lower plateaus in the discharge curve. The evolution of electrolyte resistance follows closely to the Li$^+$ concentration variation during discharge, as shown in Figure 8.4. The Ohmic resistance of Li–S cells varies with the electrolyte concentration, which is controlled by the dissolution and precipitation of polysulfides.

8.3.2 Anode Potential

The inclusion of the anodic reaction Eq. (8.19) brings additional potential variations at the anode. The reaction potential at the anode is, according to the Nernst equation,

$$E_{\text{anode}} = \frac{RT}{F} \ln\left(\frac{c_{\text{Li}}}{1000}\right) \tag{8.28}$$

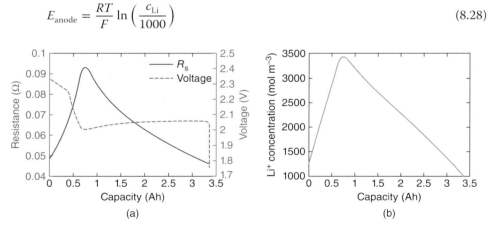

Figure 8.4 (a) Simulated discharge voltage and Ohmic resistance of a Li–S cell during a 0.34 A discharge; (b) the Li$^+$ concentration variation during the discharge.

where the standard potential for Li^+ dissolution at $1 \, mol \, l^{-1}$ is set to be zero. According to the equation, the anode potential increases with Li^+ dissolution and decreases with Li_2S precipitation. We assume that the Li^+ dissolution/deposition kinetics is fast enough so that the anodic activation overpotential is negligible compared to the activation overpotentials of the polysulfide reduction/oxidation reactions. With the inclusion of anode potential, the cell voltage can be written as

$$V = E_j - \eta_j - E_{\text{anode}} - IR_s \tag{8.29}$$

where E_j and η_j are the reaction potential and activation overpotential of any cathodic reaction j. In Eq. (8.29), we have not included the resistance of the anode SEI layer, which could contribute to an important voltage loss mechanism for aged Li–S cells.

Figure 8.5 demonstrates the modeled anode potentials during 0.1 and 0.3 C discharges. Lithium dissolution in the higher plateau causes a potential change of around 30 mV at the anode. The anode potential subsequently drops in the lower plateau as Li^+ precipitates in the form Li_2S. It is interesting to note that there is a difference in the anode potential between the 0.1 and 0.3 C discharges despite the assumption of zero anodic activation overpotential. This anode potential difference is caused by the limited precipitation rate of Li^+. Since the Li^+ precipitation rate is only concentration dependent but the rate of Li^+ dissolution is current dependent, during a faster discharge the Li^+ concentration – and therefore the anode reaction potential – is higher. This precipitation-induced overpotential is also present in the cathode where the limited rate of S^{2-} precipitation decreases the cathode reaction potentials at higher discharge currents [3]. Therefore, the precipitation bottleneck effectively manifests as a resistance that reduces the Li–S cell voltage at higher discharge rates.

8.3.3 Surface Passivation

Owing to the insulating nature of Li_2S, its precipitation during discharge reduces the conductive cathode surfaces. According to the Butler–Volmer equation, Eqs. (8.7) and (8.8), reduced cathode surface should lead to increased activation overpotential for charge transfer. To capture the surface passivation effect of precipitation, a phenomenological expression for the cathode surface area, a_r, is introduced [16]:

$$a_r = a_{r,0} \left(1 - \frac{v_{Li_2S}}{v_{Li_2S,\text{max}}} \right)^{\theta_1} \tag{8.30}$$

Figure 8.5 Simulated anode potential during 0.1 and 0.3 C discharges.

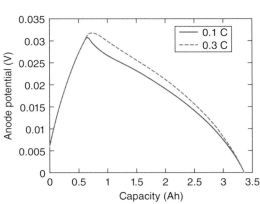

In this expression, $a_{r,0}$ is the initial cathode surface area, v_{Li_2S} is the volume fraction of precipitated Li_2S, $v_{Li_2S,max}$ is the maximum volume fraction of Li_2S, and θ_1 is a fitting constant with a value of 0.5 in our simulations. The surface area available for electrochemical reaction approaches zero as the precipitated volume of Li_2S approaches the limiting value of $v_{Li_2S,max}$. The Li_2S volume fraction can be calculated in the 0D model knowing the molar mass and molar volume of Li_2S.

With the effect of Li_2S precipitation on the conductive cathode surface area included, the 0D model reproduces the characteristic kink at the beginning of charge, as shown in Figure 8.6. This charge kink is often observed upon charging a Li–S cell from low state-of-charge (SoC) [16]. From the simulated cathode surface area and activation overpotential, it is clear that the charge kink is related to the higher activation overpotential at the beginning of charge. The activation overpotential increases during discharge as a result of reducing active cathode surface area, and reaches a peak at the end of discharge. Upon recharge, the activation overpotential gradually reduces with the dissolution of Li_2S, which frees up the covered cathode surfaces. The charge kink is therefore a result of the surface passivation effect of Li_2S precipitation, and is only noticeable when charging from a low SoC [16]. The inclusion of surface passivation has also altered the shape of the discharge curve in Figure 8.6: the gradually increasing activation overpotential allows for a gradual decrease in cell potential toward the end of discharge, as opposed to a sharp drop in voltage as seen in Figure 8.5.

8.3.4 Transport Limitation

Mass transfer becomes a limiting factor for the discharge capacity of Li–S cells at high currents. Transport limitation could arise at the cell level due to low ionic diffusion rates [13], as well as at the particle level due to choking of particle pores by precipitates [8, 18]. A recent study on the Li–S impedance at different discharge rates suggests that mass transport may be the main limiting factor for the rate capability of Li–S cells during discharge [13].

The 0D model does not resolve any spatial dimension; therefore, it is unable to include the transport of ions. However, it is possible to model the effect of

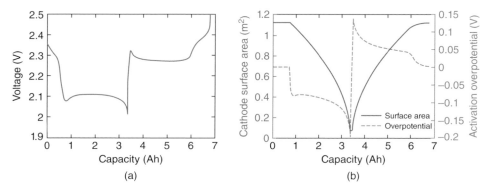

Figure 8.6 (a) Simulated cell voltage during a 0.2 C discharge and 0.1 C charge; (b) simulated active surface area in the cathode and the activation overpotential for the reaction in Eq. (8.24).

transport limitation empirically via the transport-limited form of the Butler–Volmer equation [16]:

$$\frac{i_j}{a_r i_{j,0}} = \left(1 - \frac{I_j}{I^{\text{lim}}}\right) \exp\left(\frac{-F\eta_j}{2RT}\right) - \left(1 + \frac{I_j}{I^{\text{lim}}}\right) \exp\left(\frac{F\eta}{2RT}\right) \tag{8.31}$$

Here, i_j is the current of cathodic reaction j, $i_{j,0}$ is the exchange current density of reaction j, η_j is the overpotential of reaction j, and I^{lim} is the limiting current due to mass transfer, assumed here to be the same for all reactions. The limiting current represents the maximum current carried by the transport of reactants toward the electrode surface. Since the precipitation of Li_2S alters the pore volume in the cathode available for ionic transport, the limiting current should depend on the amount of Li_2S formed in the cathode. We assume I^{lim} is a function of the Li_2S volume in a form similar to Eq. (8.30):

$$I^{\text{lim}} = I_0^{\text{lim}} \left(1 - \frac{V_{Li_2S}}{V_{Li_2S,\text{max}}}\right)^{\theta_2} \tag{8.32}$$

where I_0^{lim} is the initial limiting current with no Li_2S present and θ_2 is another fitting constant with an assumed value of 1.5.

The simulated discharge curves at 0.2 and 0.6 C are shown in Figure 8.7 using the 0D model with transport-limited reaction currents. At a higher discharge current, the model predicts a reduction in the lower plateau capacity while the higher plateau capacity remains the same. This predicted behavior agrees with experiments [16]. Figure 8.7 also shows the ratio between the reaction current for the reduction of S_2^{2-} (Eq. (8.24)) and the limiting current. In the lower plateau, the limiting current is reduced by the increasing amount of precipitated Li_2S, leading to an increase in the transport overpotential. The discharge terminates when the reaction current reaches the limiting current ($I_j/I^{\text{lim}} \to 1$). For a faster discharge, the limiting current is reached earlier so the discharge capacity is reduced. While the 0D model captures the effect of transport limitation with the phenomenological limiting current in Eq. (8.32), it does

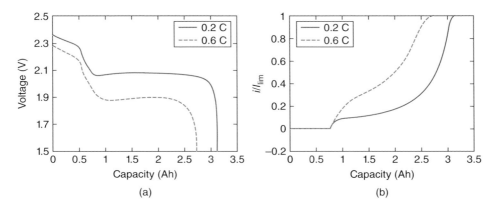

Figure 8.7 (a) Simulated discharge curves at 0.2 and 0.6 C with the transport limitation effect described by Eqs. (8.31) and (8.32) ; (b) simulated ratio between the reaction current for Eq. (8.24) and the limiting current during the discharges.

not resolve concentration gradients in the cell. To examine the spatial distributions of ionic species during Li–S cell charge/discharge, higher dimensional models are required.

8.4 Higher Dimensional Models

8.4.1 One-Dimensional Models

As explained in the last section, the 0D Li–S model does not resolve the material properties, ionic concentrations, and potentials spatially. Most of the existing Li–S models in the literature are 1D models that discretize the spatial dimension between the current collectors. The first 1D Li–S model was introduced by White and coworkers based on the dilute-solution theory [10]. The White model employs the Nernst–Plank equations to describe self-diffusion and migration of ionic species in a Li–S cell, assuming that the electrolyte is dilute. The model also calculates the volume of precipitated polysulfide formed during discharge, thereby capturing the porosity distributions in the cathode and the separator. While the White model qualitatively reproduces the discharge curve of Li–S cells, it also requires more parameters than the 0D model to describe ionic transport. To understand the sensitivity of the White model with respect to its large number of physical and phenomenological parameters, Chen and coworker performed a parametric analysis and revealed that the White model is highly sensitive to the parameters governing the precipitation/dissolution of polysulfides. In particular, it was shown that the White model fails to model charge because the low saturation concentration for Li_2S severely limits its re-dissolution upon charge. Chen and coworker [18] also reported that the White model is insensitive to ionic transport properties, with transport limitation only becoming important when the diffusion coefficients were reduced by more than 2 orders of magnitude to 10^{-12} m^2 s^{-1}.

Following Chen's sensitivity analysis [18], Zhang et al. further analyzed the effect of transport limitation on the discharge behaviors of Li–S cells using the White model [13]. It was found that with low ionic diffusion coefficients on the order of 10^{-12} m^2 s^{-1}, a significant Li$^+$ concentration gradient is developed in the separator during discharge. The Li$^+$ concentration gradient forces a large amount of polysulfides to migrate into the separator in order to maintain charge neutrality, preventing their further reductions in the cathode. These trapped polysulfides thus represent inaccessible capacity during a fast discharge. Interestingly, the trapped polysulfides could diffuse back into the cathode upon removing load, thereby allowing the cell to restore capacity. This capacity recovery effect was experimentally confirmed. As shown in Figure 8.8, after a 1 C discharge a Li–S cell is able to be discharged again after one hour of rest. The simulated average S_2^{2-} concentration in the separator during the discharge–rest–discharge process is shown in Figure 8.8. The S_2^{2-} anions accumulate in the separator during the 1 C discharge periods, and diffuse back into the cathode during the rest. The cell-level transport limitation shown here could arise in Li–S cells with lean electrolyte and high-viscosity solvents such as sulfolane.

By including ionic transport, 1D Li–S models can describe polysulfide shuttle as a transport-reaction process rather than the simplified reaction-only process in the 0D

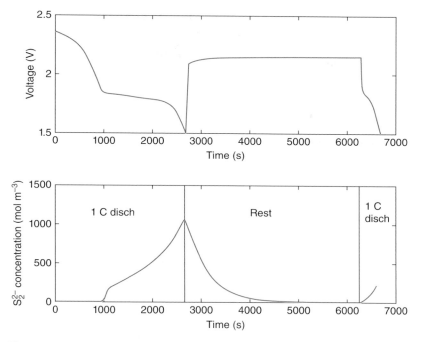

Figure 8.8 Top: Measured voltage of an OXIS Energy 3.4 Ah Li–S pouch during a 1 C–rest–1 C discharge procedure; Bottom: simulated average S_2^- concentration in the separator during the 1 C–rest–1 C discharge procedure.

model. In their 1D model, Bessler and coworkers modeled shuttle as the diffusion of S_8 anion to the anode and their subsequent reduction to S_4^{2-} [11]. This diffusion rate of S_8 toward the anode was shown to balance the diffusion rate of S_4^{2-} back toward the cathode, leading to the typical "infinite charging" behavior toward the end of a low-rate charge. Bessler's model also captures the accumulation of precipitated Li_2S in the anode and its effect on the anode resistance and cell capacity fade. Yoo et al. replaced the single shuttle reaction between S_8 and S_4^{2-} with a series of polysulfide reduction reactions at the anode in their 1D Li–S model [15]. With typical model parameters, it was shown that reducing shuttle reaction rates at the anode (e.g. by protecting anode with $LiNO_3$ additive) is more effective in suppressing shuttle than limiting the diffusion rates of polysulfides.

So far, all published 1D Li–S models employ the Nernst–Plank equations to describe ionic transport as the diffusion and migration of isolated ionic species in a dilute solution. The dilute-solution theory, however, is a poor assumption for Li–S cells as their ionic concentration can reach up to 4–5 M from the dissolution of polysulfides, as has been shown with the 0D model (Figure 8.4). On the other hand, modeling Li–S cells with concentrated-solution theory is difficult, not only because of the complexity of formulating the transport of multiple polysulfide species with the Maxwell–Stefan equations but also due to lack of measured transport properties required for the concentrated-solution formulation. Recently, Nazar and coworkers developed the theoretical framework for describing diffusion in a ternary system containing Li_2S_6/Li_2S_4, and lithium bis(trifluoromethanesulfonyl)imide (LiTFSI) dissolved in dioxolane (DOL)/dimethoxyethane (DME), and further measured the six required transport properties using a concentration cell, restricted diffusion, and transference polarization experiments [19]. It was found that the polysulfide diffusion rate is highly dependent on the salt concentration, and the transference number of Li$^+$ and salt is also strongly affected by the polysulfide concentration. Nazar's study thus highlights the importance of cross-term diffusion coefficients, which are not considered in existing 1D Li–S models based on the dilute-solution theory. Extending Nazar's formulation to model a working Li–S cell is still difficult since its charge/discharge involves many more dissolved polysulfide species in the three studied by Nazar and coworkers. The lack of a practical modeling framework for multicomponent ionic transport in Li–S cells is the main limitation of 1D Li–S models.

8.4.2 Multi-Scale Models

Recently, Li–S models that resolve the electrode microstructure have been proposed to study the effect of cathode morphology on cell performance. Danner et al. introduced a pseudo-2D model that discretizes both the cell-scale dimension between the electrodes and the microscopic dimension within the carbon/sulfur particles [7]. The transport of Li$^+$ and salt is modeled with concentrated-solution theory at the cell scale, whereas the transport of polysulfides is modeled with dilute-solution theory at the particle scale. It was found that the confinement of polysulfides within carbon/sulfur particles causes the micropores to be blocked at the particle surface due to Li$_2$S precipitation. Furthermore, polysulfide confinement induces a transport overpotential at the particle surface due to the need for Li$^+$ to go against its concentration gradient from the relatively dilute bulk electrolyte into the highly concentrated particles. Such transport overpotential represents a strong driving force for the dissolution of polysulfides into the bulk electrolyte, which makes the complete encapsulation of sulfur in particles difficult in practice. The transport overpotential also limits the sulfur utilization of Li–S cells with high sulfur loadings.

Thangavel et al. introduced a microstructurally resolved Li–S model that distinguishes between the interparticular porosity of the carbon/sulfur particles and the mesoporosity within the particles (Figure 8.9) [8]. The model does not resolve the particle-scale dimension but rather adopts a lumped approach to describe mass balance inside the particles. A phenomenological expression is used to describe the ionic flux between the bulk electrolyte and the particles as a function of the thickness of Li$_2$S layer deposited on the particles. Thangavel et al. found that with small carbon/sulfur particles, a Li–S cell's discharge capacity is mainly limited by the reduced interparticular porosity from precipitations near the cathode/separator interface, which gives rise to cell-level transport limitation. With large carbon/sulfur particles, however, the limiting mechanism is changed to the complete chocking of mesopores inside the particles caused by thick Li$_2$S deposition layers over the particle surfaces. Thangavel's model can therefore be used to predict the optimal particle radius, carbon/sulfur ratio, and porosity distributions for a Li–S cell.

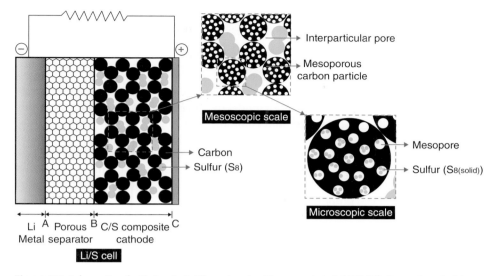

Figure 8.9 Schematic of a Li–S cell at different scales. Thangavel et al. 2016 [8]). Reproduced with permission from Electrochemical Society.

8.5 Summary

In this chapter, we have described the basic formulation for modeling the basic charge/discharge behaviors of Li–S cells with the introduction of a 0D Li–S model. By capturing the characteristic features of the Li–S chemistry such as multi-step electrochemical reactions, Li_2S precipitation/dissolution, and polysulfide shuttle, the 0D model provides mechanistic insights into the operational behaviors of Li–S cells. For example, the charge/discharge voltage of a Li–S cell is shown to be largely controlled by the thermodynamics and kinetics of Li_2S precipitation. The precipitation and dissolution of Li_2S controls the concentration of dissolved polysulfides in the electrolyte, which in turn determines the reaction potential and the Ohmic resistance of a Li–S cell during charge/discharge. The surface passivation and pore-blocking effects of the insulating precipitates further give rise to increased charge-transfer resistance and transport overpotential toward the end of a Li–S cell discharge.

Based on the 0D model, the cycling behaviors of a Li–S cell reflect the competition between the electrochemical processes of sulfur reduction/oxidation and the chemical processes of dissolution and shuttle. The 0D model explains the reduction of a Li–S cell's charge capacity at high charge currents with the depletion of dissolved polysulfides for oxidation reactions as a result of the relatively slow dissolution of Li_2S. The "infinite charging" behavior at low charge currents, on the other hand, is explained by the model as a consequence of the rate of shuttle being comparable to the rate of S_4^{2-} oxidation. The interplay between electrochemical reactions and the processes of shuttle and Li_2S precipitation/dissolution renders a portion the cell capacity temporarily irretrievable during cycling of Li–S cells. This "dormant" capacity represents a reversible degradation mechanism, which is shown to be recoverable by slowly charging the cell to its maximum voltage periodically [17]. By tracing the progress of both electrochemical and chemical processes during cycling, a mechanistic model such as the 0D model can predict the true SoC and state-of-health (SoH) of the Li–S cell.

While existing Li–S models such as the 0D model allow some insights into the complex working mechanisms and limiting processes of a Li–S cell, significant further developments are still required from both modeling and experimental characterizations to provide a complete quantitative understanding of Li–S behaviors. Capturing the effect of ionic transport and electrode morphology in Li–S cells requires 1D and higher dimensional Li–S models, the development of which is currently hindered by the lack of measured transport properties of polysulfides and the absence of in situ images of the electrode microstructure variations during charge/discharge. Modeling ionic transport in Li–S cells is further complicated by the need to utilize the concentrated-solution theory, which has not been applied to model an electrolyte that contains multiple reacting polysulfides. Further efforts are also necessary to obtain the parameters governing the thermodynamics and kinetics of sulfur oxidation/reduction reactions, to model the Li–S degradation processes including Li dendrite and SEI layer growth at the anode, and to include the temperature dependency of model parameters.

References

1 Xu, R., Lu, J., and Amine, K. (2015). Progress in mechanistic understanding and characterization techniques of Li–S batteries. *Advanced Energy Materials* 5: 1500408. https://doi.org/10.1002/aenm.201500408.
2 Wild, M., O'Neill, L., Zhang, T. et al. (2015). Lithium sulfur batteries, a mechanistic review. *Energy and Environmental Science* 8: 3477–3494.
3 Zhang, T., Marinescu, M., O'Neill, L. et al. (2015). Modeling the voltage loss mechanisms in lithium–sulfur cells: the importance of electrolyte resistance and precipitation kinetics. *Physical Chemistry Chemical Physics* 17: 22581–22586.
4 Ren, Y.X., Zhao, T.S., Liu, M. et al. (2016). Modeling of lithium–sulfur batteries incorporating the effect of Li_2S precipitation. *Journal of Power Sources* 336: 115–125.
5 Mikhaylik, Y.V. and Akridge, J.R. (2004). Polysulfide shuttle study in the Li/S battery system. *Journal of the Electrochemical Society* 151: A1969.
6 Marinescu, M., Zhang, T., and Offer, G.J. (2015). A zero dimensional model of lithium–sulfur batteries during charge and discharge. *Physical Chemistry Chemical Physics* 18: 584–593.
7 Danner, T., Zhu, G., Hofmann, A.F., and Latz, A. (2015). Modeling of nano-structured cathodes for improved lithium–sulfur batteries. *Electrochimica Acta* 184: 124–133.
8 Thangavel, V., Xue, K.-H., Mammeri, Y. et al. (2016). A microstructurally resolved model for Li–S batteries assessing the impact of the cathode design on the discharge performance. *Journal of the Electrochemical Society* 163: A2817–A2829.
9 Cheng, X.-B., Huang, J.-Q., and Zhang, Q. (2018). Review—Li metal anode in working lithium–sulfur batteries. *Journal of the Electrochemical Society* 165: A6058–A6072.

10 Kumaresan, K., Mikhaylik, Y., and White, R.E. (2008). A mathematical model for a lithium–sulfur cell. *Journal of the Electrochemical Society* 155: A576.

11 Hofmann, A.F., Fronczek, D.N., and Bessler, W.G. (2014). Mechanistic modeling of polysulfide shuttle and capacity loss in lithium–sulfur batteries. *Journal of Power Sources* 259: 300–310.

12 Ghaznavi, M. and Chen, P. (2014). Sensitivity analysis of a mathematical model of lithium–sulfur cells. Part I: applied discharge current and cathode conductivity. *Journal of Power Sources* 257: 394–401.

13 Zhang, T., Marinescu, M., Walus, S., and Offer, G.J. (2016). Modelling transport-limited discharge capacity of lithium–sulfur cells. *Electrochimica Acta* 219: 502–508.

14 Barai, P., Mistry, A., and Mukherjee, P.P. (2016). Poromechanical effect in the lithium–sulfur battery cathode. *Extreme Mechanics Letters* 9.

15 Yoo, K., Song, M.-K., Cairns, E.J., and Dutta, P. (2016). Numerical and experimental investigation of performance characteristics of lithium/sulfur cells. *Electrochimica Acta* 213: 174–185.

16 Zhang, T., Marinescu, M., Walus, S. et al. (2018). What limits the rate capability of Li–S batteries during discharge: charge transfer or mass transfer? *Journal of the Electrochemical Society* 165: A6001–A6004.

17 Marinescu, M., O'Neill, L., Zhang, T. et al. (2018). Irreversible vs. reversible capacity fade of lithium–sulfur batteries during cycling: the effects of precipitation and shuttle. *Journal of the Electrochemical Society* 165: A6107–A6118.

18 Ghaznavi, M. and Chen, P. (2014). Analysis of a mathematical model of lithium–sulfur cells. Part III: electrochemical reaction kinetics, transport properties and charging. *Electrochimica Acta* 137: 575–585.

19 Safari, M., Kwok, C.Y., and Nazar, L.F. (2016). Transport properties of polysulfide species in lithium–sulfur battery electrolytes: coupling of experiment and theory. *ACS Central Science* 2: 560–568.

9

Battery Management Systems – State Estimation for Lithium–Sulfur Batteries

Daniel J. Auger, Abbas Fotouhi, Karsten Propp and Stefano Longo

Cranfield University, School of Aerospace, Transport and Manufacturing, Cranfield MK43 0AL, UK

9.1 Motivation

Most batteries are vulnerable to being overdischarged; many are also vulnerable to overcharging. In lithium–sulfur, while overcharging is not seen as a "safety" issue [1], the high charge state "shuttle" phenomenon is associated with capacity fade [2]. Consequently, it is important to avoid extremes of charge state. As illustrated in Figure 9.1, our ability to access the full "usable capacity" depends on the ability to reliably estimate the state of charge at these extremes: uncertainty in state measurements translates directly into a reduced window of operation. This underlines the importance of reliable state estimation techniques: the effective capacity of a battery depends not only on the chemistry but also on the accuracy of the state estimation algorithms. Without these, an otherwise good battery technology can be rendered useless.

There are well-established techniques for state estimation in lithium ion, and these are well described in the literature [3–5]. Unfortunately, the two most straightforward techniques, open-circuit voltage measurement and "Coulomb counting," do not transfer directly to lithium–sulfur. The reasons for this are explored further. Before this, it is helpful to formally define state of charge and other relevant quantities.

9.1.1 Capacity

For practical reasons, lithium–sulfur cells are typically operated within a restricted voltage window to prolong lifetime, such as between 2.45 V corresponding to "fully charged" and 1.5 V corresponding to "fully discharged," in line with manufacturer recommendations [6]. The high voltage corresponds to the onset of significant shuttle effect [7] and the low voltage corresponds to the point of onset of significant power limitations caused by the depletion of polysulfide species and the buildup of precipitates [8].

Given a fully charged battery or cell with zero self-discharge, the capacity is defined by the equation

$$Q = \int_{t_0}^{t_1} I_{\text{dis}}(\tau) \ d\tau \tag{9.1}$$

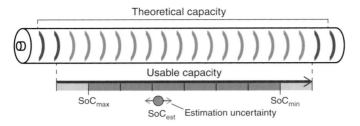

Figure 9.1 Theoretical capacity and usable capacity. At extremes of state of charge, there is a risk of damage, avoided by operating within a restricted "usable capacity" range. Any uncertainty in the state estimate reduces the effective usable capacity even further. Source: Propp 2017 [38]. Reproduced with permission from Cranfield University.

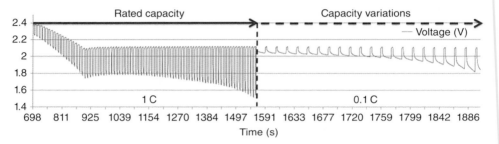

Figure 9.2 Example of variation in capacity with voltage-based discharge end points for a lithium–sulfur cell. In the first part of the discharge cycle, which takes place at 1 C, the 1.5 V limit is hit shortly after 1500 seconds, and any further attempt to discharge at 1 C will result in the terminal voltage falling below 1.5 V. If, however, the discharge rate is reduced to 0.1 C, the cell can be discharged further without breaching the voltage cutoff limit. Source: Propp 2017 [38]. Reproduced with permission from Cranfield University.

where Q is the capacity (ampere-seconds), t_0 is the time at which discharging starts (seconds), t_1 is the time at which discharging ends (seconds), and $I_{dis}(t)$ is the time varying discharge current (amperes).

In practice, cell performance varies with temperature, current, and usage profile. A common technique for verifying capacity is to discharge cells slowly over approximately 10 hours, i.e. at a nominal C-rate of 0.1 [9] in a carefully controlled environment, stopping when the terminal voltage hits the minimum value of 1.5 V; unless degraded cells are specifically being characterized, only fresh cells are used. Manufacturers' data sheets usually detail behavior for several different C-rates and temperatures, but Li–S batteries are particularly sensitive to changes in current profile, age, and temperature [10].

However, the use of the dynamic terminal voltage as the end point can cause unusual behaviors, as illustrated in Figure 9.2, illustrating a pulsed-current discharge initially at a rate of 1 C. This results in the cutoff voltage being reached very early, but by reducing the current rate to 0.1 C, it is possible to discharge the cell further. It would be possible to maximize capacity by repeatedly reducing current, although this will have limited applicability to a practical application where current demands are dictated externally.

9.1.2 State of Charge (SoC)

In terms of battery management, *state of charge (SoC)* is usually interpreted as the fraction of the usable capacity that currently remains accessible within the cell. A value of 1 (100%) indicates a fully charged battery, and a value of 0 (0%) represents a fully discharged one. In mathematical terms

$$(SoC)(t) = \frac{q(t)}{Q} \tag{9.2}$$

where $q(t)$ is the remaining accessible charge within the cell (ampere-seconds) and Q is the (usable) capacity of the cell (also in ampere-seconds). In an ideal cell, one would expect to find that

$$q(t) = Q - \int_0^t I_{dis} \, d\tau \tag{9.3}$$

Unfortunately, due to the history effects observed in lithium–sulfur, this is not always true. A more rigorous definition might be

$$q(t) = E\left[\int_t^\infty I_{dis}(\tau) \, d\tau\right] \tag{9.4}$$

where $E[X]$ denotes the expectation of the random variable X: essentially $q(t)$ is the amount of charge we expect to be able to extract from the cell or battery until total discharge *from the current point in time onwards.*

The definitions of state of charge used in this chapter may well differ from those used by electrochemists: however, they are practical and useful for battery management as they describe the "remaining capacity" at any given point in time.

9.1.3 State of Health (SoH)

There is no universally accepted definition of "state of health," (SoH) but typically, definitions focus on either one of two properties:

- Cell *capacity*, which typically degrades as the battery ages. Often a lithium ion battery is said to have reached the end of its useful life when the usable capacity falls to, say, *80%* of its original value.
- Increases in cell *internal resistance*, which typically increases as the battery ages. Often a battery is said to have reached the end of its useful life when its internal resistance reaches *double* its original value.

To a certain extent these properties are actually "surrogate outcomes": in an automotive application, the real consumer concerns are vehicle range and consistent powertrain performance [11]. However, they are widely used and easily understood.

Given a cell whose original capacity is Q_{fresh} ampere-seconds and whose degraded capacity is $Q_{aged}(t)$, one state of health metric, $\eta_Q(t)$, can be defined by

$$\eta_Q(t) = 1 - \frac{Q_{fresh} - Q_{aged}(t)}{0.2Q_{fresh}} = 5\left(\frac{Q_{aged}(t)}{Q_{fresh}}\right) - 4 \tag{9.5}$$

The capacity at any given instant can be expressed as

$$Q_{aged}(t) = 0.8Q_{fresh} + (0.2Q_{fresh}) \times \eta_Q(t) \tag{9.6}$$

thus giving us 100% of original capacity when $\eta_Q(t)$ equals unity, and 80% of the original capacity when $\eta_Q(t)$ is zero.

Given a cell whose original internal resistance is R_{fresh} ohms and whose degraded resistance is $R_{aged}(t)$, another state of health metric, $\eta_R(t)$, can be defined by

$$\eta_R(t) = \frac{(2R_{fresh}) - R_{aged}(t)}{R_{fresh}} \tag{9.7}$$

The internal resistance at any given instant can be expressed as

$$R_{aged}(t) = 2R_{fresh} - (R_{fresh}) \times \eta_R(t) \tag{9.8}$$

thus giving us the original internal resistance, R_{fresh}, when $\eta_R(t)$ equals unity, and twice this value when $\eta_R(t)$ is zero.

To date, published work on state estimation in lithium–sulfur has concentrated on state-of-charge estimation rather than on state of health.

9.1.4 Limitations of Existing Battery State Estimation Techniques

As discussed at the start of Section 9.1, there are many techniques for battery state estimation; some of these are more sophisticated than others. However, it is probably fair to say that for lithium ion batteries, it is possible to get good results with two conceptually simple methods, open-circuit voltage and "Coulomb counting." Many industrial engineers have a high degree of confidence in these methods and it is obviously good sense to avoid unnecessary complexity when a simple algorithm will already do.

In the following sections, the applicability of these methods is discussed. Unfortunately, there are big obstacles to the use of either technique and hence the need for the more sophisticated techniques introduced later in this chapter.

9.1.4.1 SoC Estimation from "Coulomb Counting"

The state of charge of an ideal battery with capacity Q ampere-seconds can be expressed in terms of an integral relationship [3]:

$$X(t) = X_0 + \frac{1}{Q} \int_0^t (I_{load}(t) - I_{loss}(t)) \, d\tau \tag{9.9}$$

where $X(t)$ is the present state of charge, X_0 is the state of charge at time zero, $I_{load}(t)$ is the instantaneous external load current (amperes), and $I_{loss}(t)$ is the instantaneous loss current.

To work effectively, this method requires knowledge of several variables. It also assumes that the capacity Q is independent of the duty cycle and that it remains constant. (It may, of course, be affected unpredictably by degradation and aging.)

In a laboratory setting, it is often possible to get accurate measurements of I_{load} and an approximate value of X_0. However, in lithium–sulfur cell Q depends on the duty cycle.

I_{load} and the precise value of X_0 are hard to find because of chemistry's behavior [2, 7]. In practical applications, there is even greater uncertainty around the initial state and the duty cycle. There may be specialist applications where this limitation is not a concern, but in general terms, Coulomb counting is a poor fit for lithium–sulfur.

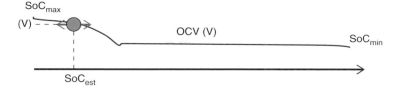

Figure 9.3 Determining the state of charge from open-circuit voltage (OCV). The OCV–state of charge curve for lithium–sulfur has distinct features, including a downward-sloped "high plateau" and a flat "low plateau." At high states of charge, the OCV can be used to infer state of charge as there is a unique mapping between the two, but at lower states of charge, the curve has a flat "low plateau," which makes state-of-charge estimation using this method unreliable. Source: Propp 2017 [38]. Reproduced with permission from Cranfield University.

9.1.4.2 SoC Estimation from Open-Circuit Voltage (OCV)

The fundamental idea behind the open-circuit voltage (OCV) method for state-of-charge estimation is that a battery's state of charge is highly correlated to its resting OCV. Thus, if one reads off the resting OCV, one can easily convert this into the state of charge. The basic idea is that for all states of charge $X \in [0,1]$,

$$V_{OC} = h_{OC}(X) \tag{9.10}$$

where $h_{OC}(X)$ is a strictly monotonic function of X – in other words,

$$\frac{dh_{OC}}{dX} > 0 \text{ for all } X \in [0, 1] \tag{9.11}$$

If this is true, then h_{OC} is invertible on $[0, 1]$ and given a measured OCV value $\hat{V}_{OC}$, the state-of-charge estimate $\hat{X}$ is given by

$$\hat{X} = h_{OC}^{-1}(\hat{V}_{OC}) \tag{9.12}$$

This method obviously relies on being able to measure the OCV, which requires the battery to be in a resting state often enough to give a useful measurement. A potential advantage is that the state of charge can be determined more or less instantaneously, without the need for initial conditions or long measurements.

Unfortunately, the open-circuit voltage method does not work well for lithium–sulfur. Figure 9.3 shows the shape of a typical OCV vs state-of-charge curve. The first part of this, representing "high" states of charge, has a monotonic gradient. This part of the graph is commonly referred to as the "high plateau," and the reactions here are dominated by high-order polysulfide species such as Li_2S_8. Within this range, OCV measurements could be expected to work well, as the strict monotonicity condition of Eq. (9.11) holds in this region. However, at lower states of charge, the curve enters the "low plateau," which is often relatively flat, and the strict monotonicity condition is violated.

9.1.5 Direction of Current Work

In the previous sections, the two most convenient methods for determining state of charge "in application" have been considered and have been found to be unsuitable, at least in an unmodified form. OCV methods do not work well except at high states of charge; Coulomb counting is highly vulnerable to uncertainties caused by self-discharge, variable short-term capacity changes brought about by the duty-cycle-induced "history" effects, and long-term aging and degradation.

Essentially, the OCV method uses measurements, whereas Coulomb counting uses a simple predictive model. One solution to this problem might be to try to "stitch" these techniques together and form an *ad hoc* estimator combining features of both. Fortunately, there is a wealth of formal techniques that can be used to do this in a rigorous and systematic manner; there are two main sources of these: control theory and computer science. These will be explored in Sections 9.3 and 9.4. First, we will describe the experimental setup that has been used to test, develop, and explore these techniques.

9.2 Experimental Environment for Li–S Algorithm Development

The development of any battery management algorithm requires a suitable test environment. Models are only approximations of real cells, and it is important that any algorithms are proved with the real thing. The lithium–sulfur cells considered in much of the authors' work to date have been supplied by OXIS Energy [6]. The majority of the cells have been "long-life" cells with 3.4 Ah nominal capacity and other specifications presented in Table 9.1, although the use of any cell is of course possible.

Kepco power source/sink devices are used for the battery experiments. The battery tester can apply a current as an "input" and measure the voltage as an "output" (or vice versa). The cell under test is contained inside an aluminum test box, which is placed inside a thermal chamber as shown in Figure 9.4.

Table 9.1 Specifications of OXIS lithium–sulfur cell [6].

Type	Rechargeable lithium–sulfur pouch cell
	Remarks: Li Metal Anode
Nominal dimension	145 mm × 78 mm × 5.6 mm
Applications	Recommended discharge current: 680 mA
Nominal voltage	2.05 V
Capacity	Typical: 3400 mAh when discharged at 680 mA to 1.5 V at 30 °C
Charging condition	340 mA to 2.45 V at 30 °C
Recommended charging condition in applications	340 mA constant current (C/10)
	Charge termination control recommended:
	CC stop at 2.45 V or 11 h max charge time
Clamped charging voltage	2.45 ± 0.05 V
Service life	>95 cycles at 100% depth of discharge
	>150 cycles at 80% depth of discharge
Weight	Approximately 50.7 g
Ambient temperature range	Charge/discharge: 5 –80 °C Storage (1 year): −27 to 30 °C

Figure 9.4 Li–S cell test equipment; Kepco BOP device and Li–S cell inside the thermal chamber.

9.2.1 Pulse Discharge Tests

A common approach of battery testing is called "pulse testing" in which consecutive current pulses are applied to a battery and its terminal voltage is recorded. Figure 9.5 demonstrates a pulse discharge test on a Li–S cell including load current (input) and terminal voltage (output), which are recorded at 25 °C. The test starts from a fully charged state (i.e. 2.45 V) and continues until the terminal voltage drops below the cutoff voltage (i.e. 1.5 V). Accordingly, cell parameterization is possible at different SoC levels. Between the consecutive pulses, a "relaxation time" is allowed, during which the current is zero. Data is collected in the time domain with a sampling rate of 1 Hz. The measurements available include time, temperature, current, and terminal voltage as shown in Figure 9.5.

9.2.2 Driving Cycle Tests

In a more realistic scenario for electric vehicle (EV) application, another type of battery testing was conducted based on EV power demand on a driving cycle. The urban dynamometer driving schedule (UDDS) [12] was used. To get enough information for this type of battery test, a typical EV (i.e. Nissan LEAF) was simulated on UDDS driving cycle. More details of EV modeling and simulation can be found in [13, 14]. The power demand signal was then scaled down to be applied to a single cell. Average power of

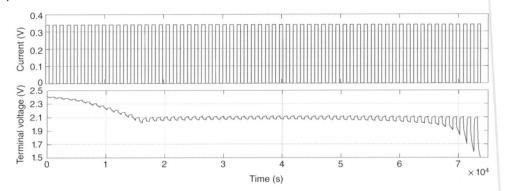

Figure 9.5 Load current and terminal voltage during a pulse discharge test.

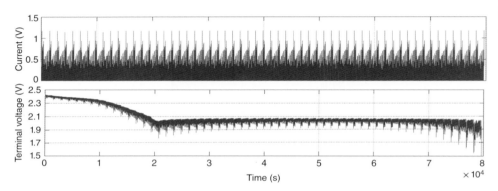

Figure 9.6 Li–S cell discharge test based on UDDS driving cycle.

a single Li–S cell that is used in this study is about 5 W. Assuming a maximum power demand of 60 kW in a typical EV to move according to UDDS, 12 000 of these proto-type Li–S cells are needed. Based on this calculation, the power demand from a single cell is obtained by dividing the EV power by the number of cells. The Li–S cells were then tested using scaled-down current profiles obtained from the EV simulation. Like the pulse discharge test explained in the previous section, current and terminal voltage were recorded at a sampling rate of 1 Hz. Figure 9.6 demonstrates a Li–S cell discharge test based on UDDS. The test is performed by repeating the UDDS cycle from a fully charged state to a depleted state.

9.3 State Estimation Techniques from Control Theory

So far, each of the basic methods for SoC estimation faces issues when applied to Li–S batteries. While Coulomb counting is accurate for short periods, it needs proper initial conditions and drifts with time. In contrast, the OCV method can estimate the SoC without prior information but is not applicable in a wide range of the cells' performance window. As a method to combine both advantages the idea of an observer is introduced from control engineering. Here, a mathematical model runs parallel to the real system, predicting the output, i.e. the terminal battery voltage. By using the error between the

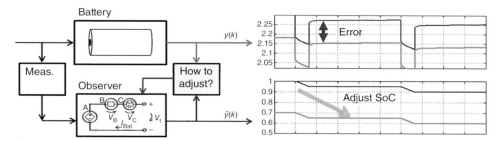

Figure 9.7 Principle of model-based estimation. Source: Propp 2017 [38]. Reproduced with permission from Cranfield University.

prediction and the real measurements the estimation of the hidden states of a system can be improved. Here, the fact that the parameters of the model and also the behavior of the battery change over the discharge range is exploited. As a rule of thumb we can say that as the parameter variations over the SoC become more pronounced, the ability to make a precise estimation gets better. The principle of the model-based estimation is also presented in Figure 9.7. The simple example shows that the voltage prediction of the observer is above the measurements. To correct for this, now the SoC is adjusted to fit the prediction.

However, a battery model capable of meeting the requirements is needed. There are many battery models spanning from the reproduction of the inner cell reactions to the more simple imitation of the voltage curve with current flow [15].

9.3.1 Electrochemical Models

The principle of electrochemical modeling is to reproduce the behavior and performance of the battery by the inner cell reactions. Since they are based on chemical reactions, these models are seen as the most accurate of the various modeling methods and due to their high depth, they are particularly useful in enhancing the understanding of the cell [16]. However, they also require a large number of physical and chemical parameters that are partly difficult to obtain. Furthermore, they need a large amount of computational power to run. For the complex reduction of sulfur, which is still a matter of current research [17], simplified assumptions of a two-step reaction led to good results representing the main properties of the battery [7] (Figure 9.8).

Efforts are on to produce electrochemical models that are simple enough to implement within a battery management system (BMS): in [18] a model is presented, which

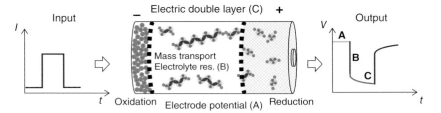

Figure 9.8 Input–output relationship of the mechanistic model. Source: Propp 2017 [38]. Reproduced with permission from Cranfield University.

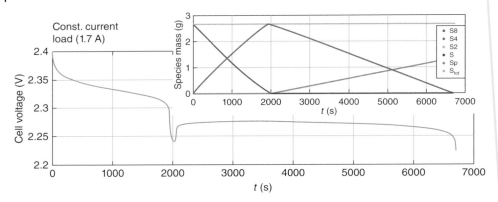

Figure 9.9 Terminal voltage predicted by mechanistic model. Source: Propp 2017 [38]. Reproduced with permission from Cranfield University.

is able to reproduce the cell's equilibrium voltage, its dynamic behavior with current, the polysulfide shuttle, and precipitation. Figure 9.9 shows that the development of the species governs the output voltage, with the example of a constant current of 1.7 A. The use of this model for state estimation is an active area of ongoing work. However, the general situation is still that most electrochemical models are too complex to be implemented in real-time software on the type of computer typically available to the BMS designer.

9.3.2 Equivalent Circuit Network (ECN) Models

Instead of reproducing the chemical reactions of the battery, equivalent circuit models only represent the input–output (input: current, output: terminal voltage) relationship of the cell with basic electrical components such as resistors, capacitors, and voltage sources. They are mostly used for SoC and SoH estimation since they are computationally inexpensive and relatively simple to parameterize. An acceptable trade-off between accuracy and complexity is seen in the Thevenin Model for Li ion batteries [19], shown in Figure 9.10.

Parameterization is usually done with optimization and fitting methods after the battery has been discharged with a pulse pattern (shown in Figure 9.10). Thereby, the pulses help assign typical battery behavior to the individual components of the model. While the voltage before the pulse is seen as the OCV the immediate response to current is assigned to be the internal resistance. The slow voltage drop afterwards is determined

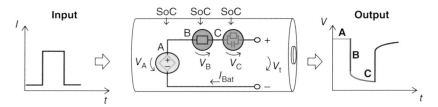

Figure 9.10 Input–output relationship of the ECN model. Source: Propp 2017 [38]. Reproduced with permission from Cranfield University.

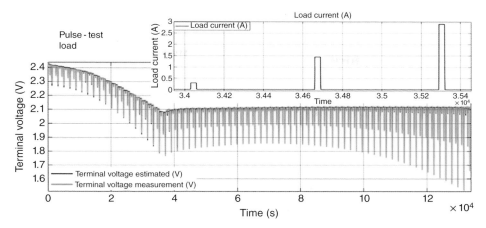

Figure 9.11 Terminal voltage predicted by the ECN model. Source: Propp et al. 2017 [23]. Reproduced with permission from Elsevier.

by the double layer capacitance of the electrodes and is represented by a parallel RC circuit [20]. In standard state space form the system can be described as

$$S \begin{cases} \dot{x}(t) = Ax(t) + Bu(t) \\ y(t) = Cx(t) + Du(t) \end{cases} \tag{9.13}$$

Equivalent circuit network (ECN) models are presented in [21, 22]. As shown in Figure 9.11, the terminal voltage can be represented well when each of the model parameters changes over the discharge range.

9.3.3 Kalman Filters and Their Derivatives

As introduced before, batteries are complex systems with many small variations occurring, even for their simple definitions for capacity and SoC. Furthermore the model itself, irrespective of its high or low fidelity, always contains some level of abstraction and simplification. Methods that have been found to be robust against these imperfections are recursive filters that assign a stochastic variable to the observations y and states x respectively. Thereby, the framework of a hidden Markov model (HMM) of sequential data [24] is applied, where the observations are influenced by unpredictable noise while the evolution of the hidden states and their probabilities is predictable (Figure 9.12a).

For the discrete propagation of the probabilities the Bayes rule is used to get a posterior probability density from the state transition probability $(P_{(xt|xt-1,ut)})$ and the measurement probability $(P_{(yt|xt)})$ from an initial value $P_{(x0)}$. The algorithm possesses two main steps [25].

Prediction:

$$\overline{\mathrm{bel}}(x_t) = \int p(x_t|x_{t-1},u_t) \; p(x_{t-1}|y_{1:t-1},u_{1:t-1})$$

where the actual probability density is calculated from past measurements, the system model, and the control input u_t. Thereafter, the posterior probability density $\mathrm{bel}(x_t)$ is calculated by employing the actual observation.

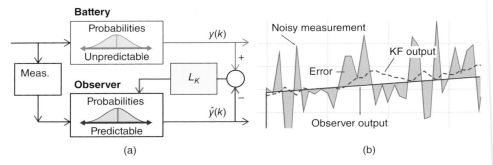

Figure 9.12 Principle of Kalman filter.

Update:

$$\text{bel}(x_t) = \eta^{-1} p(y_t | x_t) \; \overline{\text{bel}}(x_t)$$

where η is a normalization factor $p(z_t | z_{1:t, u_{1:t}})$, standing for certainty of all observations. An analytical solution of the Bayes filter for linear systems and Gaussian distributions is the Kalman filter (KF), minimizing the error variance between the true and estimated states (Figure 9.12b). Practically, the introduced uncertainty allows the user to decide how trustworthy a parameter or measurement is for the state estimation. As only a Thevenin ECN model has been used for state estimation yet, we introduce the parameterization and results for this example.

Simply speaking, the ECN-model-based KF uses three pieces of information. The first one is the measurement, influenced by Gaussian white noise with zero mean. The second one includes both system states (SoC, V_{RC}), which can be adjusted, and the third one includes the model parameters (V_{oc}, R_i, R_p, C_p) over the discharge range. Since the parameters of the model depend on the SoC, ideally a wrongly estimated SoC leads to an imprecise terminal voltage prediction. Since the algorithm is recursive, the continuous system has to be discretized as

$$x_k = A x_{k-1} + B u_{k-1}$$
$$y(t) = C x_k + D u_k \qquad (9.14)$$

The prediction step uses the input current and the past states for the estimation of the present state and probabilities.

Time update equations:

$$\widehat{x}_k^- = A \widehat{x}_{k-1}^+ + B u_{k-1} \qquad (9.15)$$

$$P_k^- = A P_{k-1}^+ A^T + Q \qquad (9.16)$$

Now an estimation of the predicted terminal voltage is possible. The error between the prediction and measurement is then multiplied by the Kalman gain to correct the states.

Measurement update equations:

$$L_k = P_k^- C^T (C P_k^- C^T + R)^{-1} \qquad (9.17)$$

$$\widehat{x}_k^+ = \widehat{x}_k^- + L_k (y_k - C \widehat{x}_k^-) \qquad (9.18)$$

$$P_k^+ = (I - L_k C) P_k^- \qquad (9.19)$$

The Kalman gain dynamically weights the estimation between the model prediction and the measurements. (For the mathematical background of the minimization the reader is guided to [26].) However, the estimation covariances develop depends on the user choices for the measurement noise variance R and the system noise variance Q. While R is relatively easy to determine by evaluating sensors or measurements, Q is guessed most of the time. However, parameterization follows patterns. As mentioned before, Li–S batteries have relatively constant battery parameters in the middle of the low plateau [21, 22]. Here, the voltage measurement gives only limited information about the SoC so the state estimation relies heavily on Coulomb counting. However, the actual capacity of Li–S may vary, which would impede a precise estimation. As shown in Figure 9.13, experiments showed that the capacity of the cell varies from 10561 to 9072 As with the same current profile but different gains [23]. Therefore, a state estimation also needs to adjust the SoC sufficiently. As shown in [23] this is solved reasonably well by basing the estimation on the Coulomb counting state but leaving enough uncertainty to correct the state if an error between prediction and measurement arises. Since the Li–S battery contains nonlinear dynamics, derivatives of the KF (extended Kalman filter (EKF), unscented Kalman filter (UKF)), and the particle filter (PF) are used in the study.

9.4 State Estimation Techniques from Computer Science

The Kalman filter methods described in the previous section use largely "known" – or at least assumed – models relating parameters (and therefore behavior) to state. There is an alternative to using a fixed model structure, namely "black-box modeling."

Black-box modeling is a technique to model the behavior of a dynamical system without the need to understanding the cause of the behavior. The reason for using such models in comparison to the conventional mathematical models, such as differential equations, is that sometimes a system cannot be defined well due to lack of information or the high uncertainty in the system. A black-box model is constructed based on the system's input–output relationship only. In this case, knowing what is inside the "box" is not of interest; the only important thing is building a proper mapping between inputs and outputs.

A schematic of black-box modeling of a battery is illustrated in Figure 9.14. In this approach, the black-box should behave exactly like the real battery, which means having the same inputs, and the outputs should be the same as well. To build such a model, experimental tests on the real battery are needed to provide enough data for a proper input–output mapping.

9.4.1 ANFIS as a Modeling Tool

In the previous section, a proper mapping between inputs and outputs of a dynamical system were discussed. Different tools can be used for this purpose; one is called adaptive neuro-fuzzy inference system (ANFIS). ANFIS is a powerful tool to construct a proper input–output mapping of a system based on (i) human knowledge and reasoning and (ii) input–output data [27, 28]. Before explaining ANFIS structure and the algorithms used for training an ANFIS structure, the two main parts of ANFIS need to be clarified; human knowledge and input–output data. Since explanation of the human knowledge and reasoning in a mathematical way needs more space, a separate

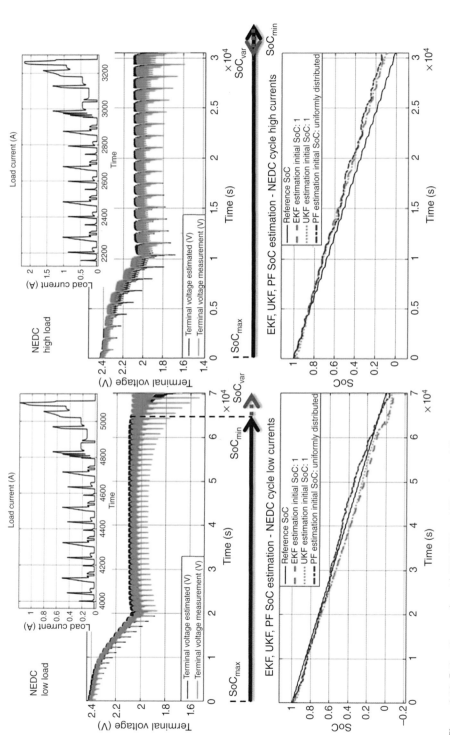

Figure 9.13 Estimation results with EKF, UKF, and PF and two different current densities. Source: Propp et al. 2017 [23]. Reproduced with permission from Elsevier.

Figure 9.14 Modeling of a battery using the black-box technique.

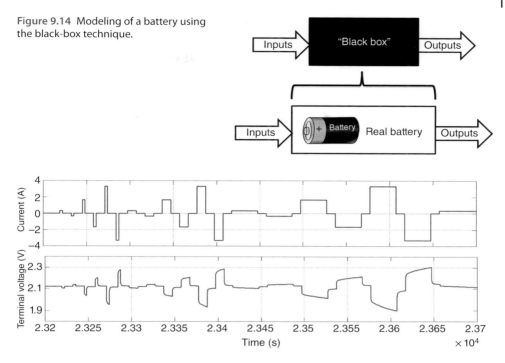

Figure 9.15 Input (current) – output (voltage) recording during a Li–S cell test.

section is allocated to this. The second part, input–output data, can be summarized as follows.

The concept of input–output data for a dynamical system can be explained as the system's response (output) to an excitation (input). To be able to model a system, such input–output data is needed. Usually, the data is provided by testing the system under various conditions and recording both inputs and outputs. For example, Figure 9.15 demonstrates input–output recording of a Li–S cell where the input is the cell's current and the output is the cell's terminal voltage, which is a single-input single-output (SISO) system. Using this data, a black-box model can be extracted that would be able to give battery terminal voltage if it knows the current as input.

9.4.2 Human Knowledge and Fuzzy Inference Systems (FIS)

A fuzzy inference system (FIS) is a tool for modeling the qualitative aspects of human knowledge and reasoning processes without employing precise quantitative analyses. An FIS gets human knowledge in the form of a number of rules called "fuzzy if–then rules" or fuzzy conditional statements. In an expression such as "if A, then B," the premise part is "if A," and the consequent part is "then B." A and B are labels of fuzzy sets characterized by membership functions (MFs) [29]. For example, "if battery OCV is high, then battery SoC is high." In this case, the word "high" is called a "linguistic label," which is defined quantitatively by an MF, and battery OCV and SoC are called "linguistic variables" [30].

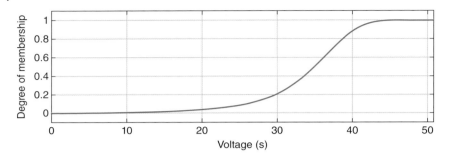

Figure 9.16 High voltage membership function.

When one person says, for example, "the voltage is high," the word "high" seems vague because no one can estimate even the range of the voltage. Assume that we know that the voltage value, which was called "high," is 50 V. Then, what can we expect 40 V to be called by that person? Should it be called "high" as well? What about 30, 20 V, or less voltage values? Let us assume that a voltage value more than 40 V is called a "high" voltage by that person. Does it mean that 39.99 V is not a high voltage? An MF in an FIS helps us define a linguistic label (e.g. "high") in a proper way using human knowledge. Figure 9.16 demonstrates a sample MF for this example. It is assumed that the voltage changes between zero and 50 V. Using the fuzzy concept, a degree of membership is defined for every value of a linguistic variable (e.g. voltage). In this example, the degree of membership of 50 V to the group of "high" voltages is 1, which means that this number is 100% inside the group. From Figure 9.16, the degree of membership of 40 V to the high voltages group is around 0.9 and so on. MFs provide a smooth transition between linguistic labels in an FIS.

Another example is presented in Figure 9.17 where three variables are plotted against the cell's SoC. Figure 9.17a illustrates a Li–S cell's OCV calculated in the whole range of SoC. From this plot, one can say "at very high SoC, OCV is high" or "if OCV is high, then SoC is very high." The second sentence can be used in a case where we do not know the SoC value (e.g. SoC estimation problem). Similarly, one can say "if OCV is medium, then SoC is high." In the case of low OCV, we can just say the SoC is not very high or high (i.e. medium or low or very low).

If a second variable such as ohmic resistance is used as shown in Figure 9.17b, it might be possible to say something more about SoC, especially when OCV is low. From the figure, when ohmic resistance is low, the SoC is very high or high, depending on the location of the dashed line (this will be discussed later). When ohmic resistance is medium, we can just say that the SoC is not very high or very low. Finally, when ohmic resistance is high, we know that the SoC would be very low.

Even after using the second variable, still more information is needed especially in the middle range of the SoC. Taking a look at the derivative of ohmic resistance against SoC in Figure 9.17c, it is clear that there is a change in the sign of the gradient in the middle SoC around 45%. So use of the sign of the ohmic resistance gradient against SoC as a third variable can be helpful in this example. When the gradient is negative (ohmic resistance is decreasing), SoC would be medium.

All the dashed lines in Figure 9.17 can be shifted up or down depending on the definitions of the linguistic labels (very high, high, medium, low, and very low). These linguistic

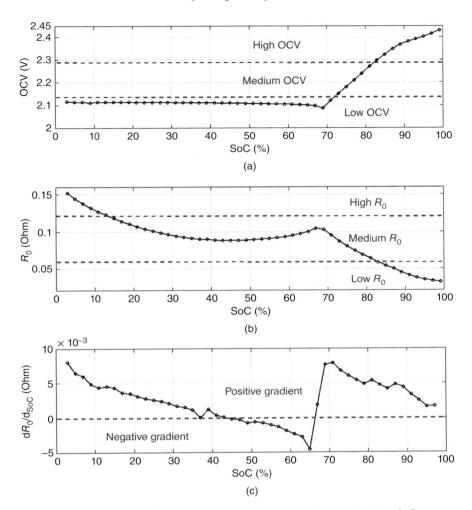

Figure 9.17 Li–S cell's (a) OCV, (b) ohmic resistance, and (c) its derivative against SoC.

labels are quantitatively defined using MFs as discussed before. However, the MFs need to be tuned properly in order to maximize a performance index.

If we want to summarize the above information (knowledge) by a number of if–then rules in an FIS for the Li–S cell's SoC estimation, the general rules would be as follows:

- If OCV is high, then SoC is very high (regardless of ohmic resistance and its derivative).
- If OCV is medium, then SoC is high (regardless of ohmic resistance and its derivative).
- If OCV is low and ohmic resistance is high, then SoC is very low (regardless of the derivative).
- If OCV is low and ohmic resistance is medium and the derivative is negative, then SoC is medium.
- If OCV is low and ohmic resistance is medium and the derivative is positive, then SoC is low.

Table 9.2 Rule base of a FIS for Li–S cell SoC estimation.

Rule number	OCV	Ohmic resistance	Derivative of resistance	SoC
1	High	High	Positive	Very high
2	High	Medium	Positive	Very high
3	High	Low	Positive	Very high
4	High	High	Negative	Very high
5	High	Medium	Negative	Very high
6	High	Low	Negative	Very high
7	Medium	High	Positive	High
8	Medium	Medium	Positive	High
9	Medium	Low	Positive	High
10	Medium	High	Negative	High
11	Medium	Medium	Negative	High
12	Medium	Low	Negative	High
13	Low	High	Positive	Very low
14	Low	Medium	Positive	Low
15	Low	Low	Positive	Medium (does not fire)
16	Low	High	Negative	Very low
17	Low	Medium	Negative	Medium
18	Low	Low	Negative	Medium (does not fire)

All possible scenarios should be considered in an FIS; the total number of rules is 18 in this example as presented in Table 9.2. Such a table is called a "rule base" of the fuzzy system. It should be noted that even the rules that may not fire in a real condition are included in the table. In this case, rules 15 and 18 may not fire since both OCV and ohmic resistance cannot be low at the same time. The conclusion part of these two rules can be anything (we just put "medium" in the table).

9.4.3 Adaptive Neuro-Fuzzy Inference Systems

Based on the idea developed by Zadeh in [29, 30], another form of fuzzy modeling technique was proposed by Takagi and Sugeno [31, 32] in which the fuzzy sets are used only in the premise part. An example of this type of FIS is the resistant force on a moving object:

If the velocity is high, then

$$\text{force} = k \times (\text{velocity})^2 \tag{9.20}$$

where the premise part is the same as before, but the consequent part is in the form of a non-fuzzy equation of the input variable, velocity [27].

ANFIS is a Takagi–Sugeno type FIS with more capabilities due to the use of neural networks (NNs) in addition to the fuzzy sets. ANFIS was first introduced by Jang in 1993 [27]. ANFIS is a powerful tool to construct a proper input–output mapping in a system based on (i) human knowledge and reasoning and (ii) input–output data. The first part is handled by using an FIS as discussed in the previous section for the example of Li–S cell SoC estimation. The second part, i.e. the ability to use the input–output data, distinguishes ANFIS from an ordinary FIS. As discussed in [27], there are two main limitations for an FIS to be applied for a real application:

Firstly, it is a demanding task to transform the existing knowledge or experience into a number of if–then rules in an FIS. Such a transformation was done for the Li–S cell SoC estimation example in the previous section, but it was not straightforward. Indeed, a deep understanding of the problem is essential to extract such effective rules.

Secondly, there is a need for tuning the MFs of an FIS in order to maximize a performance index. The tuning of MFs can be simply defined as finding an optimal shape for each MF. For example, Figure 9.18 illustrates a generalized bell-shaped MF. To tune this MF, we can play with three tunable parameters, x_1, x_2, and x_3 as shown in Figure 9.18. Since this MF is symmetric, no more variable parameter is needed. If this MF is used for the Li–S cell SoC estimation example (which had 8 MFs in total), the total number of tunable parameters of the FIS becomes 24. However, optimization of all the parameters is a demanding task.

ANFIS is able to solve the abovementioned problems by using neural networks in addition to the FIS. In this structure, human knowledge or experience should be provided in the form of input–output data as discussed before. ANFIS parameters including NN weights and FIS MF parameters are then tuned using the input–output data (i.e. training process). In other words, the rules are constructed automatically inside the ANFIS structure during the training process. After training, ANFIS can be used for another batch of input–output data (i.e. testing process) or it can be used in real time. A hybrid learning approach is used in the ANFIS training process, combining the gradient descent method with the least squares estimate. The learning algorithms of the ANFIS are explained in detail in [27, 28].

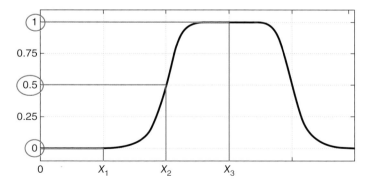

Figure 9.18 Generalized bell-shaped MF tunable parameters.

9.4.4 State-of-Charge Estimation Using ANFIS

The Li–S cell's SoC estimation is more challenging than that of other types of battery because of its SoC observability issues [33]. For this reason, a generic framework was developed in [34] (Figure 9.19), that can cope with different types of batteries including Li–S. The idea was to identify battery parameters in real time and use them as an indicator of battery SoC. So, a mapping function between the identification results and the SoC needs to be constructed, such as the function g in Equation 9.21:

$$\text{SoC} = g(P_1, P_2, \dots, P_n) \tag{9.21}$$

The whole framework is presented in Figure 9.19 including measurement, identification, and estimation units. In the measurement unit, battery current and terminal voltage are recorded to be used for parameter identification. In the identification unit, a fitting algorithm is used to find a battery model's parameters in real time. More details about battery parameterization algorithms can be found in [35]. In the estimation unit, an ANFIS is used that estimates battery SoC regarding the updated values of the battery parameters. The type and number of the battery parameters that are needed to estimate SoC depends on the chemistry of the battery as discussed in [33]. In our case, i.e. Li–S cell, SoC can be estimated using the three inputs as discussed in Section 9.4.1:

$$\text{SoC} = g\left(V_{\text{OC}}, R_0, \frac{dR_0}{d\text{SoC}}\right) \tag{9.22}$$

In this case, the inputs are V_{OC}, R_0, and dR_0/d_{SoC}, and the output is the SoC. Other specifications of the ANFIS are provided in [36]. The number of MFs used for V_{OC}, R_0, and dR_0/d_{SoC} are 5, 3, and 2 respectively. So, there are 10 MFs and 30 rules in total. More MFs can also be used; however, the results demonstrate that it would just increase the complexity of the system without further improvement in the accuracy.

The estimation results are not demonstrated here because it is an ongoing research. At the present time, the proposed estimation method can estimate Li–S cell's SoC with average and maximum errors of 5% and 14% respectively for a test data that ANFIS has not seen before.

In conclusion, Li–S cell SoC estimation via ANFIS is not straightforward due to many reasons. For example, prediction of the exact location of the break point between the high plateau and the low plateau in a Li–S cell is a demanding task since it depends on many factors. This plateau change happens around 80% SoC with 10% tolerance, which is a challenge for SoC estimation. In addition, the total capacity of Li–S cells can change

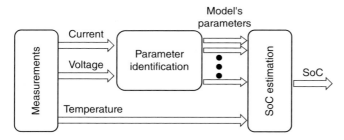

Figure 9.19 Battery measurement, identification, and SoC estimation.

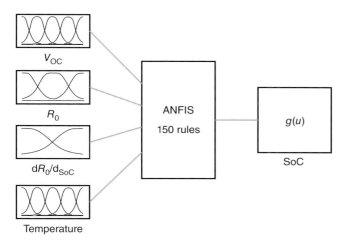

Figure 9.20 ANFIS structure including 4 inputs, 15 MFs, and 150 rules.

from one cell to another owing to manufacturing tolerances. Even for the same cell, the total capacity could change a lot under various working conditions due to variation in temperature, discharge rate, and aging, which makes Li–S cell's SoC estimation a challenging task.

Various solutions are under consideration to overcome the abovementioned challenges and research is going on in this area. For example, to consider the effect of temperature, more than one ANFIS structure can be used where each one is trained at a particular range of temperature. Alternatively, a more complex ANFIS structure can be trained to consider the temperature as the fourth input as shown in Figure 9.20. For the aging effect, an SoH estimator can be trained such as the work done for SoC estimation. For example, ohmic resistance can be monitored as an indicator of SoH. So, ohmic resistance might be used as an input of an ANFIS that is trained for SoH estimation. As mentioned before, all these ideas are under consideration and research is going on in this area.

9.5 Conclusions and Further Directions

This chapter has introduced the motivations for developing chemistry-specific state-of-charge estimation techniques for lithium–sulfur: essentially, OCV methods fail because of the shape of the OCV–SoC characteristic, and Coulomb counting fails because of uncertainties in short- and long-term capacity arising from lithium–sulfur's chemistry. To address this, two methods have been explored: a model-based technique using ECN models and Kalman-type filters. These have been tested in a laboratory bench environment that gives a reasonable representation of a target automotive application, but there is still some work to be done before they can be considered "mature."

There are several areas of work that are the subject of active research:

- *State estimation using simplified electrochemical models*: As discussed in Section 9.3.1, electrochemical models have traditionally been too computationally expensive to run on practical BMS hardware. Recently published "zero-order" electrochemical models offer scope for improved state estimation with a greater physical basis than ECN models [18]. In the longer term, there is also scope to apply techniques that have been developed to simplify spatially distributed models in lithium ion batteries [37], although this will depend on the availability of suitably mature spatially distributed lithium–sulfur models as a starting point.

- *State-of-health estimation*: There is potential to apply Kalman-type estimators to models of degradation of the form presented in Section 9.1. Again, fast-order electrochemical models will open up the potential for further analyses of degradation and aging.
- *Development of tools for rapid model/algorithm tuning*: At present, the techniques developed in this chapter require specialist knowledge and expertise to apply. This is a high "barrier to entry," and tools that would be a better fit for an industrial test and commissioning environment would be useful.
- *Widening the application base*: Different target environments are likely to present subtly different challenges, particularly concerning thermal management and duty cycles.

Acknowledgments

This work was funded by the UK's Engineering and Physical Sciences Research Council (EPSRC) under the grant EP/L505286/1 "Revolutionary Electric Vehicle Battery (REVB) – design and integration of novel state estimation/control algorithms and system optimization techniques." Much of the work described in this chapter has appeared as Open Access articles in journals [22, 23, 34]. In particular, the mathematics of the Kalman filter has borrowed heavily from [23]. The authors have endeavored to ensure that work is appropriately referenced and apologize if any small instances have escaped our notice. Research data access statements and links to any data that is available are included in the journal articles themselves.

The authors would particularly like to thank their collaborators, in particular Tom Cleaver, Mark Wild, Rajlakshmi Purkayastha, and their present and former colleagues at OXIS Energy, Greg Offer, Monica Marinescu, Teng Zhang, and the rest of the team at Imperial College London, John Bailey and his team at Ricardo UK, Francis Assadian at the University of California Davis, and Vaclav Knap at Aalborg University, Denmark. The authors' thanks also go to the many support staff at Cranfield and to the many MSc students who have contributed to this work.

References

1 Manthiram, A., Fu, Y., Chung, S.H. et al. (2014). Rechargeable lithium–sulfur batteries. *Chemical Reviews* 114 (23): 11 751–11 787.

2 Diao, Y., Xie, K., Xiong, S., and Hong, X. (2013). Shuttle phenomenon–the irreversible oxidation mechanism of sulfur active material in Li-S battery. *Journal of Power Sources* 235: 181–186.

3 Piller, S., Perrin, M., and Jossen, A. (2001). Methods for state-of-charge determination and their applications. *Journal of Power Sources* 96: 113–120.

4 Chang, W.Y. (2013). The state of charge estimating methods for battery: a review. *ISRN Applied Mathematics.*

5 Cuma, M.U. and Koroglu, T. (2015). A comprehensive review on estimation strategies used in hybrid and battery electric vehicles. *Renewable and Sustainable Energy Reviews* 42: 517–531.

6 OXIS Energy – Next Generation Battery Technology (company home page). http://www.oxisenergy.com (accessed 3 February 2016).

7 Mikhaylik, Y.V. and Akridge, J.R. (2004). Polysulfide shuttle study in the li/s battery system. *Journal of the Electrochemical Society* 151 (11): A1969–A1976.

8 Ryu, H.S., Guo, Z., Ahn, H.J. et al. (2009). Investigation of discharge reaction mechanism of lithium| liquid electrolyte| sulfur battery. *Journal of Power Sources* 189 (2): 1179–1183.

9 Sauer, D.U., Bopp, G., Jossen, A. et al. (1999). State of charge–what do we really speak about. *Proc. 21st Int. Telecom. Energy Conf.*, pp. 6–9.

10 Busche, M.R., Adelhelm, P., Sommer, H. et al. (2014). Systematical electrochemical study on the parasitic shuttle-effect in lithium–sulfur-cells at different temperatures and different rates. *Journal of Power Sources* 259: 289–299.

11 Fotouhi, A., Auger, D.J., Propp, K., and Longo, S. (2014). Simulation for prediction of vehicle efficiency, performance, range and lifetime: a review of current techniques and their applicability to current and future testing standards. *Proc. 5th Hybrid Elec. Veh. Conf. (HEVC2014)*, London. https://doi.org/10.1049/cp.2014.0959.

12 DieselNet, Emission test cycles: FTP-72 (UDDS). https://www.dieselnet.com/standards/cycles/ftp72.php (accessed 11 February 2017).

13 Fotouhi, A., Propp, K., and Auger, D.J. (2015). Electric vehicle battery model identification and state of charge estimation in real world driving cycles. *Proc. 7th Comput. Sci. Electron. Eng. Conf. (CEEC)*, Colchester, pp. 243–248.

14 Fotouhi, A., Shateri, N., Auger, D.J. et al. (2016). A matlab graphical user interface for battery design and simulation: from cell test data to real-world automotive simulation. *Proc.13th Conf. Synth. Model. Anal., Simulat. Methods, Appl. Circuit Design (SMACD)*, Lisbon.

15 Ramadesigan, V., Northrop, P.W.C., De, S. et al. (2012). Modeling and simulation of lithium-ion batteries from a systems engineering perspective. *Journal of the Electrochemical Society* 159 (3): R31–R45.

16 Fotouhi, A., Auger, D.J., Propp, K. et al. (2016). A review on electric vehicle battery modelling: From lithium-ion toward lithium–sulphur. *Renewable and Sustainable Energy Reviews* 56: 1008–1021.

17 Wild, M., O'Neill, L., Zhang, T. et al. (2015). Lithium sulfur batteries, a mechanistic review. *Energy & Environmental Science* 8 (12): 3477–3494.

18 Marinescu, M., Zhang, T., and Offer, G.J. (2016). A zero dimensional model of lithium–sulfur batteries during charge and discharge. *Physical Chemistry Chemical Physics* 18 (1): 584–593.

19 Hu, X., Li, S., and Peng, H. (2012). A comparative study of equivalent circuit models for li-ion batteries. *Journal of Power Sources* 198: 359–367.

20 Jossen, A. (2006). Fundamentals of battery dynamics. *Journal of Power Sources* 154 (2): 530–538.

21 Knap, V., Stroe, D.I., Teodorescu, R. et al. (2015). Electrical circuit models for performance modeling of lithium–sulfur batteries. *Proc. Energy Conv. Congress and Expos. (ECCE)*, pp. 1375–1381.

22 Propp, K., Marinescu, M., Auger, D.J. et al. (2016). Multi-temperature state-dependent equivalent circuit discharge model for lithium–sulfur batteries. *Journal of Power Sources* 328: 289–299.

23 Propp, K., Auger, D.J., Fotouhi, A. et al. (2017). Kalman-variant estimators for state of charge in lithium–sulfur batteries. *Journal of Power Sources* 343: 254–267.

24 Bengio, Y. (1999). Markovian models for sequential data. *Neural Computing Surveys* 2 (1049): 129–162.

25 Thrun, S., Burgard, W., and Fox, D. (2005). *Probabilistic Robotics*. Cambridge, MA: MIT Press.

26 Gelb, A. (ed.) (1984). *Applied Optimal Estimation*. Cambridge, MA: MIT Press.

27 Jang, J.S.R. (1993). ANFIS: adaptive-network-based fuzzy inference system. *IEEE Transactions on Systems, Man, and Cybernetics* 23 (3): 665–685.

28 Jang, J.S.R., Sun, C.T., and Mizutani, E. (1997). *Neuro-Fuzzy and Soft Computing: A Computational Approach to Learning and Machine Intelligence*. Englewood Cliffs, NJ: Prentice-Hall.

29 Zadeh, L.A. (1965). Fuzzy sets. *Information and Control* 8: 338–353.

30 Zadeh, L.A. (1973). Outline of a new approach to the analysis of complex systems and decision processes. *IEEE Transactions on Systems, Man, and Cybernetics* 3 (1): 28–44.

31 Takagi, T. and Sugeno, M. (1983). Derivation of fuzzy control rules from human operator's control actions. *Proc. IFAC Sym. Fuzzy Inf. Knowl. Repres., Decis. Anal.*, pp. 55–60.

32 Takagi, T. and Sugeno, M. (1985). Fuzzy identification of systems and its applications to modeling and control. *IEEE Transactions on Systems, Man, and Cybernetics: Systems* 15 (1): https://doi.org/10.1109/TSMC.1985.6313399.

33 Fotouhi, A., Auger, D.J., Propp, K., and Longo, S. (2016). Electric vehicle battery parameter identification and SoC observability analysis: NiMH and Li-S case studies. *Proc. IET Conf. Power Electron. Mach., Drives (PEMD)*, Glasgow. https://doi.org/10.1049/cp.2016.0142.

34 Fotouhi, A., Auger, D.J., Propp, K., and Longo, S. (2016). Accuracy versus simplicity in online battery model identification. *IEEE Transactions on Systems, Man, and Cybernetics: Systems* https://doi.org/10.1109/TSMC.2016.2599281.

35 Fotouhi, A., Auger, D.J., Propp, K. et al. (2017). Lithium–sulfur cell equivalent circuit network model parameterization and sensitivity analysis. *IEEE Transactions on Vehicular Technology* 66 (9): 7711–7721.

36 Fotouhi, A., Auger, D.J., Propp, K., and Longo, S. (2018). Lithium–sulfur battery state-of-charge observability analysis and estimation. *IEEE Transactions of Power Electronics* 33 (7): 5847–5859.

37 Bizeray, A.M., Zhao, S., Duncan, S.R., and Howey, D.A. (2015). Lithium-ion battery thermal-electrical model-based state estimation using orthogonal collocation and a modified extended kalman filter. *Journal of Power Sources* 296: 400–412. https://doi.org/10.1016/j.powsour.2015.07.019.

38 Propp, K. (2017). Advanced state estimation for lithium-sulfur batteries. PhD thesis. Cranfield University.

Part IV

Application

This part focuses on the practical application and commercial exploitation of lithium–sulfur battery technology.

In Chapter 10 a detailed market analysis details the strengths and limitations of lithium–sulfur technology, and the battery requirements for a range of relevant markets are summarized. The aim is to develop a better understanding of the needs of commercial applications engineers for battery technologies and to focus research efforts to deliver those solutions. It is apparent that today Li–S batteries are no longer a one size fits all academic curiosity and that the knowledge exists to tailor the technology to meet market demand such that a meaningful market share can be realized. This can be achieved through knowledge of the working mechanism (Part II), materials research (Part I), and improved control systems (Part III).

Chapter 11 briefly outlines the significant differences to be taken into account by battery engineers designing battery packs for next-generation lithium–sulfur technology that might not otherwise be obvious.

Chapter 12 provides a real-world case study where lithium–sulfur batteries enabled the record-breaking high-altitude, long-endurance unmanned aerial vehicle (HALE UAV) to be developed in the United Kingdom for space-based surveillance.

10

Commercial Markets for Li–S

Mark Crittenden

OXIS Energy, 20 Nuffield Way, Abingdon, Oxfordshire, OX14 1RL, UK

10.1 Technology Strengths Meet Market Needs

Commercial opportunities for lithium–sulfur (Li–S) are enormous – the lithium ion (Li ion) "family" of technologies, undoubtedly the dominant performance battery technology over the last decade, has forecasted sales of around \$100 billion per annum for 2025. As the performance characteristics of Li–S have been increasingly demonstrated and harnessed over the last decade, this had led to significant investment in Li–S to exploit the commercial potential into this enormous market.

Key for the exploitation of any emerging technology is to understand which markets are most suited to the technology – the so-called "low hanging fruit" – and initially focus the commercial effort in these areas where companies will have the most success. Thus we will consider aspects that in which Li–S is strong compared to any competing technology, particularly with respect to Li ion, and then in which applications that strength is critical for success – this is where most commercialization activities are seen today.

10.1.1 Weight

Li–S has a theoretical gravimetric energy density (GED) of around $2700\,Wh\,kg^{-1}$, which is five times that of Li ion. Beyond the theory, from an engineering and customer perspective, what really matters is what has been achieved already and will be achieved in the near term, and how this can be integrated into batteries for specific applications.

This is where Li–S really stands out from the crowd. Li–S has already achieved over $400\,Wh\,kg^{-1}$ in commercial-size pouch cells. Figures are climbing fast and predictions are that they will reach $500\,Wh\,kg^{-1}$ by 2019 and $600\,Wh\,kg^{-1}$ by 2025. With small increases each year, Li ion on the other hand is seeing diminishing returns on investment and is expected to achieve $300\,Wh\,kg^{-1}$ in the next five years and perhaps $350\,Wh\,kg^{-1}$ in the long term.

Thus, batteries using Li–S are already significantly lighter than those using Li ion, with the likelihood of being half the weight in the next 10 years. This is clearly a major advantage for any application where the batteries have to be carried and should be seen as its key strength driving commercial uptake.

It should be noted that there are several variants of Li–S. The highest GED cells, achieving in excess of 400 Wh kg^{-1}, have a lower cycle life – today up to 200 cycles, depending on conditions.[1] A different variant has achieved in excess of 200 Wh kg^{-1} and is capable today of around 1500 cycles.

10.1.2 Safety

In recent years, there have been numerous high-profile accidents when using Li ion batteries. These include battery fires in the Boeing 787 Dreamliner, which was subsequently grounded for three months by the Federal Aviation Authority. Samsung recently recalled all Galaxy 7 Note smartphones globally due to fires with huge reputational and financial impact. The list goes on to Tesla vehicle battery fires, to hover-boards, and other applications.

Both the public and manufacturers are therefore increasingly demanding safer battery technology.

Of course, no battery technology is completely safe, but tests on Li–S cells have shown that they have strong safety benefits due to two key mechanisms, a ceramic lithium sulfide passivation layer and the choice of electrolyte. The Li–S cells have been shown to withstand many abusive conditions such as nail penetration, short-circuit, and even bullet penetration, with no adverse reaction.

10.1.3 Cost

Li–S has two fundamental advantages in terms of costs. Firstly, the high GED (e.g. Wh kg^{-1}) equates to less mass of material for a given energy storage, directly leading to lower cost per watt-hour. Secondly, sulfur is a key component in the cell, which is a very cheap commodity, particularly when compared to costly materials in Li ion such as cobalt. Therefore long-term costs for Li–S are very favorable.

In fact, the problem that needs to be overcome today is a problem that many emerging technologies face and that is to achieve high production quantities. Economies of scale are a critical part of any cell costs, as demonstrated by the ambition of Elon Musk and Tesla, building a Gigafactory in Nevada, USA. Tesla expects that the first Gigafactory will produce up to 150 GWh of Li ion battery packs per year for electric vehicles and energy storage, driving costs and pricing down.

Commercial scale-up of Li–S is already underway. A year ago, all Li–S cells were manually produced by skilled operators assembling cells from the constituent components; but now automated cathode coating lines and semiautomated cell assembly lines are operational and producing cells for batteries. Cell costs are therefore reducing accordingly and this trend will undoubtedly continue as the demand drives the production investment.

10.1.4 Temperature Tolerance

Temperature tolerance is generally good, during operation, for storage, and indeed during unexpected extremes. At the high end, cells operate at temperatures as high as 80 °C, albeit impacting cycle life, and have successfully completed safety tests where cells are

---

1 For example, measured at 80% depth of discharge and 60% beginning of life.

held at 150 °C without incident, although at this extreme, the cells are damaged. At the lower end, variants of the chemistry have been shown to operate below −60 °C, and cells are generally unaffected when frozen.

10.1.5   Shipment and Storage

A common problem with Li ion batteries is that they often need regular recharge during storage to prevent damage, which can lead to warranty issues. An example of this is that many electric scooters and bikes are kept in storage over winter periods, and consumers often do not realize that they need to periodically recharge the batteries – once the weather improves they find that the batteries fail and contact the manufacturer for replacement. Li–S cells have been shown to have a long shelf-life without causing damage, only requiring charging, or in extreme cases a few cycles, before being useful again.

This ability to fully cope without stress in the discharged state is also beneficial for shipment. Transportation of batteries is an increasing problem for both the industry and the public when boarding the aircraft.

The internationally recognized authority for the testing, packaging, and shipping of cells and batteries is the United Nations and in particular the UN Manual of Tests and Criteria, section 38.3 entitled "Lithium Metal and Lithium Ion Batteries." The manual details the following series of tests to be completed prior to shipment:

Test	Title	Purpose
T1	Altitude simulation	Simulates air transport under low-pressure conditions
T2	Thermal test	Assesses the cell and battery seal integrity and internal electrical connections. The test is conducted using rapid and extreme temperature changes
T3	Vibration	Simulates vibration during transport
T4	Shock	Assesses the robustness of cells and batteries against cumulative shocks
T5	External short circuit	Simulates an external short circuit
T6	Impact/crush	Simulates mechanical abuse from an impact or crush that may result in an internal short circuit
T7	Overcharge	Evaluates the ability of a rechargeable battery or single cell rechargeable battery to withstand an overcharge condition
T8	Forced discharge	Evaluates the ability of a primary or a rechargeable cell to withstand a forced discharge condition

Many Li–S cell designs have already passed this standard. Furthermore, shipping cells in the discharged state, coupled with the safety benefits discussed above, reduces the safety risk when shipping. UN transportation regulations are in the process of being updated to reflect new technologies that had not been considered when they were written.

10.1.6   Power Characteristics

Li–S must be considered a high-energy, rather than high-power, chemistry. Applications that require discharge time longer than an hour are generally the most suitable. However, depending on the discharge profile hybridization systems are now being considered, combining Li–S with a higher power technology. Similarly, maximum charge rates are currently two to three hours, so applications that require fast charging are not currently suitable. This nevertheless continues to be an area of research.

10.1.7 Environmentally Friendly Technology (Clean Tech)

The chemistry of Li–S is considered to have less environmental impact when compared to other technologies such as Li ion. The Li–S cell utilizes sulfur in place of heavy metals such as cobalt, which have a significant environmental impact if they enter the food chain in concentrations that would otherwise not occur naturally. In contrast, both lithium and sulfur occur naturally: indeed sulfur is found in large concentrations around volcanoes and deposits in the United States. For manufacturing, it can be obtained as a recycled by-product from the oil industry. Lithium or its salts are abundant in rocks and spring waters. Studies have shown how the lithium can be removed from Li–S cells at the end of life for recycling purposes.

10.1.8 Pressure Tolerance

In tests performed by the National Oceanography Centre, a leading marine research establishment based in the United Kingdom, Li–S cells were shown to continue to operate under a pressure of 660 bar, which is equivalent to sea water depths of approximately 6600 m.

10.1.9 Control

More complex chemical and electrochemical reactions within the cell necessitate the use of advanced control algorithms in the battery management system (BMS) in order to accurately predict state of charge (SoC) and state of health (SoH).

Understanding the strengths and limitations of the technology is essential for ensuring that the correct applications are being developed. Applications development includes land-based electric vehicles (cars, trucks, buses, and scooters), electric aircraft, marine applications, and energy storage. OXIS Energy is currently involved in the development of batteries in all the applications listed in this chapter. Sion Power is known to have developed batteries for the unmanned aerial vehicle (UAV) sector.

10.2 Electric Aircraft

Clearly, the energy required to lift a battery and keep it in the air is significant, so weight is the most important factor in the choice of battery technology for this application. Chapter 12 provides a case study of a fixed-wing UAV, showing the advantages that lightweight technology brings to this application, and so will not be discussed in detail in this chapter.

There has also been a very strong demand for small electric aircraft carrying a pilot and/or passengers. For example, EasyJet is working with Wright Electric to develop a London to Paris electric aircraft within 10 years. Airbus had initially been developing a pure-electric two-seater aircraft, called the E-Fan, and now has plans for a hybrid-electric aircraft, the E-Fan X, in collaboration with Rolls-Royce and Siemens.

The benefits of switching to electric aircraft are seen as reduced costs and reduced noise pollution. To be successful in pure electric aircraft, the GED has to be high enough that the battery capacity is sufficient to power the aircraft for the flight duration, including a safety reserve. To understand this relationship, let us consider a small two-seater aircraft to see how the flight duration is affected by gravimetric energy and battery weight.

Li ion batteries have already been implemented in such electric aircrafts, but are unable to achieve commercially viable flight times. By increasing the battery weight, the flight time can be increased, but there are of course limits to this weight – such as airframe considerations and motor power considerations, and ultimately the battery becomes too heavy to lift. Practically the limit on a two-seater aircraft is around one hour using Li ion batteries.

The following graph shows the expected flight duration for a given battery weight and for different GEDs.

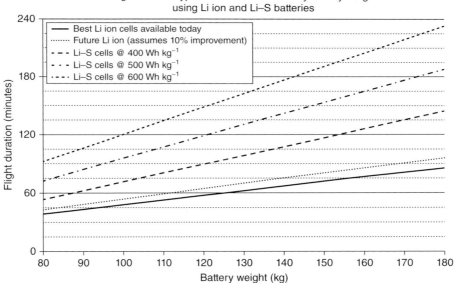

Flight time for typical two-seater aircraft by battery weight using Li ion and Li–S batteries

As Li–S energy densities improve, it can be seen that with the weight restriction in such aircraft, the flight duration is expected to extend to around four hours by 2025. Using modular battery packs allows quick swap on the ground, so that planes will not have to sit on the tarmac while batteries are being recharged.

Similar calculations can be performed on larger aircraft showing how EasyJet's aspirations to have a larger electric passenger aircraft flying from London to Paris within 10 years are possible.

Furthermore, in aircraft, safety is paramount – the impact of a mid-flight fire can clearly have catastrophic consequences. Li–S offers the strong advantage of reducing this likelihood.

Looking beyond fixed wing aircraft, in the last few years there has been an enormous growth in small drones not only for leisure, but also for commercial activities such as inspection, surveillance, and filming. Historically, the power requirements were beyond the abilities of Li–S, but as energy density improves, flight times increase and discharge rates[2] reduce. Li–S is now reaching the turning point where flight times will be improved compared to Li ion due to these less stressful discharge rates. A rapid uptake of Li–S in this application should be expected over the next few years.

10.3   Satellites

The leader in research and development for the implementation of Li–S batteries in space applications batteries is Airbus Defence and Space, who have several projects in this area.

There are many reasons why Li–S is attractive to Airbus and its competitors in this sector and clearly the reduction in battery mass is the main advantage. In a typical satellite, the rechargeable battery has a mass of 100–200 kg, with launch costs of €10–20 000 per kilogram per launch. Conservative estimates are that upgrading to Li–S will save between 30% and 50% of the battery weight, saving around €1 million per launch.

Furthermore, thermal management of batteries is currently required to maintain Li ion batteries between 10 and 30 °C. Li–S offers the opportunity of widening this operation window and simplifying the thermal control, reducing associated power requirements, weight, and thus costs.

Safety is again seen as a strong benefit due to the reduced risk of catastrophic failure on extremely high value equipment, which have very long lead times to replace. Other benefits include the reduced toxicity of Li–S and reduced maintenance.

Cycle life needs vary significantly across the different space applications, from launchers to low earth orbit (LEO) satellites to geostationary (GEO) satellites – accordingly requirements vary from tens of cycles to tens of thousands of cycles. Therefore, Li–S is only suitable to a subset of space applications today, but this subset will grow as cycle life improves.

10.4   Cars

Electric vehicles (EVs) have been around for a surprisingly long time. The first demonstration EVs were produced in the 1830s, but were only commercially produced from the late nineteenth century with the introduction of mass-produced rechargeable batteries. In fact, the first person to break the mile-a-minute barrier was Gaston de Chasseloup-Laubat in the Jeantaud Duc EV in 1898. By the 1920s several hundred thousand EVs had been produced. However, the internal combustion engine (ICE) soon began to dominate due to its superior GED. Petrol has a GED of over 12 000 Wh kg^{-1} compared to lead acid at 30 Wh kg^{-1}, so with little consideration to the environmental impact, the ICE became the clear winner.

---

2 Here we mean "C" rates: power divided by capacity.

The key barrier to the widespread uptake of electric vehicles is known as "range anxiety." This is essentially the concern that a vehicle has insufficient range to reach the desired destination and the occupants of the vehicle will be left stranded. Thus as the vehicle range increases, anxiety decreases, and more electric vehicles are purchased. For conventional petrol and diesel cars to be replaced, the range of electric vehicles needs to be similar to the ICE equivalents, and at a competitive price.

However, EVs powered today by Li ion batteries are now beginning to gain acceptance among consumers. Strong growth is being seen and is predicted to continue for the new generation of EVs that offer a 500 km range to the user, such as the Tesla Model S and Model X, the Audi Q6 e-tron, and Porsche Mission E. These vehicles compete directly with ICE vehicles in terms of performance and range, but are relatively high cost and aimed at the premium market.

There is now a concerted effort to provide affordable electric vehicles with a large range that will significantly increase EV adoption as long as reliability, safety, and life-time can be guaranteed. Tesla Motors and Chevrolet are now targeting this gap with the Model 3 and the Bolt respectively.

Safety is an increasingly emotive subject for operators, manufacturers, governments, and the public. The safety of ICE vehicles is well understood, and such vehicles are designed to protect fuel tanks as escaped fuel can ignite relatively easily.

Li ion by contrast has a poor reputation for safety due to the coverage given to the small number of failures that occur and may relate to the difficulty of extinguishing lithium fires. Similarly to ICE vehicles, many design decisions have therefore been made to electrically and physically protect the Li ion batteries from an event occurring and to minimize the impact of any such event. However, when these protection mechanisms are overcome by a freak occurrence or when the battery is punctured, fires still occur. Li–S can offer an extra layer of protection here, where Li ion cannot.

The revolutionary electric vehicle battery (REVB) project was an Innovate UK-funded collaboration between OXIS Energy, Cranfield University, Imperial College London, and Ricardo. The project developed automotive Li–S cells and battery modules and analyzed the key requirements for a typical "D-segment" car. "D-segment" is a European market car classification defined by the European Commission – examples include Audi A4, BMW 3 Series, Jaguar XE, Mercedes-Benz C-class, Volvo S60, Ford Mondeo, Honda Accord, and Vauxhall (Opel) Insignia. The REVB analysis of battery requirements is summarized in the following table:

	D-segment	Unit
Specific energy	400	$Wh\,kg^{-1}$
Volumetric energy	400	$Wh\,l^{-1}$
Maximum discharge	2	C-rate
Maximum charge	1.6	C-rate
Cycle life	330	Cycles
Operating temperature	−20 to +60	°C
Cell cost	125	$/kWh
Cell capacity	30+	Ah

Except for cost, all of these parameters have already been met individually in different cells. The key to success is to develop a single cell design that can deliver all these characteristics. Many argue that the greatest challenge is to combine the high GED with the cycle life. This should all be within the capability of Li–S within a few years.

It is also worth noting that volumetric energy density tends to be more challenging for smaller vehicles, and hence larger vehicles are more likely to adopt Li–S.

Although the cost requirement has not yet been achieved, as discussed above, the problem is essentially one of scale-up for these figures to be met.

Most of the major car manufactures are now investigating the merits of Li–S to power electric vehicles. This includes two Horizon 2020 projects, funded by the European Commission:

The "Advanced Lithium–Sulfur Batteries for Hybrid Electric Vehicles" (ALISE) project, to develop a battery for a SEAT hybrid electric vehicle, containing both a conventional ICE and an electric motor system.

The "High Energy Lithium–Sulfur Cells and Batteries" (HELIS) project to develop cells that are to be tested for automotive use, with the participation of PSA Peugeot Citroen. Also considering that the integration and testing process for such new technologies in cars can take 5–10 years, we would expect to see the release and uptake of Li–S around 2025–2030.

10.5 Buses

Much progress has been made in electric cars but the most polluting vehicles on the road are heavy vehicles such as buses, trucks, and vans. These vehicles are considered harder to develop because the cost of design is high, unit cost is high, and the market is smaller in terms of the number of vehicles, leading to a more cautious development approach from original equipment manufacturers (OEMs).

By switching from an ICE vehicle to a battery vehicle, emissions at the point of use drop to zero. This will provide a clear benefit in terms of air quality and overall health of citizens, and help combat climate change. Electric buses are now being seen, but there are several challenges to overcome before the widespread adoption of electric buses.

The key advantages for Li–S in buses when compared to using Li ion are weight saving and safety. The range of the system is defined by the capacity of the battery, which directly affects the mass. Here, Li–S can offer a significant advantage over Li ion.

Consider the example of an electric bus battery with a capacity of 476 kWh. Below is a comparison of battery weights using Li–S and lithium–iron phosphate (LFP). Weight saving allows more passengers to be carried in the bus, as shown below.

Battery	Weight (tons)	Saving (%)	Passenger increase
LFP	4	–	–
Li–S (300 Wh kg^{-1})	1.6	60	34 European/41 Asian passengers
Li–S (400 Wh kg^{-1})	1.2	70	39 European/48 Asian passengers
Li–S (500 Wh kg^{-1})	1.0	76	43 European/52 Asian passengers

Instead of reducing weight, the bus manufacturer has the opportunity of increasing range. For example, the range of a battery using Li–S (at 400 Wh kg^{-1}) would be 233% more than that of an LFP battery of the same weight. Of course, the manufacturer also has the option of both reducing the weight and increasing the range, balancing both these advantages.

From a safety perspective, with a battery that is an order of magnitude greater than in a car, battery risks are clearly higher. Thus Li–S offers a major advantage.

The lifetime cost of the system must also be low enough to undercut the ICE vehicles; this is possible because of the economies of scale now being seen in the battery industry. Li ion battery chemistries are mature and readily available today and so costs are already relatively low. As previously discussed, the cost of Li–S cells is predicted to undercut that of Li ion in the long term, as most of the production processes are the same, but the raw materials are cheaper. The result of this is that Li–S batteries will eventually undercut ICE and Li ion powertrains and therefore further increase the adoption of electric buses.

Cycle life is of course a large consideration for electric buses and indeed forms part of the lifetime cost calculation. Thus, to be successful in the electric bus market, higher cycle life must be achieved.

Generally, overnight charging is required for electric buses, which fits with Li–S capabilities. In some cases, there may be opportunities for top-up charges through the day, thus reducing the battery size.

As already mentioned, Li–S is best suited to lower power applications. However, when power rates are very low, this can have an impact on both performance and lifetime. One solution for this problem is to have banks (or strings) of battery packs. If power is needed all banks are allowed to discharge, but when power rates are very low, a subset (or even a single bank) is discharged to increase the power rate on those selected banks. Bank selection can be rotated to ensure banks remain at the same state of charge, voltage, and lifetime.

10.6 Trucks

Vans and trucks often carry out the same delivery and collection tasks, but the truck can carry almost 10 times the payload, which lowers operating costs and emissions. Therefore, operators prefer trucks but cannot always use them due to vehicle size constraints; this is why smaller vans are also used. Tractors and trailers are used for very large customers with their own depots as well as for long distance travel between user depots.

As an example, let us first understand the operation of vehicles in a typical logistics depot and how the associated trucks and vans work over a typical weekday – this is shown in the diagram below, noting that the time of the day is indicated at the top.

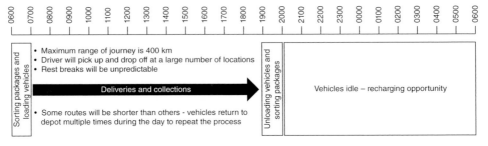

Key points to add are as follows:

- The total distance covered on most days is less than 400 km, but operators need all vehicles to be capable of this to avoid the situation where deliveries and collections need to be made, but no vehicles are available.
- The locations of delivery and collection vary each time.
- Vehicles often leave a depot with only 80% of the stops planned and the fleet manager will add stops during the day for further deliveries and collections.
- Vehicles are only idle for a very short time.
- Rest stops are unpredictable in terms of location and duration, being at the discretion of the driver, while keeping within agreed working rules. Indeed, many rest stops can be in a layby where there is no grid infrastructure for a charger.

Thus the opportunity to charge in the day is extremely limited and overnight charging with an intelligent charging strategy is the best solution.

By charging overnight, the power required to charge the vehicles is small enough to use standard, off-the-shelf chargers. This will also allow easier integration to the grid, reducing cost, and will mean that standard connections are used, allowing other vehicles to be charged. By integrating the Fleet Management System with the chargers and vehicles, the loads can be planned in to gently ramp up and down to prevent significant spikes and system destabilization.

In this situation, the advantages of Li–S are thus the following:

- The reduction of weight using Li–S allows batteries with greater capacity, and thus lasting all day, enabling overnight charging.
- When comparing with Li ion, Li–S battery chemistry offers the advantage of a higher payload due to a lighter battery.
- Similarly to other vehicles discussed, Li–S offers better safety in the event of an accident.
- Li–S involves lower long-term cost.

10.7 Electric Scooter and Electric Bikes

There are more electric scooters and bikes in the world than all other electric vehicles combined. China is dominant, both in production and sales. Historically electric scooters and bikes were powered by lead acid batteries but there is an ever-increasing demand for Li ion batteries partly due to the demand on vehicle range and in part due to environmental considerations of the heavily polluting lead acid batteries. Batteries typically have a capacity of 1–2 kWh.

In this market, volumetric energy density tends to be more important than GED as weight saving of a few kilograms has minimal impact on the overall weight of the bike and rider. However, many of the issues of space on scooters can be overcome through good design as there is generally unused space under the seat.

Power requirements can be high, particularly when carrying a heavy rider up a long incline, but with larger capacity batteries, power limitation can be overcome.

Scooter batteries are already under development and test. However, price is predominantly the key driver for battery sales, particularly in the Chinese market, and once pricing drops, there should be a strong demand in this area.

10.8 Marine

The term "marine" encompasses many different applications, for small leisure boats, hotel loads (powering air conditioning, lighting, and entertainment systems) on superyachts, large transport vehicles, and underwater vehicles. Owing to the ability of Li–S cells to handle pressure, the relatively small size of the batteries, and the high price of these batteries, underwater vehicles is the main area of focus for Li–S in the marine sector.

As part of an Innovate UK-funded project, Steatite Batteries has been working with the National Oceanography Centre (NOC), OXIS, and MSubs to develop a pressure-tolerant Li–S battery capable of withstanding the harsh pressures of deep sea exploration. The NOC has tested cells and shown that they operate at a pressure equivalent to depths of 6600 m.

Thus rather than having to use expensive housing, batteries have been developed that can directly withstand this pressure.

Furthermore, current lithium polymer (LiPo) batteries require expensive syntactic foam to ensure neutral buoyancy of an underwater vehicle. As Li–S has a mass density similar to that of sea water, the foam is not required, saving on weight and cost.

The table below shows a comparison between LiPo and Li–S. Net battery energy density (NBED) considers the overall effect of the cells within a battery, plus any buoyancy foam – a significant improvement is seen compared to LiPo.

Depth rating (m)	NBED (%)
0	+212
1000	+218
4500	+244
6000	+257

The initial drive has been for autonomous underwater vehicles, but this solution can equally be applied to manned vehicles, particularly when the added benefit of safety is considered. Looking more widely in marine, there is significant potential to grow into surface vessel applications such as hotel loads, where high-end customers demand power with the noise of noisy generators, and motive applications including pleasure boats, which have requirements similar to cars.

10.9 Energy Storage

Energy storage has become a rapidly growing market for batteries, driven by an increasing share of renewable energy generation. This changing mix of energy production has created new problems to ensure grid supply and minimize electricity bills. Energy storage provides an effective solution for grid stabilization, power generation management, and residential storage. Industry and consumers want a cost-effective solution that is safe and easy to install with minimal maintenance.

Although being lightweight has some small advantages for installation, the vast majority of this market is primarily driven by the battery cost and the cost per cycle. The improvement in costs has already been covered in this chapter; improvements in cycle life will further drive down the cost per cycle such that Li–S would be strong in this market within 10 years.

The early adoption of Li–S will be in those specific applications where safety and weight play an important factor.

10.10 Low-Temperature Applications

Although niche, there is a demand for very low temperature applications. For example, the British Antarctic Survey (BAS) currently uses lead acid batteries that slowly discharge their energy over the winter months to power electrical equipment. Temperatures are very much at the limit of lead acid capabilities and achieve 20–25 Wh kg^{-1}. Batteries are usually airlifted to the location, so transport costs to both deliver batteries to the site and later remove them are very high. Reducing weight will therefore have a significant cost saving.

As mentioned previously there is not one single Li–S chemistry, and variants have been developed. One such cell is capable of temperatures below −60 °C, albeit at a reduced GED of 100 Wh kg^{-1}. This is nevertheless four times better than the lead acid batteries used today and has the potential to save 75% in transportation costs.

10.11 Defense

The defense sector has a keen interest in lightweight batteries not only in bespoke versions of the applications already discussed but also in man-portable batteries. On the battlefield, soldiers frequently carry a heavy burden due to the assortment of electronics systems, ammunitions, food, and water. With battery loads on a soldier sometimes as high as 10 kg, reducing this burden is a key priority for the defense industry.

Furthermore, safety is equally important. Being able to withstand bullet penetration with no adverse reaction and no significant temperature rise, Li–S is perfectly suited to offer critical power to soldiers having to operate in extreme temperature and harsh conditions.

10.12 Looking Ahead

As mentioned at the start of this chapter, for the successful commercialization of Li–S it is important to focus on where the technology is strongest. The challenge for the technology is the volumetric energy density, but this is of course only a disadvantage when "size matters." Volume is of course a consideration in all of the applications listed above, but these applications are driven by another factor, usually weight, and the volume characteristics are sufficient.

There are many applications where volume is critical, and the biggest market by far is consumer electronics such as mobile phones (or "cell phones" to US readers). For Li–S to replace Li ion in this application, volumetric energy must reach $600\text{--}1000\,\mathrm{Wh\,l^{-1}}$. This is several years away, but certainly within the realms of the possible.

10.13   Conclusion

There are many applications suited to Li–S today, particular those playing to the key strengths of high GED. We now see increasing activity producing commercial cells and developing batteries for commercial release, such that in the next decade it is possible that Li–S could take a significant share of the battery market.

11

Battery Engineering

Gregory J. Offer

Imperial College London, Exhibition Road, London SW7 2A, UK

11.1 Mechanical Considerations

As described in the section on degradation, the volume and therefore the thickness of a lithium–sulfur pouch cell can change significantly during operation [1]. Upon a normal charge–discharge cycle a lithium–sulfur cell can vary by 5% or more, known as "breathing." This change is driven by both the increase in the number of species dissolved in the electrolyte as the long chain polysulfides become oxidized to lower chain polysulfides, and the plating and dissolution of the lithium electrode that seems to contribute the most to irreversible thickness increase. This irreversible thickness increase as the cell degrades (i.e. loses usable capacity) can lead to a 30–40% increase in thickness at the end of life. Cells will also undergo thermal expansion as the temperature changes, which is comparable to the "breathing" caused by cycling.

From a systems engineering perspective, this presents a challenge, as space must be left to allow cells to both "breathe" during normal cycling or to account for changes in temperature, and irreversibly expand as they age. This can impact on the achievable volumetric energy density for the pack even at the beginning of life.

11.2 Thermal and Electrical Considerations

These two considerations are so intrinsically intertwined that it is not recommended to consider them in isolation. Previous work on lithium ion batteries has shown that thermal gradients across a cell can significantly affect the performance [2] and also have a significant effect on accelerating degradation [3]. In summary, if the large surfaces of a pouch cell made of multiple layers are used for thermal management, then this causes the individual layers to be at different temperatures. The layers are thermally in series. As the layers are connected electrically in parallel and their impedance is a strong function of temperature, this will lead to significant current inhomogeneities between layers. For example, at the beginning of discharge, the middle layers will get hot and stay hot, their impedance will be lower, and they will discharge first. As they become depleted, the layers producing the most current will move outwards, until at the end of discharge, a few layers, with high impedance because they are cold, will produce most of

Lithium–Sulfur Batteries, First Edition. Edited by Mark Wild and Gregory J. Offer.
© 2019 John Wiley & Sons Ltd. Published 2019 by John Wiley & Sons Ltd.

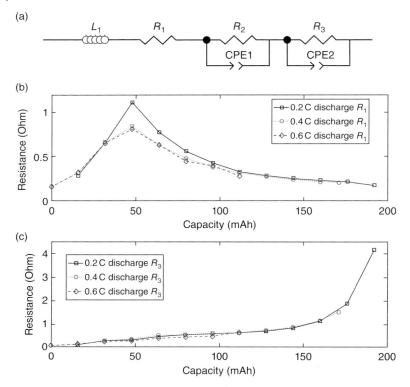

Figure 11.1 The classic "volcano"-shaped high-frequency resistance R_1 caused by the concentration of species in solution in plot (b) and the low frequency resistance R_3 caused by precipitation reducing the active surface area in the cathode in plot (c) and the equivalent circuit used to extract the parameters (a). Source: Zhang et al. 2018 [4]. Reproduced with permission from Electrochemical Society.

the current. This will lead to the voltage cutoff being reached early, leaving considerable capacity behind. The accelerated degradation over many cycles is caused by the higher local C rates on each layer, as they discharge one after another, compared to the constant average C rate for the cell as a whole.

This effect has been reproduced for lithium–sulfur, and shown to suffer from additional positive feedback effects caused by the impedance of the lithium–sulfur cell increasing in the high plateau and decreasing in the low plateau as the cell is discharged, and the inverse on charging.

This is again caused by the increase in the number of species dissolved in the electrolyte as the long chain polysulfides become oxidized to lower chain polysulfides, which increases the viscosity and therefore decreases the conductivity of the electrolyte, as shown in Figure 11.1. The consequence of this on a multilayer cell behavior is that any layer that makes it over the peak of the volcano before another will race ahead of the others; as the impedance decreases it produces more current, discharging faster and racing ahead even more. This is further compounded by the temperature increase and further impedance decrease caused by the heat generated from this higher current. All this

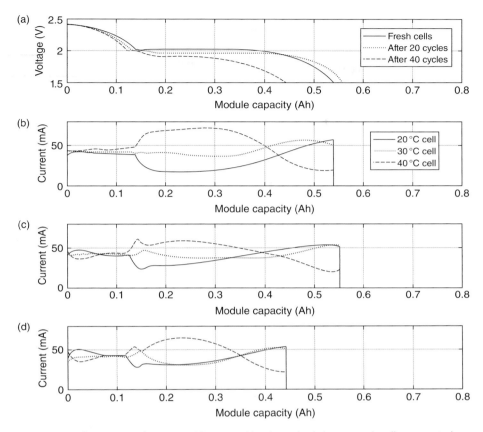

Figure 11.2 The current inhomogeneities caused by three single layer pouch cells connected electrically in parallel but artificially held at different temperatures during discharge. Source: Hunt et al. 2018 [5]. Reproduced with permission from Electrochemical Society.

contributes toward significant positive feedback. During charging, exactly the same phenomenon occurs and could lead to a small number of layers in the middle of the cell becoming fully charged before the others and entering the shuttle well before the cell would have done if it was behaving uniformly (Figure 11.2).

From a systems engineering perspective, this means that surface cooling of pouch cells for lithium–sulfur batteries is not recommended for any application other than very slow charge/discharge rates where the rate of heat generation is not significant enough to lead to a thermal gradient of more than 1 or 2 °C. In any application with the risk of significant thermal gradients, tab cooling should be considered as this in theory cools each layer by the same amount and therefore should not contribute toward current inhomogeneities within the cell. Considering how and if the cell has been designed for this type of cooling should therefore be a significant concern of the pack engineer; it is not sufficient to just assume the cell designer has designed the tab and tab welds for good thermal management as well as good electrical connection.

References

1 Waluś, S., Offer, G., Hunt, I. et al. (2018). Volumetric expansion of lithium–sulfur cell during operation – fundamental insight into applicable characteristics. *Energy Storage Materials* 10 (May): 233–245.

2 Troxler, Y., Wu, B., Marinescu, M. et al. (2014). The effect of thermal gradients on the performance of lithium-ion batteries. *Journal of Power Sources* 247: 1018–1025.

3 Hunt, I.A., Zhao, Y., Patel, Y., and Offer, J. (2016). Surface cooling causes accelerated degradation compared to tab cooling for lithium-ion pouch cells. *Journal of the Electrochemical Society* 163 (9): A1846–A1852.

4 Zhang, T., Marinescu, M., Walus, S. et al. (2018). What limits the rate capability of Li–S batteries during discharge: charge transfer or mass transfer? *Journal of the Electrochemical Society* 165 (1): A6001–A6004.

5 Hunt, I., Zhang, T., Patel, Y. et al. (2018). The effect of current inhomogeneity on the performance and degradation of Li–S batteries. *Journal of the Electrochemical Society* 165 (1): A6073–A6080.

12

Case Study

High-Altitude, Long-Endurance Unmanned Aerial Vehicles (HALE UAVs) – A Perfect
First Application for Lithium–Sulfur Batteries
Paul Brooks

Prismatic Ltd, Cody Technology Park, Farnborough, Hants GU14 0LX, UK

12.1   Introduction

Flying forever using only the power of the sun – this is the vision for solar-powered,
high-altitude, long-endurance, unmanned aerial vehicles (HALE UAVs). It is a vision
that has been around since the early 1980s and, despite the attention of some of the
great names in aerospace and innovation, remains as yet an unfulfilled promise.

The concept is simple. Fly using electric motors and propellers, with photovoltaic
(solar) cells providing the power to fly during the day and to charge batteries to maintain flight overnight. There are two variants to the concept – an aeroplane that flies
using aerodynamic lift and an airship (lighter than air, LTA) that uses a lifting gas (usually helium) to maintain altitude and electric propulsion to maintain location. In both
cases, the need for a continuous source of power results in the use of photovoltaic (PV)
solar cells and regenerative electric energy storage. This case study takes the highest level
interest and requirements in such systems – focusing on aircraft that have been more
successful to date – and considering the technical and business drivers on the systems
shows how regenerative electrical storage, specifically high specific capacity batteries, is
both the critical element in the technology and an attractive first application of any new
battery technology.

We start with a potted and relatively unhappy history of HALE UAV and LTA activity
and use this to highlight the technical and business demands. The experience gained by
the author through the implementation of the only successful solar HALE UAV to date
(Zephyr) is then used to detail more precise requirements on battery technology and to
leave a clear, short, valuable and, in some ways, relaxed objective for the experts in the
lithium–sulfur field.

12.2   A Potted History of Eternal Solar Flight

When discussing the history and progress of HALE UAV, I will start by putting the
emphasis on the H – High (altitude). Flying high to enable persistent operation is essential both for staying above the clouds, which would prevent recharging of the battery,
and for staying above the worst of the winds (the jet streams), which would prevent the

vehicle from staying where the user wants it. Why is this different from normal aircraft? Conventional, modern aircraft tend to fly much faster, so while the wind speed affects the time to get where they are wanted it is unusual for an aircraft not to be able to defeat the wind. The problem for solar-powered vehicles is that it takes a lot of power to fly fast (increasing with the cube of the airspeed) and as will be seen later, solar electric aircraft must minimize power usage, so must fly as slow as possible. This means that the combination of requiring to fly slow, requiring to stay in sunlight, and requiring to stay in one location needs an "eternal aircraft" to fly high, above the jet streams that extend up to 65 000 ft in parts of the world. We will therefore quickly move on to those programs that attempted to fly high and long on solar power. This is a theme that will be returned to in the chapter – it is much simpler to fly at low altitude than high altitude and it is much simpler to fly using solar power in the summer than it is in the winter.

Early attempts at solar flight paralleled the improvements in solar PV technology, both in terms of areal power ($W\,m^{-2}$) and specific power ($W\,kg^{-1}$). The early model flights of Robert Boucher [1] were followed by Paul Macready's (manned) Gossamer Penguin and Solar Challenger in the early days of AeroVironment [2]. These flights were taken up by the US government in the 1980s and were then brought into the NASA Environmental Research Aircraft and Sensor Technology (ERAST) program in 1994, which led to the high-altitude flights of Pathfinder, Centurion, and eventually Helios, which set a record altitude of 98 000 ft in 2003. While the ERAST program effectively stopped development of solar-powered HALE UAV following the loss of Helios in 2003, its success encouraged a number of other programs including the QinetiQ Zephyr [3] and the DARPA Vulture [4]. While the DARPA program never progressed beyond subsystem technology development, the Zephyr program resulted in the successful demonstration of high-altitude, long-duration flight of a solar-powered aircraft with a 14-day flight in 2010 [5] and a further 11-day flight under winter conditions (nights longer than the days) in 2014 [6] (Figure 12.1).

In parallel with these HALE UAV programs, other groups were investigating the use of solar-powered airships for high-altitude, long-endurance flight. The most significant

Figure 12.1 Zephyr 7 in flight. Source: Courtesy of Airbus.

progress was made by US Department of Defence programs through the HALE-D, Hi Sentinel, and LEMV programs [7]. While vehicles were built and flown and the Hi-Sentinel ascended to 74 000 ft in 2005 for a few hours, none of these programs, to date, have resulted in a flight of more than a few hours and only one has achieved a useful altitude.

The latest phase of intent for persistent high-altitude flight started in 2013. Airbus purchased the Zephyr program from QinetiQ to complement its satellite-based services and both Facebook and Google announced plans to develop high-altitude platforms. This latest phase has followed past history to date, with the Zephyr successfully flying, resulting in the procurement of three vehicles by the UK Ministry of Defence [8] and no other program resulting in any flight of altitude or endurance.

Two thoughts might be forming as you read through this potted history. The first thought may be, "Is it a little conceited to talk of the Zephyr as the only successful HALE UAV to date?" The fact is that despite 30 years of development, hundreds of millions of dollars, and the world's best aerospace and innovation companies' efforts, to date, only the Zephyr has managed to fly for more than one day at an altitude suitable for persistent flight. This is not to say that others will not succeed and succeed soon. There is no magic to the Zephyr. The program was developed in a very focused environment and was also born out of a satellite engineering team and the end result is a very light, very efficient aircraft.

The second thought might be "Why has it been so difficult?" This will be addressed in the next section and be related specifically to the challenge for the battery industry. Before moving on to that section we need to ask the more specific question, "Why has *what* been so difficult?"

The key issue around HALE flight is the ability to stay high. We will see from the analysis in this paper that it is much easier to fly at low altitude than high altitude (requiring three times less power). It is also easier to fly during the summer than during the winter (as the sun is higher, the days are longer, and the winds less demanding). It is also easier to fly high during the day – as we can get 24 times more energy out of a solar cell than we can out of a battery. So the answer to the question "Why has what been so difficult?" is another question "Why is it so difficult to stay at high altitude, overnight in the winter?"

The author fully expects that other aircraft and airships will fly using solar power – this has been undertaken frequently since the 1970s. It is also fully expected that solar aircraft and airships will fly at high altitudes for long durations soon, breaking the seven-year-old record of Zephyr 7. To determine how close any group is getting to achieving the goal of year-round persistent flight, two questions need to be asked:

"What was the lowest altitude descended to during the night?"
"How long was the night compared with the day?"

12.3 Why Has It Been So Difficult?

While there are many issues to be dealt with in delivering a long duration flight in the harsh stratospheric environment, many of these have already been successfully dealt with through space programs – specifically the build of high-reliability electronic systems that can cope with low pressures, extreme temperatures, and an ionizing radiation environment. This experience from the space business was one of the critical elements to the success of the Zephyr – our team was a space technology team who had previously designed and built spacecraft and systems that had gone as far as Mars. Putting aside the challenges of environment and reliability, it all comes down to energy, and as we will see, specifically, the energy stored per kilogram by the batteries.

For an aircraft to fly, two equations govern the energy balance.

To fly requires sufficient lift to balance the weight of the aircraft

$$F_{\text{lift}} = \frac{1}{2}\rho A C_{\text{L}} V^2 \tag{12.1}$$

where

F_{lift} is the lift force, which must equal the weight of the aircraft,
ρ is the air density,
A is the lifting area of the aircraft (the wing area),
C_{L} is the coefficient of lift (typically 0.7–1.0), and
V is the velocity of the craft through the air.

The movement through the air incurs a drag that must be balanced to maintain the flight speed. This drag force is given by

$$F_{\text{drag}} = \frac{1}{2}\rho A C_{\text{D}} V^2 \tag{12.2}$$

where

F_{drag} is the drag force, which must be balanced by the thrust produced by the aircraft to keep flying at the given speed,
ρ is the air density,
A is the lifting area of the aircraft (the wing area),
C_{D} is the coefficient of drag, comprising of a fixed element and an element induced by the forcing of the air down to produce the lift, and
V is the velocity of the craft through the air.

The power required to keep the aircraft flying is then

$$\text{Power} = F_{\text{drag}} \cdot V \tag{12.3}$$

A quick substitution in these equations shows that the power to fly an aircraft is inversely proportional to the density of the air (ignoring Reynold's number effects on aerodynamic efficiency). At 65 000 ft the air density is approximately 8% of the density at sea level, so the power required to fly at high altitude is nearly four times that required to fly at sea level. This is the first indicator of how close a group is to HALE flight – if all they have done is fly near the ground, they are a factor of four away just in terms of energy, ignoring all the other issues of high-altitude flight such as temperature and radiation.

These equations will be used in the context of a practical HALE UAV following a similar discussion of an airship.

For an airship the lift is produced by the difference in density of the lifting gas and the surrounding air and the propulsive power is dictated by the power needed to offset the force of the wind on the balloon. Airships have the inherent joy of not needing power to stay aloft, but the large volumes required to float at high altitude mean that the aerodynamic drag on these large envelopes can be very large and rapidly exceed the power required by an aircraft to fly at the same speed.

The lifting force, F_{lift}, achieved by a balloon of volume, V, is

$$F_{\text{lift}} = V \cdot (\rho_{\text{air}} - \rho_{\text{gas}})$$ (12.4)

where

ρ_{air} = the density of air at the float altitude and
ρ_{gas} = the density of the lifting gas at the float altitude.

For an airship using helium at 65 000 ft the lift generated is 80 gm^{-3}. Note again the difficulty of flying at high altitude as at sea level the same balloon provides 1 kg of lift per cubic meter.

Even the smallest proposed demonstrators of such airships required volumes of 150 000 m^3 to lift a nominal 30 kg of payload [1]. Even with a simple single envelope, this volume requires an airship that is some 20 m in diameter and 80 m in length.

The drag of an airship uses the same equation as for the aircraft, with the reference area being the cross-sectional, frontal area rather than the wing area.

$$F_{\text{drag}} = \frac{1}{2}\rho A C_{\text{D}} V^2$$ (12.5)

where A is the cross-sectional area of the airship and the other symbols retain their meaning. Typical C_{D} values for airships are in the range 0.025–0.045 [9].

The power equation remains the same, with the relevant velocity now being the velocity of wind that the aircraft needs to counter to maintain station rather than the velocity needed to maintain sufficient lift to fly. Some airships do allow for aerodynamic lift to be used to optimize the need to fly forward and provide greater flexibility on the tricky issue of buoyancy control [10].

The power requirement for the airship therefore again follows a cube relationship with airspeed and as a worked example for a simple system of similar size to the HALE D demonstrator,

$$\text{Power} = \frac{1}{2}\rho A C_{\text{D}} V^3$$ (12.6)

which to counter a typical wind of 15 m s^{-1} at 65 000 ft requires some 2 kW of electric propulsion. As will be seen, this power requirement is similar to the power required to fly an aircraft with a similar payload capacity.

There are many arguments as to whether an aircraft or airship is a better approach to provide high-altitude persistence, beyond the scope of the current chapter. What will become clear is that both approaches are critically dependent on the ability to efficiently store electrical energy to maintain propulsion at night and the degree of this criticality will be considered in a worked example after the objectives for such system are clarified.

12.4 Objectives of HALE UAV

It has already been set out that any HALE capability – aircraft or airship – needs to maintain an altitude above the clouds and above the most significant winds. It has also been shown that for an aircraft it takes considerably more power to fly at higher altitudes

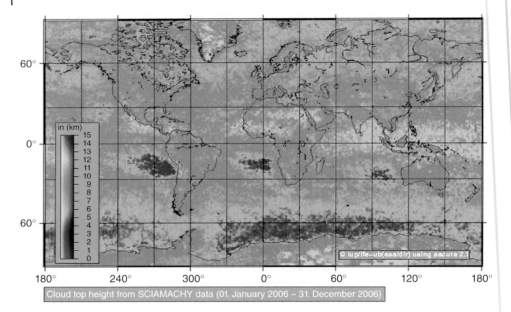

Figure 12.2 Cloud top height statistics [11].

and thinner air and for an airship the volume of envelope required to float also increases with altitude. In short, staying high is difficult so the objective is to maintain flight above the *minimum* sustainable altitude that enables the platform to stay in one location.

12.4.1 Stay Above the Cloud

The simpler requirement is to stay above the clouds. While there are high cirrus clouds at altitudes above 50 000 ft, in general most clouds top out below 15 km (50 000 ft) as shown in [11] (Figure 12.2).

12.4.2 Stay Above the Wind

The most significant wind speeds are associated with the jet streams, which largely occur at medium latitudes between 15° and 55° as indicated in Figure 12.3.

For more specific examples consider the wind speeds as a function of altitude and time of year for San Francisco (38° latitude) (Figure 12.4) and Bogota (5° latitude) (Figure 12.5).

The effect of the jet stream can be seen clearly, with average wind speeds in excess of 60 km present for large parts of the year at altitudes around 40 000 ft over San Francisco, with no such winds apparent at any time of year over Bogota.

This effect means that the minimum altitude required to be maintained by a slow-moving, high-efficiency platform is a function of both latitude and time of year and this has been summarized by Prismatic as Figure 12.6.

While 65 000 ft remains the general goal for sustainable altitude, a year-round capability is viable nearer the Equator by maintaining altitude above 53 000 ft, which, as has already been shown, makes the task significantly easier.

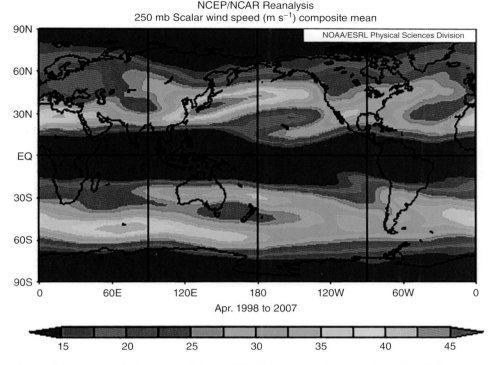

Figure 12.3 Average wind speeds at 250 mbar (40 000 ft). Source: Courtesy of NOAA/ESRL Physical Sciences Division.

While considering the effect of wind, take another look at Figure 12.4. This highlights the further benefit of demonstrating long-endurance flying during the summer at most latitudes and of also considering those applications that are most essential during the summer months. The significant jet stream winds are both lower in altitude and slower.

12.4.3 Stay in the Sun

The next requirement is for there to be enough sun to enable the solar charge/battery discharge cycle to be sustainable. Figure 12.7 shows the energy available from a 30% efficient solar array on the aircraft model used as the later worked example and compares that with the total energy required to maintain flight (noting the variation due to the minimum required altitude indicated in Figure 12.6) as a function of latitude at winter solstice.

As expected, there is no solution at latitudes above 67° latitude – as there is no sun above the Arctic Circle at winter solstice. Of particular interest is the crossover point, as this indicates the latitude at which, regardless of battery technology, persistent solar operations become nonviable with the current generation of 30% efficient arrays on the main horizontal surfaces. This is a simple picture but the rapid reduction in available solar energy with latitude at solstice strongly indicates that such persistent, year-round systems will not be viable above latitudes of 40°–50°, regardless of how good the batteries get. The phrase "horizontal surfaces" is deliberately noted in the section above as

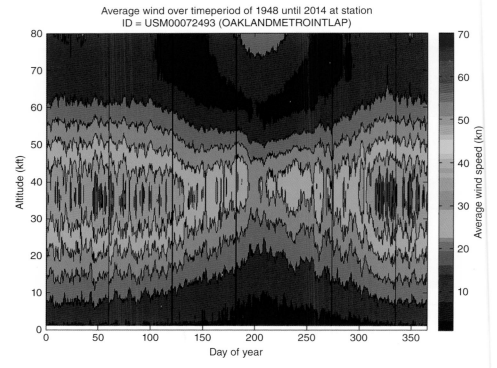

Figure 12.4 Average wind speed as a function of time of year and altitude at San Francisco. Source: Courtesy of Airbus.

any immediate review of concepts for solar HALE UAV will bring out concepts of tilting arrays to follow the sun and reconfigurable wing forms to make the most of low elevations. Certainly for polar summer missions where the sun is up most of the time and always low, near the horizon, there will be a benefit in considering vertical surfaces and possibly even surfaces that can rotate. In general, however, our experience is that the aircraft will usually be orientated to fly into the prevailing wind, not to orientate itself to the sun, so the potential benefit of such surfaces will not outweigh the mass of their implementation.

This brings us to the last and most important aspect of availability for these systems – where is the market? Where do HALE UAV systems need to provide these year-round services? And where else can valuable services be provided that may not need year-round coverage?

12.4.4 Year-Round Markets

Figure 12.8 shows the distribution of population as a function of latitude. The coverage generally proposed as a target for year-round availability of solar-powered HALE systems is ±40°. This target covers approximately 87% of the world's population and perhaps more importantly it covers the areas of greatest economic growth and the poorest levels of infrastructure – the ideal market conditions for a flexible, movable platform such as HALE UAV. This 40° target is therefore a sound target from both a technical and a business perspective.

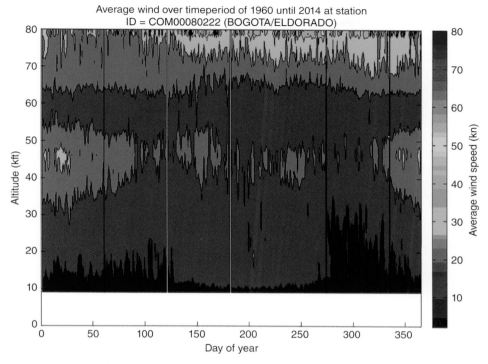

Figure 12.5 Average wind speed as a function of time of year and altitude at Bogota. Source: Courtesy of Airbus.

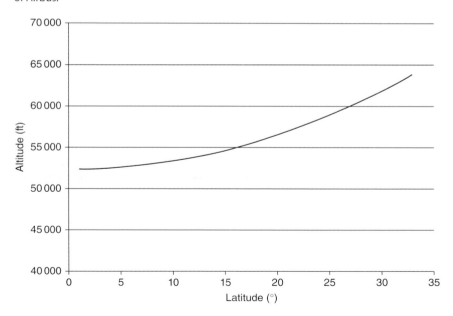

Figure 12.6 Minimum sustained altitude required as a function of latitude.

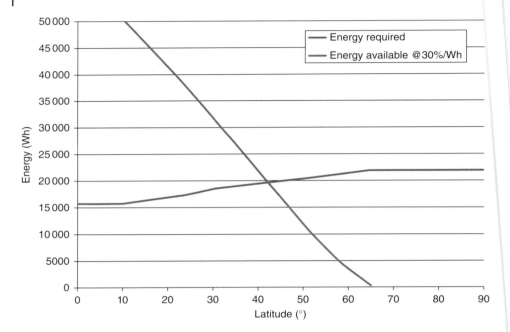

Figure 12.7 Energy available and energy required at winter solstice.

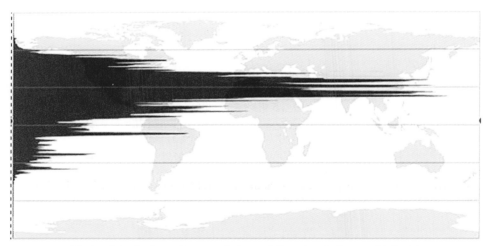

(Horizontal axis shows the sum of all population at each degree of latitude)

Figure 12.8 Population density as a function of latitude. Source: Courtesy of Radical Cartography.

12.4.5   Seasonal Markets

In addition to this year-round market, there are other niche markets for which year-round availability may not be essential. Precision agriculture, polar science, forest fires, and major events are all related to more significant interest during local summer conditions. As has already been discussed, it is much simpler to operate a solar-powered vehicle in the summer when the days are long, the sun is high, and most importantly, the nights are short.

12.4.6   How Valuable Are These Markets and What Does That Mean for the Battery?

This chapter is not intended to set out the business case for HALE UAV. It is, however, intended to highlight the value of improving battery technology, specifically in terms of reducing the mass of a given energy capacity. To that end the following precise, if slightly simple, analysis is provided.

One of the major markets that has reawakened a wider interest in solar-powered HALE UAV is the provision of internet services, championed by the Facebook [10] and Google [12] programs. As an open, well-publicized market this allows a simple indication to be made of the marginal benefit of improving performance.

Consider the following. If we were to reduce the mass of the aircraft by 1 g, how much bigger a market could be served?

A 1 g reduction in mass is equivalent to an improvement in specific energy (on a 20 kg battery mass) from 400 to 400.02 Wh kg^{-1}.

If the aircraft is 1 g lighter it is capable of maintaining 65 000 ft at a slightly higher latitude at winter solstice than before. At a nominal 30° latitude this increase in addressable latitude is 0.0004°.

From Figure 12.8, at 30° N latitude the population density is 183 million people per degree.

The 1 g reduction in mass therefore opens up a market of $0.0004 \times 183\,000\,000$ = 73 000 people.

Facebook's accounts for 2014 show [13], as a global average, an annual revenue per user (RPU) of \$11 and a user uptake of 1.3 billion of a global population of 7.2 billion − 18%.

Therefore, if Facebook can reach a further 73 000 population, it is reasonable to expect that their revenues can increase by 18% of 73 000 people developing \$11 each.

So − saving 1 g on the aircraft or increasing specific energy density of the battery by 0.005% can lead to annual revenue increase of \$145 000.

Amortized over a three-year life the value of mass loss on a HALE UAV is \$435 000/g, compared with gold at \$15/g, diamond at \$55 000/g, and californium 252 at \$27M/g.

12.5   Worked Example – HALE UAV

Having looked generally at both airplane and airship approaches to persistent HALE capability, and the value of improving the specific energy density of batteries, we will now put the challenge for new battery technology in a real context; we can consider an aircraft of the same class as the world-leading Zephyr 7, which set the record for the longest duration flight of any aircraft at 14 days 30 minutes and 8 seconds in July 2010.

This is not to say that this is the only approach that will work in the future, but as it is the only approach that has worked to date, it provides the soundest foundation for a discussion on battery technology. It should also be noted that what is presented here is a case example similar to the Zephyr 7. While the Zephyr 7 set its world record flight about seven years ago and Airbus are now completing the development of a much improved Zephyr 8 vehicle, which is about 60% more efficient [14], the Zephyr 7 performance remains far ahead of any similar flown system so it would be inappropriate to use more information about how Zephyr 7 achieves this performance than already in the public domain [15, 16].

We will start with the simplest and most optimistic scenario where we consider only the power to fly the aircraft and charge a perfect battery: No payload, no avionics, no heaters, an excellent (and achievable) 75% efficient propulsion systems, an excellent (and achievable) 85% charge efficiency, and attempting to fly at the Equinox not the much more demanding case of higher latitude winters.

The power required to fly a 55-kg, 23-m span aircraft similar to the Zephyr 7 at our target altitude of 65 000 ft, using typical drag and lift coefficients of appropriate airfoils, can be estimated at 600 W from Eq. (12.3).

Taking the simple case of Equinox where the days are as long as the nights, the solar power system needs to produce enough power over 12 hours to fly the aircraft,

$$\text{Daylight flight energy (Equinox)} = 12 \text{ hours at } 600 \text{ W} = 7200 \text{ Wh}$$

and enough power to charge a battery to keep the aircraft flying for the 12 hours of night,

$$\text{Night flight energy (Equinox)} = 12 \text{ hours at } 600 \text{ W} = 7200 \text{ Wh}$$

which, at 85% charge efficiency = 8500 Wh.

Therefore, the solar cells need to produce 15 700 Wh, an average of 1300 W over 12 hours of daylight – noting, of course, that the sun is not directly above the aircraft but is illuminating the solar panel on the wing obliquely as a function of the altitude, time of day, and time of year. To specify the solar array requirement this integrated energy requirement over the sunlit period of the day is translated into the area efficiency conversion required by the solar panel when fully illuminated by the overhead sun at the top of the atmosphere, conditions termed as AM0.

For the case of Equinox at 40° latitude this means that the solar array needs to be able to produce a power equivalent to 4600 W if illuminated under these standard conditions. There are two limits to the solar array's ability to do this – the area of wing available to collect sunlight and the mass of array that can be allocated.

The photovoltaic industry has seen huge improvements in the efficiency of solar cells (watts per square meter) driven by the push for renewable sources of energy, with triple junction gallium arsenide cells producing over 30% conversion efficiency. This means that in this example, the aircraft with 30 m^2 of wing area can produce 12 kW of electric power from a 30% efficient array if fully populated. As seen above, it only need produce 4600 W equivalent normal power, meaning that the wing need not be fully populated to maintain power balance at Equinox. As it turns out, a fully populated wing is perfectly matched to providing the required energy to fly at 40° latitude at winter solstice – which means that the available 30% *area* efficiency array is already sufficient to enable year-round flight at 40° latitude.

For the mass, we have worked closely with Microlink devices to minimize the mass of the cells and the infrastructure required to integrate the cells into panels on the aircraft, and this has resulted in cells that can produce over $2000\,W\,kg^{-1}$ and integrated panels that can produce $1500\,W\,kg^{-1}$, resulting in a solar array of mass 3 kg to support this Equinox operation [17, 18].

Let us now consider the battery.

For an Equinox case, the battery needs to store a minimum of 7200 Wh of energy. Using the best, standard, and commercially available technology (lithium polymer) with a nominal specific capacity of $260\,Wh\,kg^{-1}$ requires a mass of battery cells of 28 kg – over half the mass of the aircraft and well above the nominal cell mass allocated on the Zephyr 7 aircraft at 20 kg. So, lithium polymer technology does not work for this design, even under the most optimistic considerations and this is the reason that Zephyr has worked over the past 10 years with Sion Power Limited Liability Company (LLC) to successfully and reliably utilize the higher specific capacity of lithium–sulfur technology [19].

Of course, the aircraft can be made stronger to carry more battery mass and thus enable the use of lower Watt hours per kilograms. The structural mass then increases, which in turn increases the mass of battery required. The aircraft can also be made bigger such that the overhead mass of avionics is diluted and the induced drag is reduced through a higher aspect ratio. Many such arguments are put forward in the range of paper studies proposed for HALE UAV and again, I will come back to the point that the Zephyr 7 is the only vehicle to have worked. So, before we get too excited about how we might work around this issue let us look at the practical implications of HALE UAV flight on the battery requirements.

12.6 Cells, Batteries, and Real Life

The discussion above is a simple and basic analysis of the issues. We are now going to throw some real-life experience into the problem to make the challenge harder and in two aspects, a little easier. Let us start with two aspects of HALE UAV that actually make the battery issue easier – cycle life and rate. Enjoy the respite.

12.6.1 Cycle Life, Charge, and Discharge Rates

The Zephyr activity has always been based on a clear business case for both military and commercial applications. In making such a case both the capital expenditure of designing, developing, and building the aircraft and the operating expenditure of operating and flying the vehicles must be considered. A number of the enthusiastic paper studies for HALE UAV have promoted five-year flight lives but our business case shows that the benefit of flying for much longer than one year is marginal. As we only charge and discharge the batteries once per day, this means that a valuable battery need only provide 365 cycles at charge and discharge rates under 0.2 C. The target for the LiS community for HALE UAV is therefore to provide the highest Watt hours per kilograms over a 400-cycle life at (low) 0.2 C charge and discharge rates.

Given this clear target, which I will return to at the end of the chapter, let me get back to making life a little more difficult.

12.6.2 Payload

The aircraft needs to fly a payload. For many applications this payload needs to be operated both day and night. Our satellite background has served us well here. Take the most obvious payload, an optical camera. With our satellites we are used to designing cameras to provide imagery from 600 km above the Earth's surface. With HALE UAV we are a mere 20 km above the Earth. Optics works on a square law so this 30 times reduction in distance equates to a 900 times reduction in difficulty. If a payload needs to be 1 tonne on a satellite it may need only be 1 kg on a HALE UAV. On a more practical basis we have proved significantly valuable surveillance and communications payloads with masses under 3 kg and power drains of under 50 W – so let us take 50 W as the payload power requirement.

12.6.3 Avionics

The aircraft needs to fly – to measure parameters, process information, activate sensors, and deal with commands and telemetry. Our satellite background again allows for very efficient processing techniques so let us assign 20 W to flying overhead.

12.6.4 Temperature

Our experience with Zephyr has shown some significantly lower temperatures than standard atmosphere models, with a lowest recorded ambient temperature of $-83\,°C$ being recorded during our 11-day winter flight in 2014. The (poor) behavior of rechargeable batteries at low temperatures is well known, which means that we have to spend considerable time, money, and mass on protecting the batteries from these temperatures.

Both passive (insulation) and active (heating) approaches are used to provide a viable environment for the batteries and the cost of this protection has already been emphasized in the "cost per gram" argument outlined previously. The battery supplier needs to consider the potential Watt hours per kilograms of the battery with all the measures required to provide a useable energy rather than the cell capacity and this calculation will be exercised at the end of this section.

12.6.5 End-of-Life Performance

The business case for a one-year flight does require a high degree of reliability, not a concern from the satellite background we brought to bear for Zephyr, nor for the automotive industry in which batteries have established a footing. However, the difference between a car and an aircraft is that if the range of the car drops by 20% over the life of the battery, it is just an issue of range that can be managed. If the aircraft battery is known to drop 20% over the year of operation the system must be designed to work at that 80% point, because if it does not it will descend into the Jetstream and will be lost.

12.6.6 Protection

Battery technology shows a similar development path as any other. The initial performance is achieved and this is iterated to provide reliability of performance and then consumer practicality. For the battery, the impact of this ongoing development is that the early examples require more careful selection, monitoring, and protection, and monitoring and protection involve adding mass to the battery in the form of sensors, controllers, actuators, and protection devices such as fuses. All such measures increase the mass of the battery and decrease the effective Watt hours per kilograms.

12.6.7 Balancing – Useful Capacity

Related to the issue of protection is that of balancing. The aircraft will use multiple, parallel strings of cells to provide the capacity at the preferred voltage. Of specific concern are the tight limits on overvoltage and undervoltage that apply to individual cells. While there are multiple ways with different levels of complexity that can be applied to this problem the same criterion as to the end of life discussion above needs to be applied – the useful capacity of the battery is that at the end of its nominal life accounting for any expected losses and the practical management of all the cells within their limits. Therefore, if a simple control is used whereby as soon as the end of discharge (EoD) limit is reached for any one cell in a series string the whole string is stopped from discharging then any remaining capacity in the cells is unusable and the effective capacity of n cells is n times the worst capacity of any cell in the string. If a more effective means is employed of maximizing charge to each cell, by shunting or even individual charge circuitry, then the effective specific capacity must include the mass and efficiency loss of this circuitry.

12.6.8 Summary of Real-World Issues

The impact of the factors noted above can all be translated into mass and power demands on the system and reductions in the effective specific energy available from any battery technology.

We start with the simple energy required to fly for 12 hours, 7200 Wh.

Payload: this adds 50 W to the required power, so the energy storage needs to increase by 12×50 W $= 600$ Wh.

Avionics: this adds 20 W to the required power, so the energy storage needs to increase by 12×20 W $= 240$ Wh.

End of life: the design must cater for an 80% effective capacity to still operate at end of life.

Balancing: assuming a reasonably tight spread (at end of life) of $\pm 5\%$ on cell capacity gives an effective 95% capacity with no incurred mass of extra control circuitry.

Instead of the nominal 7200 Wh of energy required from the batteries overnight, the batteries need to provide $7200 + 600 + 240 = 8040$ Wh.

This capacity needs to be provided at end of life, so the nominal start of life capacity needs to be $8040/0.8 = 10{,}050$ Wh.

Finally, if we assume that the series string of cells can only be depleted to the minimum voltage of the worst cell so that the realizable capacity is 95%, then $10{,}050/0.95 = 10{,}580$ Wh is the nominal capacity required, which is 47% more than the first "flight only" figure of 7200 Wh.

The protection mass, insulation mass, and heater power requirements are best considered as an effective reduction in specific energy at the battery level.

Temperature: for consideration of the effective battery mass this insulation is approximately 10% of the cell mass. Additionally, 100 W of heater power is budgeted to maintain operating temperature overnight, requiring a further 1200 Wh of energy.

Protection and control: the electronics for the charge, discharge, and protection circuitry add to the mass of the battery. This is approximately 10% of the cell mass.

Considering 20 kg of cells at 260 Wh kg^{-1} with 10% protection overhead, 10% insulation, and 100 W of heater power effectively means the specific capacity of the battery is

$$\frac{Wh}{kg} = \frac{(\text{Energy of cell} - \text{energy to heat})}{(\text{mass of cell} + \text{mass of insulation} + \text{mass of control circuitry})} \quad (12.7)$$

which for 20 kg of 260 Wh kg^{-1} cells gives

$$\frac{Wh}{kg} = \frac{(20.260 - 12.100)}{(20 + 2 + 2)}$$

167 Wh kg^{-1} – or 64% of the nominal cell specific capacity.

Combining these effects shows that the nominal capacity of the battery of the aircraft needs to be 47% more than the simple flight power calculation while the effective specific capacity of a battery considering temperature and control circuitry is 64% less than the cell specific capacity. In short, the battery problem is twice as difficult as would first be thought.

Another way of stating the battery issues is that for each kilogram of solar array we carry we can produce 9000 Wh of energy but each kilogram of battery can only store 375 Wh of energy – a factor of 24.

This is the explanation for why solar HALE UAV operations are so much easier in high-altitude summers – the sun is very powerful, solar cell technology has advanced remarkably over the past 20 years, and we are flying above the clouds.

At this point, I should be clearer as to the context of this point on specific capacity. A review of other proposed programs shows that many have been based on the use of "standard" LiPo technology for HALE flight. I am not saying that HALE flight is not possible by using specific capacities at this level, and in fact Prismatic is now working on designs that can provide year-round operations with this technology, and I am simply stating one fact and one proposition:

Fact – the only vehicle that has achieved HALE flight (Zephyr 7) required a specific cell capacity of 375 Wh kg^{-1} and even this did not allow for high-altitude flight at any time of year at any location.

Proposition – improving the specific capacity of batteries is the single most important, outstanding factor in routinely implementing high-altitude, long-duration flight.

What this clearly shows is that for what is currently the world's only successful solar-powered HALE UAV, the standard lithium ion battery technology at 260 Wh kg^{-1} is not good enough to allow year-round operations. Being even more to the point, what has been demonstrated with the Zephyr aircraft is that it is now *only* the battery technology that is preventing persistent, year-round operations at most latitudes.

12.7 A Quick Aside on Regenerative Fuel Cells

The larger HALE UAV programs (Helios, Vulture) intended to utilize regenerative fuel cell technology whereby hydrogen and oxygen are combined in a fuel cell to generate electricity during the night and then the resultant water is electrolyzed by solar cells during the day to reform the hydrogen and oxygen "fuel."

The attraction of this technology is the very high (>1000) specific energy density expected from such systems. The problems with the systems are the following:

Overhead mass: while the chemistry promises high specific energy, the mass overhead of fuel cells and gas storage means that high specific energy can only be realized with very large systems, requiring very large aircraft.

Recharge efficiency: the overall charge/discharge efficiency of the system is much lower (40%) than that achieved by batteries (85%). This results in significantly more solar array being required. For an aircraft with the limited array area provided by the wing this has led to the need for much higher (and currently unobtained) solar cell efficiencies of 50%.

Viability: to date (2018), no practical regenerative fuel cell system has been demonstrated, let alone successfully flown on a high-altitude vehicle. This is despite the fact that the Helios vehicle intended to use such fuel cells as far back as 2003.

In summary, if regenerative fuel cell systems are proved to be viable they are likely to be useful only for large stratospheric airships, which will be very expensive.

12.8 So What Do We Need from Our Battery Suppliers?

The major business for persistent flight requires year-round availability. The winter solstice case is significantly worse than for the Equinox case used above as the winds are faster, the winds are higher, the sun is lower, the days (time available to charge) are shorter, and the nights (time when we rely on the battery) are longer.

For the example vehicle discussed here with an all up mass of 55 and 20 kg of battery cells, the required specific energy density to support year-round operations as a function of latitude is shown in Figure 12.9. This shows that a specific cell capacity of 600 Wh kg^{-1}

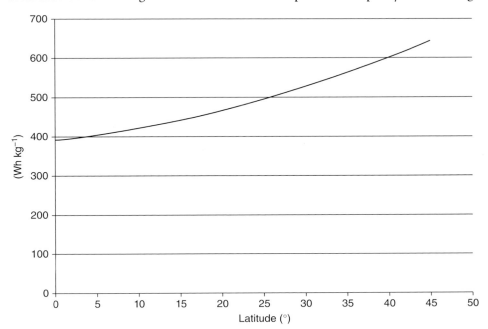

Figure 12.9 Indicative cell Wh kg^{-1} requirement for year-round operations as a function of latitude.

would enable a vehicle similar to the Zephyr 7 to operate year round at latitudes up to 40° and 400 Wh kg^{-1} is the threshold for year-round performance for this proven level of performance.

As noted earlier, this analysis is based on a vehicle similar to the Zephyr 7. Airbus have improved this vehicle significantly for Zephyr 8, and other groups, including Prismatic, are developing new, more efficient platforms but the general goal remains the same – *400–600 Wh kg^{-1} at cell level*.

12.9 The Challenges for Battery Developers

For valuable year-round operations from HALE UAVs we need

- >400 Wh kg^{-1} cell capacity
- 400 cycles
- 0.2 C charge/discharge rates
- 20 °C operating temperature.

12.10 The Answer to the Title

The reasons why HALE UAV represents such an attractive first application for lithium–sulfur technology are as follows:

- *Value*: this is a low-volume, high-value market.
- *Cycle life*: we only need 400 cycles.
- *Rates*: we cycle the batteries on a fixed and slow daily cycle.
- *Safety*: the batteries will be handled by professional engineers on the ground and then spend their entire life away from everyone and everything, 20 km above the Earth.

12.11 Summary

HALE UAV represents a viable, valuable, and growing market.

The size of the addressable market is largely dependent on the specific energy density available from rechargeable batteries.

HALE UAV and HALE airships will continue to be developed, and despite the disappointment engendered by excessive, early hype in this latest round of development, vehicles will fly that will break the Zephyr 7 record and go on to provide services all year round – at some latitudes.

It is hoped that this chapter has set out some of the challenges and wider issues that developers can now consider in producing batteries that most effectively meet the requirements of this small, growing, and valuable market for new power technology.

Acknowledgments

My thanks to Airbus Defence and Space for permission to use photographs and information on the Zephyr program.

My thoughts and thanks to Chris Kelleher, who designed the Zephyr and cofounded the program in 2003 and who sadly passed away in 2015.

References

1 Boucher, R.J. (1984). History of solar flight. In: *20th Joint Propulsion Conference, (June 11–13, 1984)*. Cincinnati, OH: American Institute of Aeronautics and Astronautics.

2 Macready, P.B., Lissaman, P.B.S., Morgan, W.R., and Burke, J.D. (1983). Sun powered aircraft designs. *Journal of Aircraft* 20.

3 BBC News (2003). Strato-plane looks skyward http://news.bbc.co.uk/1/hi/sci/tech/3016082.stm (accessed 24 June 2003).

4 DARPA. Vulture programme http://www.darpa.mil/program/vulture (accessed 13 August 2018).

5 BBC News 'Eternal' solar plane's records are confirmed (accessed 24 December 2010).

6 Daily Telegraph (2014). Fly 11 days non-stop? Now that's long-haul (31 August 2014).

7 Office of the Assistant Secretary of Defense for Research and Engineering, Rapid Reaction Technology Office (2012). Summary report of DOD funded lighter than air vehicles (June 2012).

8 UK Defence Journal Britain to purchase high altitude surveillance aircraft (10 July 2016).

9 Thompson, F.L. and Kirschbaum, H.W. (1932). The drag characteristics of several airships determined by deceleration tests. NACA Report No. 397.

10 Facebook The technology behind Aquila, Mark Zuckerberg, https://www.facebook.com/notes/mark-zuckerberg/the-technology-behind-aquila/10153916136506634/ (accessed 13 August 2018).

11 Kokhanovsky, A., Vountas, M., and Burrows, J.P. (2011). Global Distribution of Cloud Top Height as Retrieved from SCIAMACHY Onboard ENVISAT Spaceborne Observations. *Remote Sensing* 3: 836–844.

12 Google's Titan Solara 50 Business and commercial aviation (1 July 2015).

13 Facebook Q4 2014 Results.

14 QinetiQ (2010). Record breaking Zephyr offers 24/7 cost effective military surveillance and communications. Press release (23 December 2010).

15 Zephyr The high altitude pseudo-satellite https://airbusdefenceandspace.com/our-portfolio/military-aircraft/uav/zephyr/ (accessed 13 August 2018).

16 Airbus Zephyr, making the World's best better, Paul Brooks, AUVSI New Orleans, (May 2016).

17 Hybrid Air Vehicles Airlander 10 www.hybridairvehicles.com/aircraft/airlander-10 (accessed 13 August 2018).

18 Microlink Devices 2016. http://mldevices.com/index.php/news Press release (18 May 2016).

19 Sion Power® Sion Power's lithium–sulfur batteries power high altitude pseudo-satellite flight. Sion Power Press release (10 September 2014). http://www.sionpower.com/media-center.php?code=sion-powers-lithiumsulfur-batteries-power-high-alt.

Index

Lithium–Sulfur Batteries, First Edition. Edited by Mark Wild and Gregory J. Offer.
© 2019 John Wiley & Sons Ltd. Published 2019 by John Wiley & Sons Ltd.